9/04

PROCEEDINGS OF THE BRITISH ACADEMY • 106

THE SPECIATION OF MODERN *HOMO SAPIENS*

Edited by
TIM J. CROW

Published for THE BRITISH ACADEMY
by OXFORD UNIVERSITY PRESS

Oxford University Press, Great Clarendon Street, Oxford OX2 6DP

Oxford New York
Auckland Bangkok Buenos Aires Cape Town Chennai
Dar es Salaam Delhi Hong Kong Istanbul Karachi Kolkata
Kuala Lumpur Madrid Melbourne Mexico City Mumbai Nairobi
São Paulo Shanghai Singapore Taipei Tokyo Toronto

Published in the United States
by Oxford University Press Inc., New York

First published 2002
Reprinted 2003 (twice)
Paperback edition 2003

British Library Cataloguing in Publication Data
Data available

ISBN 0–19–726311–9
ISSN 0068–1202

Typeset in Times
by J&L Composition Ltd, Filey, North Yorkshire
Printed in Great Britain
on acid-free paper by
Antony Rowe Ltd
Chippenham, Wiltshire

Contents

List of Contributors

Derek Bickerton, Department of Linguistics, Moore Hall 569, 1890 East West Road, Honolulu, HI 96822, USA (e-mail bickertond@prodigy.net)

Mark Collard, Department of Anthropology, and AHRB Centre for the Evolutionary Analysis of Cultural Behaviour, University College London, Gower Street, London WC1E 6BT, UK (e-mail m.collard@ucl.ac.uk)

Norman D. Cook, Department of Informatics, Kansai University, Takatsuki, Osaka, Japan (e-mail cook@res.kutc.kansai-u.ac.jp)

Michael C. Corballis, Research Centre for Cognitive Neuroscience, Department of Psychology, University of Auckland, Private Bag 92019, Auckland, New Zealand (e-mail m.corballis@auckland.ac.nz)

Tim J. Crow, Prince of Wales Centre for Research on Schizophrenia, University Department of Psychiatry, University of Oxford, Warneford Hospital, Oxford OX3 7JX, UK (e-mail tim.crow@psychiatry.oxford.ac.uk)

Paul Mellars, Corpus Christi College, Cambridge CB2 1RH, UK

Detlev Ploog, Max-Planck Institute for Psychiatry, Kraepelinstrasse 2, D-80804, Muenchen, Germany (e-mail ploog@mpipsykl.mpg.de)

Klaus Reinhold, Institute for Advanced Study Berlin, Germany and Department of Evolution and Ecology, University of Bonn, An der Immenburg 1, 53121 Bonn, Germany (e-mail KReinhold@evolution.uni-bonn.de)

Carole A. Sargent, **Patricia Blanco** and **Nabeel A. Affara**, Human Molecular Genetics Group, University of Cambridge, Department of Pathology, Tennis Court Road, Cambridge CB2 1QP, UK (e-mail cas1001@hermes.cam.ac.uk)

James Steele, Department of Archaeology, University of Southampton, Highfield, Southampton SO17 1BJ, UK (e-mail tjms@soton.ac.uk).

Chris Stringer, Department of Palaeontology, The Natural History Museum, London SW7 5BD, UK (e-mail c.stringer@nhm.ac.uk)

Ian Tattersall, Department of Anthropology, American Museum of Natural History, New York, NY 10024–5192, USA (e-mail iant@amnh.org)

Chris Tyler-Smith, CRC Chromosome Molecular Biology Group, Department of Biochemistry, University of Oxford, South Parks Road, Oxford OX1 3QU, UK (e-mail chris.tyler-smith@bioch.ox.ac.uk)

Acknowledgements

The meeting at which these papers were presented was organised under the auspices of the Academy of Medical Sciences and the British Academy, and supported by grants from the Wenner-Gren Foundation and the Wellcome Trust. It was planned by a committee of Tim Crow, John Maddox, Paul Mellars and Chris Stringer, with the help of Rosemary Lambert of the British Academy and Greg Bock of the Novartis Foundation. Sessions at the 2-day meeting were chaired by C. Blakemore, J.R. Hurford, P.J. Lachmann, J. Maddox, S. Mithen and L. Wolpert. The proceedings and discussion were recorded and transcribed by Lynn De Lisi assisted by Carole Walton. Scripts were edited by Eva Fairnell and publication was arranged by James Rivington of the British Academy.

Introduction

TIM J. CROW

FEW, IF ANY, SYMPOSIA OR MEETINGS have addressed directly the topic of the speciation of modern *Homo sapiens*. This may be because the concept of speciation as a discrete genetic event has an ambiguous status within evolutionary theory, in part because of the difficulty we have in taking an objective view of the species of which we are members, but also because no member seems to have thought of it as a problem. Yet it is arguable that scrutiny of the origin of our species casts speciation, and maybe evolutionary theory as it now stands, in a critical light.

The meeting that took place under the auspices of the British Academy and the Academy of Medical Sciences on 28 March 2000 at the British Academy, with the subsequent workshop at the Novartis Foundation, had an informal precursor as a 1-day SANE-sponsored conference held at St Anne's College, Oxford, on 30 June 1997 (Maddox, 1997). That meeting introduced the themes of language and lateralisation as putative species characteristics and X–Y homologous genes as a class of potential mediators of change. Several speakers (Derek Bickerton, Nabeel Affara, Chris McManus, James Steele and myself) participated in both events. As reflected in these *Proceedings*, the second meeting added the dimensions of palaeontology (Chris Stringer, Ian Tattersall and Mark Collard), archaeology (Paul Mellars) and neuroanatomy (Detlev Ploog and Norman Cook) to those of linguistics, psychology (Michael Corballis), genetics (Chris Tyler-Smith) and evolutionary theory (Klaus Reinhold).

MAN'S PLACE IN EVOLUTIONARY THEORY—T. H. HUXLEY, C. DARWIN AND A. R. WALLACE

The present *Proceedings* can be placed in the longer time–course of evolutionary debates on 'man's origins'. In *Evidence of Man's Place in Nature*, Huxley (1863) summarised the anatomical and palaeontological case that Darwin (1859) had been reluctant to spell out in the *Origin of Species*, that man has a descent from the great apes. In 1871, Darwin himself approached the issue, which was now the subject of widespread debate, in *The Descent of Man*. One interest of this

Proceedings of the British Academy, **106**, 1–20, © The British Academy 2002.

publication is that it linked in a single volume, *The Descent of Man* and *Selection in Relation to Sex*, the anatomical and palaeontological case with Darwin's theory of sexual selection. Was the connection fortuitous; did he, for example, have two different theses that happened to come to fruition at the same time, or were the theses related? In his introduction, Darwin writes that:

> During many years it has seemed to me highly probable that sexual selection has played an important part in differentiating the races of man; but in my *Origin of Species* I contented myself by merely alluding to this belief. When I came to apply this belief to man, I found it indispensable to treat the whole subject in full detail. (Introduction, *The Descent of Man*, pp. 4–5)

Some passages in the second part of the book indicate that Darwin considered the two arguments to be related in a more fundamental way. Thus:

> … Sexual selection has apparently acted on both the male and the female side, causing the two sexes of man to differ in body and mind … [and] has indirectly influenced the progressive development of various bodily structures and of certain mental qualities. Courage, pugnacity, perseverance, strength and size of body … have all been indirectly gained by one sex or the other, through the influence of love or jealousy, through the appreciation of the beautiful in sound, colour and form, and through the exertion of choice … (*Selection in Relation to Sex*, p. 402)

But as to the mechanism, as of course to the genetic basis of natural selection itself, Darwin was unclear. Perhaps the fact that the two arguments are in separate halves of the book reflects Darwin's own uncertainty on the relationship between sexual and natural selection.

The above passage illustrates some of the problems. If we suppose that sexual and natural selection are to some extent based upon distinct principles, do we assume that sexual selection can act separately, '… indirectly gained by one sex or the other …' in the two sexes? One can conjure up an image of almost any bodily feature coming under the influence of capricious mate choice in one sex, and of the two sexes diverging with respect to an array of independent features. But there are clearly limits to the scope of sexual selection. Moreover, A. R. Wallace denied that any separate principle was involved (Cronin, 1991). Then again, if 'Courage, pugnacity, perseverance, strength and size of body' have all been gained through 'the influence of love or jealousy [or] through the appreciation of the beautiful and … the exertion of choice', at what stage did this happen? Do we assume that the courage, pugnacity and perseverance, which are presumably characteristics of males differentially preferred by females, were selected sequentially or as a suite? When did this happen in the history of the species? If strength and size of body are features that differentiate the sexes, are such features species-specific or do the differences that are present in humans reflect a trend that was already present in a precursor hominid or great ape? In short, what is the relationship between sexual selection and speciation?

As is well known, on the topic of speciation Darwin was a gradualist. From the time he became convinced of the mutability of species in 1838, he was at pains to emphasise continuities rather than discontinuities. For natural selection to act it was necessary for there to be variation within populations. No qualitative distinction between variation within and between species was drawn. In the only figure to appear in the *Origin of Species*, population continuities are represented as dotted lines travelling up the page over successive thousands of generations. Lineages separate, and some die out. Some that are close together at the origin are connected to points that are widely separated at the finishing line.

The critical role in Darwin's concept is that of the environment. Populations separate and come under differing selective pressures. Separations are reinforced by geographical barriers. It is the chance isolation of populations which are then subject to divergent environmental selective pressures that is the core of the concept, a concept that has been formalised in the subsequent literature as the 'biological' or 'isolation' species concept (Mayr, 1963). Populations separate and diverge in their genetic complement, and some component of those divergences eventually contributes to hybrid infertility should individuals from the two populations then reassociate.

The attraction of the theory is its simplicity. No arbitrary mechanism of speciation or discontinuity in the flow of natural selection is introduced. There is no need to be overly concerned with the nature of species boundaries. Ring species with continuous clines to infertility at the extremes are well accounted for. What is critical is the environment to which the varieties of organism adapt.

The case of *H. sapiens* raises problems for the isolation species concept. What environment was critical for the transition from a prior hominid species? What is striking about the biological success of the species is its ability to survive in diverse environments (humans are everywhere; Gamble, 1993), and even to change the environment. What specific characteristics of populations (bodily habitus, skin colour, thermoregulatory capacity?) are adaptations to particular environments, and how do these relate to the core characteristic(s) of the species? Diverse hominid species have previously co-existed. Is it just time, or the lack of an isolated population in a sufficiently distinct environment, that has prevented speciation occurring again?

The more fundamental question is the nature of a species. Is there a speciation characteristic? What holds a species, for example humanity, together? The notion seems contrary to the Darwinian continuity principle. Indeed, in the case of humans Wallace had difficulty in accepting Darwin's thesis that:

> man's whole nature—physical, mental, intellectual, and moral—was developed from the lower animals by means of the same laws of variation and survival; and, as a consequence of this belief, that there was no difference in *kind* between man's nature and animal nature, but only one of degree. My view, on the other hand,

> was, and is, that there is a difference in kind, intellectually and morally, between man and other animals; and that while his body was undoubtedly developed by the continuous modification of some ancestral animal form, some different agency, analogous to that which first produced organic *life*, and then originated *consciousness*, came into play in order to develop the higher intellectual and spiritual nature of man. (Wallace, 1905, quoted in Cronin, 1991).

Wallace's unease has a significant parallel in contemporary evolutionary theory. Maynard-Smith & Szathmary (1995) identify eight major transitions—the origin of replicating molecules (RNA, DNA and protein), eukaryotes, sexual populations, cell differentiation and multicellular organisms, colonies, societies and language (which, following Bickerton, they regard as the most characteristic function of *H. sapiens* with consciousness as an epi-phenomenon)—as critical discontinuities in evolutionary history. But one can ask why these eminent evolutionary theorists have not treated the origins of language as another speciation event, analogous, say, to the separation of the New from the Old World monkeys or to the diversification of Darwin's finches on the Galapagos islands? The answer, it seems, is twofold. First, that the jump seems too big (Wallace's point). Secondly, that these authors, as inheritors of the most rigorous and 'gene-centred' version of Darwinism, have down-played the discontinuous nature of speciation events (Maynard-Smith, 1993). The outcome is that, as in Wallace's case, the explanation of the origins of *H. sapiens* acquires the status of mystery.

LANGUAGE AS THE DEFINING CHARACTERISTIC: F. M. MULLER'S CRITIQUE OF DARWINIAN THEORY

The concept that language is the defining characteristic of humanity has an ancient origin:

> In most of our abilities we differ not at all from the animals; we are in fact behind many in swiftness and strength and other resources. But because there is born in us the power to persuade each other and to show ourselves whatever we wish, we not only have escaped from living as brutes, but also by coming together have founded cities and set up laws and invented arts, and speech has helped us attain practically all of the things we have devised. (Isocrates, BC 436–338, quoted in Harris & Talbot, 1997)

Darwin can be quoted as in agreement with this view. On page 53 of *The Descent of Man* he writes that language 'has justly been considered as one of the chief distinctions between man and the lower animals' but he seems not to have regarded this as a particular difficulty. In 1873, within 2 years of the publication of *The Descent of Man*, Friedrich Max Muller (Muller, 1996), who held the chair of Philology in the University of Oxford, delivered a series of three lectures at the Royal Institution in which he drew attention to the problems that language raises for Darwin's theory:

> My object is simply to point out a strange omission, and to call attention to one kind of evidence—I mean the evidence of language—which has been most unaccountably neglected, both in studying the development of the human intellect, and determining the position which man holds in the system of the world.

Muller placed Darwin's theory in the philosophical tradition of Locke (1690) and Hume (1739) without regard for the achievements of Kant (1781, 1788). He complained that Darwin's theory that 'man being the descendant of some lower animal, the development of the human mind out of the mind of animals, or out of no mind, is a mere question of time, is certainly enough to make one a little impatient'. The problem according to Muller was that the contents of human consciousness were not merely, as maintained by Locke and Hume, those that arose from the sensations but also from the framework (the 'pure intuitions') of space and time that is intrinsic to the human mind:

> If we are to become conscious of anything ... we must place all phenomena side by side, or in space; and we can accept them only as following each other in succession, or in time. If we wanted to make it still clearer, that Time and Space are subjective, or at all events determined by the Self, we might say that there can be no There without a Here, there can be no Then without a Now, and that both Here and Now depend on us as recipients, as measurers, as perceivers.

In other words, there must be a deictic frame, and that frame is intrinsic to the capacity for language.

In the second lecture Muller addresses the problem:

> There is one difficulty which Mr Darwin has not sufficiently appreciated ... There is between the whole animal kingdom on the one side, and man, even in his lowest state, on the other, a barrier which no animal has ever crossed, and that barrier is—*Language* ... If anything has a right to the name of specific difference, it is language, as we find it in man, and in man only ... If we removed the name of *specific difference* from our philosophic dictionaries, I should still hold that nothing deserves the name of man except what is able to speak ... a speaking elephant or an elephantine speaker could never be called an elephant. Professor Schleicher, though an enthusiastic admirer of Darwin, observed once jokingly, but not without a deep meaning, 'If a pig were ever to say to me, 'I am a pig' it would ipso facto cease to be a pig'.

Muller considers how far Darwin had gone towards conceding the point: 'Articulate language is peculiar to man' (Darwin, 1871: 54), and 'It is not the mere power of articulation that distinguishes man from other animals, for, as everyone knows, parrots can talk; but it is his large power of connecting *definite sounds with definite ideas*'.

Muller writes:

> Here, then, we might again imagine that Mr Darwin admitted all that we want, viz. that some kind of language is peculiar to man, and distinguishes man from the other animals ... but, no, there follows immediately ... 'This obviously depends upon the development of the mental faculties'.

Muller asks:

> What can be the meaning of this sentence? ... If it refers to the mental faculties of man, then no doubt it may be said to be obvious. But if it is meant to refer to the mental faculties of the gorilla, then whether it be true or not, it is, at all events, so far from being obvious, that the very opposite might be called so—I mean the fact that no development of the mental faculties has ever enabled one single animal to connect one single definite idea with one single definite word.
>
> I confess that after reading again and again what Mr Darwin has written on the subject of language; I cannot understand how he could bring himself to sum up the subject as follows: 'We have seen that the faculty of articulate speech in itself does not offer any insuperable objection to the belief that man has been developed from some lower animal'.

Muller distinguishes between what he describes as emotional and as rational language. The former he relates to the 'bow-wow' or 'onomatopoeic' origins theory of language, 'the power of showing by outer signs what we feel, or, it may be, what we think'. This he regards as shared between man and other animals. Rational language, on the other hand, he relates to the power of forming and handling general concepts, and this he regards as specific to man. He draws attention to the observations of Hughlings Jackson (1868; Taylor, 1932), that the two can be separated by the effect of disease, and that disorders that impair the intellectual and rational expression of speech are, following the observations of Broca (1861), to be found with lesions that affect the anterior lobe on the left side of the brain.

The ability to form 'roots' is what Muller regards as the essence of intellectual or rational language. He describes its significance as follows:

> There is in every language a certain layer of words which may be called purely emotional. It is smaller or larger according to the genius of each nation, but it is never quite concealed by the alter strata of rational speech. Most interjections, many imitative words, belong to this class. They are perfectly clear in their origin, and it could never be maintained that they rest on general concepts. But if we deduct that inorganic substratum, all the rest of language, whether among ourselves or among the lowest barbarians, can be traced back to roots, and every one of these roots is a sign of a general concept ... Take any word you like, trace it back historically to its most primitive form, and you will find that besides the derivative elements, which can easily be separated, it contains a predicative root, and that in this predicative root rests the connotative power of the word.

Muller summarised his case:

> If the words of our language could be derived straight from imitative or interjectional sounds, such as bow-wow or pooh-pooh, then I should say that Hume was

> right against Kant, and that Mr Darwin was right in representing the change of animal into human language as a mere question of time. If, on the contrary ... after deducting the purely onomatopoeic portion of the dictionary, the real bulk of language is derived from roots, definite in their form and general in their meaning, then that period in the history of language which gave rise to these roots ... forms the frontier ... between man and beast ... That period may have been of slow growth, or it may have been an instantaneous evolution: we do not know ... These roots, which are in reality our oldest title-deeds as rational beings, still supply the living sap of the millions of words scattered over the globe, while no trace of them, or anything corresponding to them, has ever been discovered even amongst the most advanced of catarrhine apes.

Muller's notion of the linguistic root as a general concept can be compared with de Saussure's (1916) formulation of the components of the linguistic sign. According to de Saussure it is the arbitrariness of the relationship between the signifier (the sound pattern) and the signifieds (the concepts to which it is attached) that is the first principle of linguistics. Although he seems not to have considered this principle in a comparative context, he regarded arbitrariness of association as fundamental to the human capacity for language. It is necessary to the formation of abstract concepts.

DID LANGUAGE REQUIRE A MACRO-MUTATION?

Subsequent linguists have raised the problem for evolutionary theory in different forms. Chomsky's (1959) critique of Skinnerian operant theory as an explanation of verbal behaviour carried the strong implication that there were principles underlying language that were human specific. The concept of universal grammar as a defining human characteristic and of its generativity have implications for speciation theory, as well as for neuroscience, but these have not been pursued. Linguistics and evolutionary theory have remained separated [see Pinker's (1994) discussion of Chomsky's views below].

Perhaps the most incisive attempt to cross the disciplinary boundary is that of Derek Bickerton in his contributions in *Language and Species* (1990) and *Language and Human Behavior* (1995). Bickerton (1990) distinguishes protolanguage, the use of symbols without grammatical structure, the use of null elements, subcategorisation of verbs and recursiveness, from full language, and attributes the former to trained apes, children under 2, adults who have been deprived of language in their early years, and speakers of pidgin. He dates the origin of language to the origins of modern *H. sapiens* and writes (1990: 190) that:

> The evidence ... indicates that language could not have developed gradually out of protolanguage, and it suggests that no intermediate form exists. If this is so,

> then syntax must have emerged in one piece, at one time, the most likely cause being some kind of mutation that affected the organization of the brain. Since mutations are due to chance, and beneficial ones are rare, it is implausible to hypothesize more than one such mutation.
>
> Several factors suggest, indeed, that just such a single mutation gave rise to our species. It is here that problems arise. For our species is distinguished from all others not merely by syntacticized language but also by changes in the features and dimensions of the skull and by our typical supralaryngeal tract.
>
> The problem, bluntly stated, is: How could any single event, whether a point mutation or a re-shuffling of chromosomes, occasion so many and such diverse changes? The problem would be bad enough if only a single principle were required to set syntax in motion. If seven or eight were required, the situation would be still worse.

Bickerton goes on to argue that the critical event:

> could have consisted simply in the linking (or the dramatic strengthening of pre-existing links between) those areas of the brain where the lexicon was stored and those areas where the structure of actions and events was analyzed. These linkages would then have inhibited the random chaining of words and facilitated their rapid and automatic organization into the structural units described.

On the question of the nature of the event Bickerton supposes that the primary change was in the brain itself but:

> the emergence of our species does not seem to have been attended by any conspicuous enlargement, and whatever alterations in the brain's hard-wiring may have been required to initiate syntax, it seems unlikely that there should have resulted from them a change in the shape of the brain sufficient to cause such marked alterations in skull dimensions. Perhaps some additional factor could be found that would underlie all three changes. But at present there are no obvious candidates.

Other linguists have squared up to the evolutionary dilemma. Thus in a chapter entitled 'The big bang' Pinker (1994: 332) outlines the problem:

> Language is obviously as different from other animals' communication systems as the elephant's trunk is different from other animals' nostrils. Nonhuman communication systems are based upon one of three designs: a finite repertory of calls (one for warnings of predators, one for claims to territory, and so on), a continuous analog signal that registers the magnitude of some state (the livelier the dance of the bee, the richer the food source that it is telling its hivemates about), or a series of random variations on a theme (a birdsong repeated with a new twist each time …). As we have seen, human language has a very different design. The discrete combinatorial system called 'grammar' makes human language infinite (there is no limit to the number of complex words or sentences in a language), digital (this infinity is achieved by rearranging discrete elements in particular orders and combinations, not by varying some signal along a

> continuum like the mercury in a thermometer), and compositional (each of the infinite combinations has a different meaning predictable from the meanings of its parts and the rules and principles arranging them).

> Even the seat of human language in the brain is special … [see the chapter in these *Proceedings* by Detlev Ploog] The vocal calls of primates are controlled not by their cerebral cortex but by phylogenetically older neural structures in the brain stem and limbic system, structures that are heavily involved in emotion … Genuine language … is seated in the cerebral cortex, primarily the left perisylvian region. (Pinker, 1994: 334)

After an extended discussion of how attempts in the great apes fall short of demonstrating human language capacity, Pinker considers the views of other linguists:

> Elizabeth Bates, a vociferous critic of Chomskyan approaches to language, writes: 'If the basic structural principles of language cannot be learned (bottom up) or derived (top down), there are only two possible explanations for their existence: either universal grammar was endowed to us directly by the Creator [Wallace's explanation], or else our species has undergone a mutation of unprecedented magnitude, a cognitive equivalent of the Big Bang … '

Bates backs off from the Bickertonian scenario into the timid conclusion that the evidence for linguistic and neurological discontinuity cannot be so great as to require so drastic a solution:

> We have to abandon any strong version of the discontinuity claim that has characterized generative grammar for thirty years. We have to find some way to ground symbols and syntax in the mental material that we share with other species.

Pinker contrasts this with one of Chomsky's (1988) infrequent comments on the matter:

> Can the problem [the evolution of language] be addressed today? In fact, little is known about these matters. Evolutionary theory is informative about many things, but it has little to say, as of now, about questions of this nature. The answers may well lie not so much in the theory of natural selection as in molecular biology, in the study of what kinds of physical system can develop under the conditions of life on earth and why, ultimately because of physical principles. It surely cannot be assumed that every trait is specifically selected. In the case of such systems as language … it is not easy even to imagine a course of selection that might have given rise to them.

'What could [Chomsky] possibly mean?', Pinker (1994) asks. 'Could there be a language organ that evolved by a process different from the one we have always been told is responsible for the other organs?'

Pinker considers the alternatives to natural selection but comes up with nothing more striking than Gould & Lewontin's (1979) 'spandrels', the by-products of selection for other targets. He considers the Darwinian refutation of Paley's (1803) argument for design in the evolutionary history of the eye and compares this history with that of language. But the time–course of these developments and their respective scopes are quite different. Language is recent and a species characteristic. Photodetection is ancient and crosses the boundaries of species, orders and even phyla (Quiring *et al.*, 1994). Thus the evolution of the eye, even though it does not qualify as a 'major transition' in Maynard-Smith & Szathmary's (1995) sense, has greater significance in the history of life than the evolution of language. Although the difficulty in explaining the initial change may be as great in each case, the progressive modification of the eye by selection over hundreds of millions of years casts little light on the discontinuity in the evolution of language, that, in its extant form, must be dated at less than two hundred thousand.

Aspects of Pinker's argument fall foul of Bickerton's critique. Pinker (1994) writes:

> Language … is composed of many parts: syntax, with its discrete combinatorial system building words; a capricious lexicon; a revamped vocal tract; phonological rules and structures; speech perception; parsing algorithms; learning algorithms. Those parts are physically realized as intricately structured neural circuits, laid down by a cascade of precisely timed genetic events.

But if these component genetic events are sequential innovations, at what stage and in what order were they introduced? One must assume that they were each selected and that each on its own proved advantageous. But what were the advantages of the individual components unrelated to the other elements of language? What was the order in which the components were selected (Bickerton, 1995: 72)? One must assume that the whole cascade was completed in the transition between the protolanguage that was characteristic of *Homo erectus*, and the capacity for language that is present in modern *H. sapiens*.

In answer to Bates' question 'What protoform can we possibly envision that could have given birth to constraints on the extraction of noun phrases from an embedded clause? What could it conceivably mean for an organism to possess half a symbol, or three quarters of a rule?', Pinker (1994) writes that:

> Grammars of intermediate complexity are easy to imagine [contra Bickerton, 1995: 72]. In a recent book Derek Bickerton answers Bates even more concretely. He gives the term 'protolanguage' to chimp signing, pidgins, child language in the two-word stage, and the unsuccessful partial language acquired after the critical period by Genie and other wolf-children. Bickerton suggests that *H. erectus* spoke in protolanguage. Obviously there is still a huge gulf between these relatively crude systems and the modern adult language instinct and here Bickerton makes the jaw-dropping additional suggestion that a single mutation in a single woman, African Eve, simultaneously wired in syntax, resized and reshaped the skull, and reworked the vocal tract … [W]e can extend the first half of Bickerton's argument without accepting the second half, which is reminiscent of hurricanes assembling jetliners.

Thus Pinker, along with Bates, recoils from the prospect of a big bang mutation. But neither has any answer to the questions about the time–course or the nature of the transition from the protolanguage of *H. erectus* to modern *H. sapiens*. Nor does Bickerton himself offer any hypothesis about the nature of the brain change that he postulates as responsible not only for the innovation in function but also for the change in skull shape and configuration of the vocal tract. Curiously none of these authors considers the possibility raised by Louis Bolk (Bolk, 1926) that some step in the evolution of humans was taken by the process of neoteny, i.e. by a small change in the trajectory of development, for example a delay in the plateau of brain growth that resulted in a prolongation into adult life of some of the features, for example of the face, that are characteristic of infancy in the chimpanzee.

But in discussing Chomsky's views on the evolution of language, Pinker (1994) quotes him as follows:

> These skills [for example, learning a grammar] may well have arisen as a concomitant of structural properties of the brain that developed for other reasons. Suppose that there was selection for bigger brains, more cortical surface, *hemispheric specialization for analytic processing* [italics added], or many other structural properties that can be imagined. The brain that evolved might well have all sorts of special properties that are not individually selected; there would be no miracle in this, but only the normal workings of evolution. We have no idea, at present, how physical laws apply when 10^{10} neurons are placed in an object the size of a basketball, under the special conditions that arose during human evolution.

Indeed this is a curious passage. The first sentence invokes the concept of the spandrel. The second and last sentences postulate selective pressures for brain features that apparently were specific to *H. sapiens* but are unidentified. Pinker comments dismissively on the last sentence 'We may not, just as we don't know how physical laws apply under the special conditions of hurricanes sweeping through junkyards'.

But what this dismissal has overlooked is what (italicised in the passage) has been overlooked by Bickerton, and is so familiar to Pinker (see, for example, the passage quoted above from page 334 of Pinker, 1994) that he and many others have failed to identify it as a potential key to the problem. Language is lateralised, with some critical component being localised in most individuals in the left hemisphere, as we have known since the observations of Dax (1865) and Broca (1861). It is only when it is understood, as Marian Annett has consistently emphasised, that directional handedness on a population basis is specific to *H. sapiens* (McGrew & Marchant, 1997), that it becomes recognisable as the key to the evolution of language and the speciation of *H. sapiens*.

THE STATUS OF SALTATIONS IN EVOLUTIONARY THEORY

To take seriously a structural discontinuity as an explanation for a species difference, it is necessary to consider concepts of evolutionary change other than Darwinian gradualism and the 'biological' or 'isolation' species concept. Bickerton relates how some time between his first (1981) and second (1990) formulations of the evolutionary origins of language, he heard about the theory of punctuated equilibria (Eldredge & Gould, 1972). Penner's (2000) sustained attack on evolutionary theory, on the basis that the theory cannot account for the origins of language, lacks reference even to this modest departure from classical Darwinian orthodoxy. But there have long been challenges to the gradualist version (De Vries, 1901; Bateson, 1894; Goldschmidt, 1940), and some of these have had an explicit genetic basis. Thus White (1978) and King (1993) have argued strongly for a role for chromosomal change in speciation, but their arguments have not overwhelmed the established view. Against this view it is argued (e.g. Coyne & Orr, 1998) that radical rearrangements of the chromosome complement, for example chromosomal fusions, may apparently have few phenotypic effects, and in some cases alternative chromosomal configurations persist, as it were, as a polymorphism within species.

The case for saltational change in species transitions has been argued at a macro-evolutionary level (Stanley, 1998), that the amounts of change seen within species and other taxa are simply insufficient to account for the overall pattern of evolutionary change that is seen over time, and at the level of morphology (Schwartz, 1998) that the intermediate states in the transition between species that are required by the gradualist theory are absent. But all such general arguments come up against the difficulty, in terms of an actual genetic transition, that Goldschmidt's (1940) 'hopeful monster' ran into: the greater the magnitude of the saltational change the less likely it is to have survival value, and the greater the difficulty the hopeful monster will have in identifying a mate. The difficulty is particularly great if the change has the reproductive consequence of reducing fertility in the hybrid state. The possibility that the monster can identify an individual with the same mutation is clearly dependent on reproduction already having taken place, and even then the new mutation is at a severe statistical disadvantage with respect to the existing population.

But here Darwin's (1871) juxtaposition of *The Descent of Man* and the theory of sexual selection offers a way out. If there were some way in which sexual selection and speciation were interdependent this might be relevant both to the problem of the discontinuity of the change and to that of mate selection. The case of cerebral lateralisation in modern *H. sapiens* illustrates the possibility. All authorities on the genetics of lateralisation (Annett, 1985; McManus, 1985; Corballis, 1997) are agreed that there is a sex difference: females are more right-handed than males (although the adult male brain is probably more asymmetrical than that of the female; Bear *et al.*, 1986). The female brain grows faster than that of the male (Kretschmann *et al.*, 1979) and females have greater mean verbal fluency and acquire words earlier (Maccoby & Jacklin, 1975; McGlone, 1980; Halpern, 1992) than males. If language is the species characteristic and lateralisation is the process by which it evolved, these facts are related, and they tell us about the nature of the genetic mechanism. Only two explanations of the sex difference in lateralisation are conceivable, that it is hormonal in origin (Geschwind & Galaburda, 1985), or that it reflects a sex chromosomal locus (Crow, 1993, 1994), and the facts of sex chromosomal aneuploidies (XXY and XXX individuals who differ in hormonal status have similar hemispheric deviations in development) speak decisively in favour of the latter interpretation. The hypothesis that the asymmetry factor is present on both X and Y chromosomes (Crow, 1993; Corballis *et al.*, 1996) can explain the transmission of handedness within families and apparent dosage effects in the aneuploidies. That there are problems (Corballis, 1997) in accounting for persisting variation in males and females in terms of conventional polymorphisms and heterozygote advantage explanations, should not dissuade us from pursuing the line of thought. The genetic principles involved may not be those on which we have hitherto relied.

The paradigm of *H. sapiens* therefore suggests a new version of saltational speciation, that it is not chromosomal changes in general that play a role in speciation but changes on the sex chromosomes, and perhaps particularly changes in regions of X–Y homology that are involved. These regions have a special status because they can account (as in the case of lateralisation in humans) for quantitative differences in a characteristic in males and females, and such quantitative differences are a potential substrate for sexual selection. The Y chromosome itself has a unique role, because it is not necessary for survival. There are interindividual differences on the Y (reviewed here by Chris Tyler-Smith) but there are also large interspecific differences. The Y therefore can be seen as a test-bed of evolutionary change. One possibility is that the primary changes in speciation take place on the Y, but that when they are located in regions of homology with the X there is the possibility of correlated but independent change in the two sexes. Such correlated but quantitatively differing ranges of variation have the potential to explain the type of runaway sexual selection envisaged by Fisher (1958), and this may be what occurred with respect to cerebral asymmetry at some point in hominid evolution (Crow, 1998a, 1998b); the introduction of the dimension of symmetry–asymmetry allowed brain growth to equilibrate at a new point of plateau, and this equilibration took place around successive modifications on the Y and then on the X chromosome.

There is thus a potential three-way relationship between sexual selection, sex linkage and speciation, in which the pattern suggested by hominid evolution is backed up in the recent literature relating to other species. A role for sexual selection in modifying a primary change in a sexually dimorphic feature to establish a new species boundary has been argued in relation to Hawaiian Drosophilid species by Kaneshiro (1980) and Carson (1997). Similar arguments apply in the case of the prolific speciation of cichlid fishes in the lakes of East Africa (Dominey, 1984; McKaye, 1991) and may also apply in birds (Price, 1998). Some putative speciation loci, for example the Odysseus homoeobox (Ting *et al.*, 1998) and the *per* gene (Ritchie & Kyriacou, 1994) that have been identified in *Drosophila* species, are X-linked. In discussing the relationship between the X chromosome and speciation that she finds in Lepidoptera, Prowell (1998) offers three explanations: (1) that X-linked traits evolve faster, (2) that traits related to speciation tend to be sex-limited, and that sex-limited traits tend to be on the sex chromosomes, and (3) that female-limited X-linked traits undergo faster rates of evolution when, as in the case of Lepidoptera, the female is heterogametic. These explanations are not mutually exclusive. Prowell asks whether the X chromosome bias is unique to Lepidoptera and concludes that none of these explanations is likely to be limited to this order. In discussing Haldane's rule, that when, in a species hybrid, one sex is sterile or inviable it is the heterogametic sex, Coyne & Orr (1998) consider various explanations including faster evolution and recessivity of genes on the X chromosome. While each of these observations and hypotheses is consistent with the generalisation that there is a relationship between speciation and the sex chromosomes, none of the authors considers the more restrictive formulation suggested by the sequence of events (Sargent *et al.*, 1996, 2001; described in these *Proceedings* by Carole Sargent *et al.*) on the mammalian Y chromosome, that it is the interaction between the sex chromosomes, particularly the possibility of transfer of material between them, that is critical in speciation. In this volume Klaus Reinhold considers the case that sexual selection acts selectively on sexual dimorphisms that relate to sex-linked genes, as suggested by Rice (1984).

The sequence of events, including a translocation and a paracentric inversion, suggested by the work summarised by Carole Sargent *et al.* and by the X–Y hypothesis as relevant to the course of hominid evolution, carries the further implication that epigenetic modification is involved in the process of sexual selection and speciation. In mammals genes on one X chromosome are subject to the process of X inactivation, but those gene sequences that are also represented on the Y chromosome are protected from this influence. Such genes are expressed from both X and Y in males and from both Xs in females, a similar dosage thus being maintained in each sex. The mechanism by which this protection is achieved is unknown (Burgoyne & McLaren, 1985; Crow, 1991). Gene sequences that have been transferred from the X to the Y are in a new situation; whatever the mechanism a phase of epigenetic equilibration must be assumed (Jegalian & Page, 1998). If X–Y pairing in male meiosis plays a role, the orientation of the sequence on the Y is also relevant. The paracentric inversion on the Y short arm could be critical.

A magnetic resonance imaging (MRI) investigation in monozygotic twins of handedness and asymmetry of the planum temporale (Steinmetz *et al.*, 1995) indicates that there is room for an epigenetic influence in the determination of cerebral asymmetry, and this may account for the stochastic element that is incorporated in genetic theories (Annett, 1985; McManus, 1985). There is a possibility, therefore, that the genetic mechanisms underlying the development of cerebral asymmetry in humans are a paradigm for a more general interaction between genetic and epigenetic mechanisms in sexual selection and speciation. One can contemplate the hypothesis that sexual selection and natural selection are mediated by distinct but complementary genetic processes, that natural selection reflects the response that ensures the organism's survival, of any part of the genome to environmental change, whereas sexual selection reflects the sequential response of the female genome (for example the mammalian XX complement) to change on the Y chromosome, and that this process involves particularly the epigenetic modulation of genes on the X. According to this concept speciation follows the history of the non-recombining sex chromosome, in mammals the Y.

CONTRIBUTIONS TO THIS VOLUME

These *Proceedings* reflect an attempt to cross the interdisciplinary boundaries between archaeology, palaeontology, neuropsychology, neuroanatomy, genetics and evolutionary theory to approach the problem of the origins of modern *H. sapiens*. Paul Mellars outlines the archaeological evidence (see also Mellars, 1989, 1998) for the relative recency and abruptness of appearance of artefacts associated with the creativity of modern humans. Chris Stringer gives an update on the speciation of modern *H. sapiens* and the Out of Africa hypothesis, for which he has been so effective an advocate (Stringer & McKie, 1996), and Ian Tattersall argues the case that he has previously developed (Tattersall, 1998), that the speciation of modern *H. sapiens* exemplifies the principle of punctuated equilibria, i.e. it is a saltational change. Mark Collard places this discussion within the framework of the present evidence on the longer course and diversity of hominid evolution. These contributions are followed by discussions of the singularity of human language by Derek Bickerton, and of the relevance of cerebral asymmetry by Michael Corballis. In each case they outline their own concepts of the selective advantages of the component functions that they regard as critical. Both authors modify their previous formulations; interestingly neither sees the critical steps as necessitating genetic discontinuities as strongly as some other contributors. James Steele reviews the palaeontological and archaeological evidence that allows us to address the question: when did directional asymmetry enter the record of hominid evolution? The key question here (p. 154) is whether or not directional asymmetry is a derived feature, and if so at what stage?

Detlev Ploog has accumulated a body of evidence on the neuroanatomical basis of language and on the differences between this and the substrate of communicative ability in other primates. These differences require a genetic explanation, and that genetic explanation must be relevant to the nature of species transitions. In amplifying the human case, Norman Cook has built on the arguments that he first deployed in *The Brain Code* (Cook, 1986), that the facts of human brain evolution require that language is represented in both hemispheres and that callosal connectivity is critical to their integration. These arguments are echoed in the recent literature on the role of the right hemisphere in language (Coney & Evans, 2000). When these hemispheric differentiations are recognised as dependent upon language as the speciation characteristic, it will be appreciated, as Chomsky argued, that human psychology has less in common with that of other primates and mammals than is often assumed. New neuroanatomical and neurophysiological principles have been introduced at at least one hominid species boundary.

The peculiar genetic history and population characteristics of the Y chromosome are outlined by Chris Tyler-Smith, including the interaction with the X. The small size of the Y and its sex-limited transmission make it at first sight an unlikely vehicle for the determining characteristic of the species. In addition, as Corballis (1997) has argued, there are problems in assuming that a conventional polymorphism will be maintained on the Y and could account for significant and persisting variation in the population. These problems need to be overcome if the X–Y hypothesis of human origins is to be sustained. Sargent *et al.* (1996, 2001) have summarised their own evidence and that of others, that allows one to identify those regions of the Y that may have particular significance in human evolution, including the Xq21.3/Yp11.2 region of homology, following translocation and subsequent paracentric inversion. The identification within this region of protocadherinXY, a gene that codes for a cell-surface adhesion molecule that is expressed in the brain, rescues the theory from the sudden death that would have followed the demonstration that no gene, or a gene that was unexpressed in the brain, was present in the region. I have outlined the theory together with its implications as I see them. Klaus Reinhold discusses evidence and theory on sex chromosomal linkage of sexually selected traits that may be the key to a functional separation of sexual and natural selection.

References

Annett, M. (1985) *Left, Right, Hand and Brain: The Right Shift Theory.* London: Erlbaum.

Bateson, W. (1894) *Materials for the Study of Variation*. New York: Macmillan.

Bear, D.M., Schiff, D., Saver, J., Greenberg, M. & Freeman, R. (1986) Quantitative analysis of cerebral asymmetry; fronto-occipital correlation, sexual dimorphism and association with handedness. *Archives of Neurology*, **43**, 598–603.

Bickerton, D. (1981) *The Roots of Language*. Ann Arbor: Karoma.

Bickerton, D. (1990) *Language and Species*. Chicago: University of Chicago.

Bickerton, D. (1995) *Language and Human Behavior*. Seattle: University of Washington.

Bolk, L. (1926) *Das Problem der Menschwerdung.* Jena: Gustav Fischer.

Broca, P. (1861) Remarques sur la siegé de la faculté du langue. *Bulletin de la Société Anatomique de Paris (2nd Series)*, **6**, 330–57.

Burgoyne, P.S. & McLaren, A. (1985) Does X–Y pairing during male meiosis protect the paired region of the X chromosome from subsequent X-inactivation? *Human Genetics*, **70**, 82–3.

Carson, H.L. (1997) Sexual selection: a driver of genetic change in Hawaiian *Drosophila*. *Journal of Heredity*, **88**, 343–52.

Chomsky, N. (1959) A review of B.F. Skinner's verbal behavior. *Language*, **35**, 26–58.

Chomsky, N. (1988) *Language and the Problems of Knowledge: The Managua Lectures.* Massachusetts: MIT Press.

Coney, J. & Evans, K.D. (2000) Hemispheric asymmetries in the resolution of lexical ambiguity. *Neuropsychologia*, **38**, 272–82.

Cook, N.D. (1986) *The Brain Code: Mechanisms for Information Transfer and the Role of the Corpus Callosum*. London: Methuen.

Corballis, M.C. (1997) The genetics and evolution of handedness. *Psychological Review*, **104**, 714–27.

Corballis, M.C., Lee, K., McManus, I.C. & Crow, T.J. (1996) Location of the handedness gene on the X and Y chromosomes. *American Journal of Medical Genetics (Neuropsychiatric Genetics)*, **67**, 50–2.

Coyne, J.A. & Orr, H.A. (1998) The evolutionary genetics of speciation. *Philosophical Transactions of the Biological Society of London*, **353**, 287–309.

Cronin, H. (1991) *The Ant and the Peacock*. Cambridge: Cambridge University Press.

Crow, T.J. (1991) Protection from X inactivation. *Nature*, **353**, 710.

Crow, T.J. (1993) Sexual selection, Machiavellian intelligence and the origins of psychosis. *Lancet*, **342**, 594–8.

Crow, T.J. (1994) The case for an X–Y homologous determinant of cerebral asymmetry. *Cytogenetics and Cell Genetics*, **67**, 393–4.

Crow, T.J. (1998a) Sexual selection, timing and the descent of man: a genetic theory of the evolution of language. *Current Psychology of Cognition*, **17**, 1079–114.

Crow, T.J. (1998b) Why cerebral asymmetry is the key to the origin of *Homo sapiens*: how to find the gene or eliminate the theory. *Current Psychology of Cognition*, **17**, 1237–77.

Darwin, C. (1859) *On the Origin of Species by Means of Natural Selection: Or, the Preservation of Favoured Races in the Struggle for Life.* London: John Murray.

Darwin, C. (1871) *The Descent of Man and Selection in Relation to Sex.* London: John Murray [New Jersey: Princeton University Press: 1981 facsimile of original]

Dax, M. (1865) Lésions de la moitié gauche de l'éncephale coincident avec l'oubli des signes de la pensée [Read at congrés méridional at Montpelier in 1836]. *Gazette Hebdom Méd Chirurg*, **11**, 259–60.

De Vries, H. (1901) *Die Mutationstheorie*. Leipzig: Verlag von Veit.

Dominey, W.J. (1984) Effects of sexual selection and life histories on speciation: species flocks in African cichlids and Hawaiian *Drosophila*. In: *Evolution of Fish Species Flocks* (eds A. A. Echelle & I. Kornfield), pp. 231–49. Maine: Orino Press.

Eldredge, N. & Gould, S.J. (1972) Punctuated equilibria: an alternative to phyletic gradualism. In: *Models in Palaeobiology* (ed. T. M. Schopf), pp. 82–115. San Francisco: Freeman Cooper.

Fisher, R.A. (1958) *The Genetical Theory of Natural Selection*, 2nd Edn. NY: Dover.

Gamble, C. (1993) *Timewalkers: The Prehistory of Global Colonization*. Bath: Sutton.

Geschwind, N. & Galaburda, A.M. (1985) Cerebral lateralization. Biological mechanisms, associations and pathology: a hypothesis and a program for research. *Archives of Neurology*, **42**, 428–654.

Goldschmidt, R. (1940) *The Material Basis of Evolution*. New Haven: Yale University Press [reprinted 1982].

Gould, S.J. & Lewontin, R.C. (1979) The spandrels of San Marco and the Panglossian program: a critique of the adaptationist programme. *Proceedings of the Royal Society of London*, **205**, 281–8.

Halpern, D.F. (1992) *Sex Differences in Cognitive Abilities*. New Jersey: Erlbaum.

Harris, R. & Talbot, T.J. (1997) *Landmarks in Linguistic Thought*. London: Routledge.

Hughlings Jackson, J. (1868) Observations on the physiology of language. *British Medical Journal*, **ii**, 259.

Hume, D. (1739) *A Treatise on Human Nature*. 1978 Reprint. Oxford: Clarendon Press.

Huxley, T.H. (1863) *Evidence of Man's Place in Nature*. London: Williams & Norgate.

Jegalian, K. & Page, D.C. (1998) A proposed mechanism by which genes common to mammalian X and Y chromosomes evolve to become X inactivated. *Nature*, **394**, 776–80.

Kaneshiro, K.Y. (1980) Sexual isolation, speciation and the direction of evolution. *Evolution*, **34**, 437–44.

Kant, I. (1781) *Immanuel Kant's Critique of Pure Reason*. Translated by Norman Kemp Smith 1963. London: Macmillan.

Kant, I. (1788) *Critique of Practical Reason*. Translated by Lewis White Beck 1956. Indianapolis: Library of Liberal Arts.

King, M. (1993) *Species Evolution. The Role of Chromosome Change.* Cambridge: Cambridge University Press.

Kretschmann, H.F., Schleicher, A., Wingert, F., Zilles, K. & Loeblich, H.-J. (1979) Human brain growth in the 19th and 20th century. *Journal of the Neurological Sciences*, **40**, 169–88.

Locke, J. (1690) *An Essay Concerning Human Understanding*. Reprinted 1975. Oxford: OUP.

Maccoby, E.E. & Jacklin, C.N. (1975) *The Psychology of Sex Differences*. Oxford: Oxford University Press.

McGlone, J. (1980) Sex differences in human brain asymmetry: a critical survey. *Behavioral and Brain Sciences*, **3**, 215–63.

McGrew, W.C. & Marchant, L.F. (1997) On the other hand: current issues in and meta-analysis of the behavioral laterality of hand function in nonhuman primates. *Yearbook of Physical Anthropology*, **40**, 201–32.

McKaye, K.R. (1991) Sexual selection and the evolution of the cichlid fishes of Lake Malawi, Africa. In: *Behaviour, Ecology and Evolution* (ed. M. H. A. Keenleyside), pp. 241–57. London: Chapman & Hall.

McManus, I.C. (1985) Handedness, language dominance and aphasia: a genetic model. *Psychological Medicine Monograph Supplement*, **8**, 1–40.

Maddox, J. (1997) The price of language? *Nature*, **388**, 424–5.

Maynard-Smith, J. (1993) *The Theory of Evolution*. Cambridge: Canto.

Maynard-Smith, J. & Szathmary, E. (1995) *The Major Transitions in Evolution.* Oxford: W. H. Freeman.

Mayr, E. (1963) *Animal Species and Evolution.* Massachusetts: Harvard University Press.

Mellars, P. (1989) Technological changes at the middle-upper palaeolithic transition: economic, social and cognitive changes. In: *The Human Revolution: Behavioural and Biological Perspectives in the Origins of Modern Humans* (eds P. Mellars & C. Stringer), pp. 338–65. Edinburgh: Edinburgh University Press.

Mellars, P. (1998) Modern humans, language, and the 'symbolic explosion'. *Cambridge Archaeological Journal*, **8**, 88–94.

Muller, F.M. (1996) Lectures on Mr Darwin's philosophy of language in Fraser's Magazine volumes 7 and 8, 1873. In: *The Origin of Language* (ed. R. Harris), pp. 147–233. Bristol: Thoemmes Press.

Paley, W. (1803) *Natural Theology*, 5th Edn. London: R. Faulder.

Penner, J.G. (2000) *Evolution Challenged by Language and Speech*. London: Minerva.

Pinker, S. (1994) *The Language Instinct*. Allen Lane: London.

Price, T. (1998) Sexual selection and natural selection in bird speciation. *Philosophical Transactions of the Royal Society of London (Biological)*, **353**, 251–60.

Prowell, D.P. (1998) Sex linkage and speciation in Lepidoptera. In: *Endless Forms* (eds S. H. Howard & S. H. Berlocher), pp. 309–19. Oxford: Oxford University Press.

Quiring, R.U., Walldorf, U., Kloter, U. & Gehring, W.J. (1994) Homology of the eyeless gene of *Drosophila* to the small eye gene in mice and Aniridia in humans. *Science*, **265**, 785–9.

Rice, W.R. (1984) Sex chromosomes and the evolution of sexual dimorphism. *Evolution*, **38**, 735–42.

Ritchie, M.G. & Kyriacou, C.P. (1994) Reproductive isolation and the period gene of *Drosophila*. *Molecular Ecology*, **3**, 595–9.

Sargent, C.A., Boucher, C.A., Blanco, P. *et al.* (2001) Characterization of the human Xq21.3/Yp11 homology block and conservation of organization in primates. *Genomics*, **73**, 77–85.

Sargent, C.A., Briggs, H., Chalmers, I.J., Lambson, B., Walker, E. & Affara, N.A. (1996) The sequence organization of Yp/proximal Xq homologous regions of the human sex chromosomes is highly conserved. *Genomics*, **32**, 200–9.

de Saussure, F. (1916) *Course in General Linguistics*. Paris: Payot [Illinois: Open Court, translated by R. Harris, 1983].

Schwartz, J.H. (1998) *Sudden Origins: Fossils, Genes and the Emergence of Species.* New York: J. Wiley.

Stanley, S.M. (1998) *Macroevolution: Patterns and Processes*. Baltimore: J. Hopkins.

Steinmetz, H., Herzog, A., Schlaug, G., Huang, Y. & Jancke, L. (1995) Brain (a)symmetry in monozygotic twins. *Cerebral Cortex*, **5**, 296–300.

Stringer, C. & McKie, R. (1996) *African Exodus: The Origins of Modern Humanity*. London: J. Cape.

Tattersall, I. (1998) *Evolution and Human Uniqueness*. Oxford: Oxford University Press.

Taylor, J. (1932) *Selected Writings of John Hughlings Jackson, 2 Volumes*. London: Hodder & Stoughton.

Ting, C.T., Tsaur, S.C., Wu, M.L. & Wu, C.-I. (1998) A rapidly evolving homoeobox at the site of a hybrid sterility gene. *Science*, **282**, 1501–4.

White, M.J.D. (1978) *Modes of Speciation*. San Francisco: W. H. Freeman.

I

THE ORIGIN OF THE SPECIES

Speciation, migration and symbolic representation

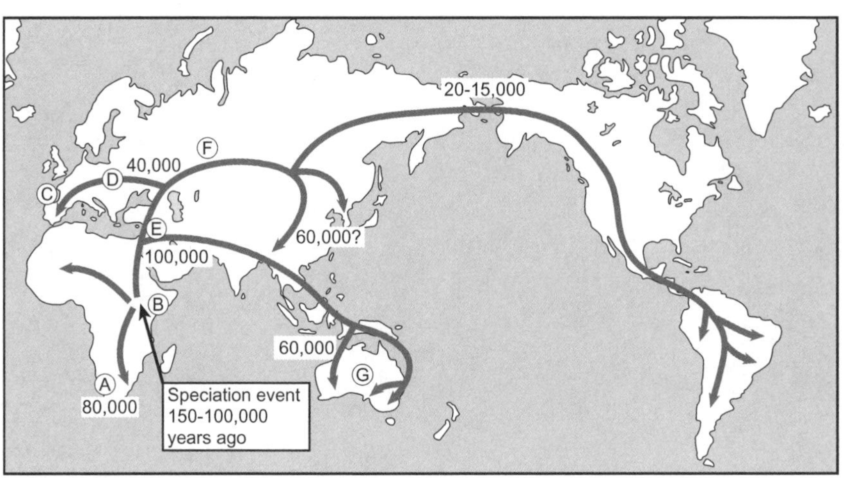

Ⓐ South Africa: Blombos cave (red ochre engraving 80,000 BP)
Klasies River (blade technology 70,000 BP)
Border cave (engraved bone and wood 36,000 BP)

Ⓑ Tanzania: Mumba (LSA industry 50,000 BP)
Kenya: Enkapune ya Muto (LSA industry 50,000 BP; perforated beads 40,000 BP)

Ⓒ France: Cromagnon (decorative grave goods 31,000 BP)
Cave paintings: Chauvet (31,000 BP), Lascaux (17,000 BP),
Pech-merle (22,000 BP)
Spain: Cueva Morin (grave goods 37,000 BP)
Altamira (cave paintings 18,000 BP)

Ⓓ Czech Republic: Mladec (ceremonial burial 18,000 BP)
Dolni V stonice (engraved mammoth tusk 26,000 BP)

Ⓔ Israel: Skhul, Qafzeh (symbolic grave goods 110-90,000 BP)

Ⓕ Russia and Ukraine: Sungir (painted ivory pendant 28,000 BP)
Kostenki (drilled beads 36,000 BP)

Ⓖ Australia: Kakadu (rock shelter art 40,000 BP)
Lake Mungo (ceremonial red ochre burials 60,000 BP)

BP = Years Before Present LSA = Late Stone Age

The Morphological and Behavioural Origins of Modern Humans

CHRIS STRINGER

Summary. The majority of the fossil and genetic evidence favours an African origin for modern humans during the later part of the Middle Pleistocene (prior to 130,000 years ago), and one or more range expansions out of Africa after that date. However, a number of uncertainties remain. If there was a speciation event at the appearance of modern humans, what was its nature? Furthermore, did the evolution of modern human behaviour occur gradually or punctuationally? In this chapter, I will examine the difficulties faced in defining what is meant by 'modern' humans, and in reconstructing the morphological and behavioural origins of our species.

INTRODUCTION

THIS IS A GOOD TIME to be writing about the morphological and behavioural origins of *Homo sapiens*. Although the Late Pleistocene fossil human record has not been greatly extended over the last few years, provocative new interpretations of it have appeared, as well as new or revised dating of important evidence. In addition, genetic data from both recent and fossil humans are allowing increasingly detailed reconstructions of early human history. The archaeological record has not expanded markedly, but key discoveries have focused debate on critical areas concerning the capabilities of Middle Palaeolithic humans, and on the concept of a behavioural 'human revolution' at the Middle–Upper Palaeolithic transition (Holden, 1998).

It is now generally accepted that modern humans had a recent African origin (Stringer, 2001a). The Multiregional Model, under which *H. sapiens* evolved across the inhabited Old World throughout the Pleistocene (Thorne & Wolpoff, 1992), has now given way to variants of Out of Africa models (Stringer, 1994). However, these have differing time-scales for the origin and dispersal of *H. sapiens*, and varying scenarios of replacement or gene flow during dispersal phase(s) outside of Africa. In the rest of this chapter I will assume that the morphological and behavioural origins of modern humans were African, but will review recent data and the current debate about the tempo and mode of that origin. In particular I will address the following issues.

1 What is meant by 'modern humans'?
2 How has the application of new dating techniques affected the debate?
3 Was speciation in humans gradual or punctuational?
4 When did modern human behaviour evolve?

WHAT IS MEANT BY 'MODERN' HUMANS?

No agreement exists among palaeoanthropologists on how to recognise ancient examples of 'modern' humans, morphologically or behaviourally. I previously favoured the use of recent skeletal variation to diagnose whether a fossil could be termed 'modern' (Stringer, 1994), but it is now apparent that modern skeletal variation is smaller than that recognised for *H. sapiens* in even the Late Pleistocene, and members of the *H. sapiens* clade in the African Late–Middle to Early–Late Pleistocene were even more distinct and diverse (Stringer, 1992; Lahr, 1996). While there seems little doubt that Aurignacian- and Gravettian-associated humans from over 25,000 years ago in Europe share enough morphological and behavioural features with recent populations to warrant the application of the term 'modern', problems soon arise as we move further back in time. The samples from Skhul and Qafzeh in Israel appear to represent a primitive form of *H. sapiens* (Stringer, 1992; Lahr, 1996). However, they are associated with Middle Palaeolithic artefacts, comparable with those made by Neanderthals, and with only disputed evidence of 'modern' symbolic behaviour. The contrast between their morphology and their inferred behaviour is sufficient for Klein (2000) to employ the term 'near-modern' for them, implying that they represent an evolutionary stage where modern anatomy was evolving *prior* to truly modern behaviour. In the case of the late Neanderthals of south western France, dating from about 35,000 years ago, an inversion of that situation has been posited by d'Errico *et al.* (1998). They argue that these Neanderthals were developing 'modern' symbolic behaviour independently of *H. sapiens*, thus producing a contrasting decoupling of modern anatomy and behaviour from that envisaged by Klein, since the Neanderthals were apparently evolving aspects of 'modern' behaviour separately from the appearance of 'modern' anatomy.

It seems to me that these different ideas, whether ultimately accurate or not, are important for the way that they highlight difficulties inherent in any absolute concept of 'modernity'. Was modernity (morphological and/or behavioural) a package that had a unique African origin in one time, place and population, or was it a composite whose elements appeared at different times and places, and were then gradually assembled to assume the form we recognise today? In the rest of this chapter I will review the evidence that leads me to favour the second alternative, one which does, however, bring with it other difficulties.

HOW HAS THE APPLICATION OF NEW DATING TECHNIQUES AFFECTED THE DEBATE?

The majority of the fossil human record can still only be relatively dated. However, over the last 15 years new or improved physical dating techniques have allowed better age estimates for sites that were previously at, or beyond, the limits of radiocarbon dating (approximately 35,000 years) (Stringer, 2001b). When I began my doctoral studies some 30 years ago, Europe and the Americas were the only continents that could be said to have reasonably accurate time-scales for the appearance of modern humans. The Neanderthal–modern interface in Europe appeared to occur at about 35,000 years, while in the Middle East it was estimated to be only slightly older. The time-scale for the appearance of modern humans was poorly known in eastern Asia and Australasia, perhaps occurring at less than 30,000 years. Africa, although central to hominid origins, was considered to have lagged behind regions such as Europe during later human evolution. Now, however, African fossils such as Florisbad, Singa and Guomde can be seen more appropriately as relics of the early evolution of the modern human clade, since they all probably date beyond 130,000 years (Grün *et al.*, 1996; McDermott *et al.*, 1996; Bräuer *et al.*, 1997). In the Middle East, the Skhul and Qafzeh samples of early modern (or near-modern for Klein, 2000) humans probably date to more than 100,000 years, while anatomically modern humans were apparently even present in south-eastern Australia prior to 60,000 years (Thorne *et al.*, 1999; Grün *et al.*, 2000). Thus Europe can no longer be seen as crucial for an understanding of the early evolution of *H. sapiens*, although it is clearly important for what it reveals about the extent of human behavioural complexity 30,000 years ago.

WAS SPECIATION IN HUMANS GRADUAL OR PUNCTUATIONAL?

There is no agreement about the number of human species that have existed through the Pleistocene. For some workers (multiregionalists) there may have been only one: *H. sapiens.* For others, there may have been at least eight. My preference lies between these extremes, and for the rest of this chapter I will use four species names: *Homo erectus*, its probable descendant *Homo heidelbergensis*, and the two probable descendant species of *heidelbergensis*, *Homo neanderthalensis* and *H. sapiens*. However, it should be recognised that any discussion of speciation processes in humans presupposes a workable species recognition concept for the fossils that will allow speciation events to be examined. An additional complication is that some workers confuse different species concepts; for example, some multiregionalists insist on applying biological species concepts in an attempt to show that *H. neanderthalensis* and *H. sapiens* must have been conspecific. However, even if we accept controversial claims for the existence of supposed Neanderthal–modern hybrids (for example Duarte *et al.*, 1999), it is well known that many closely related mammal species, including primates, can hybridise, and may even produce fertile offspring. But if this is not a widespread behaviour, it may have little or no impact on the populations that constitute the core of the different species. Thus in fossils, morphological criteria are necessarily the mode of species recognition.

However, some genetic data are now available from Neanderthal specimens (Krings *et al.*, 2000), and these can be compared with the fossil record. Both support the idea of a separation between the Neanderthal and modern human lineages during the Middle Pleistocene, and both suggest that Neanderthal–modern human differences were of the order of two or three times that found within modern humans. But even in this case, where morphological differences are clear-cut, the genetic data can be used to support either a conspecific or specific difference between Neanderthals and modern humans.

If we assume that the Neanderthal–*sapiens* separation was a specific one occurring in the middle part of the Middle Pleistocene, what can we say about the nature of the origins of these species? The European fossil record of this period can now be interpreted as showing a gradual accretion of Neanderthal characteristics. This is best exemplified in the rich fossil sample from the Sima de los Huesos at Atapuerca, dating from about 300,000 years ago (Arsuaga *et al.*, 1997). Individual specimens show mosaic *heidelbergensis* and *neanderthalensis* characters, and the sample as a whole can be interpreted as a derived form of *heidelbergensis* or a primitive form of *neanderthalensis*.

The relatively rich and well-dated European Middle Pleistocene record thus appears to demonstrate the gradual, local, nature of Neanderthal evolution. If this model of gradual, regional, evolution can also be applied to the African fossil record, an accretional mode of *sapiens* evolution would consequently be expected (Stringer, 1998). However, acceptance of a gradualistic scenario for the origin of modernity means that diagnosing 'modernity' will be dependant on the particular criteria selected. Additionally, while individual anatomical characters may be used to recognise which fossils belong to the *sapiens* clade, membership of this clade will not necessarily be synonymous with modernity as an assemblage, since this may have evolved long after the cladistic origin of *H. sapiens* (which, in my view, was at the *neanderthalensis*–*sapiens* cladogenetic event). Thus fossils such as Skhul, Qafzeh, and even those from Singa and Florisbad, probably belong to *H. sapiens* cladistically, but do not necessarily represent 'modern' humans.

WHEN DID MODERN HUMAN BEHAVIOUR EVOLVE?

If the characteristic morphology of modern humans evolved in a gradual, mosaic, fashion, what of modern human behaviour? The concept of a 'human revolution', demarcating a punctuational origin of a package of modern human behaviours, such as complex language, symbolism and specialised technologies, has been central to much archaeological debate over the last 10 years (Klein, 2000). Originally focused on apparent contrasts between the Middle and Upper Palaeolithic records in Europe, this concept has now been extended to the Middle Stone Age–Later Stone Age transition in Africa. It is argued that the major changes in the whole of human behavioural evolution occurred there about 50,000 years ago (possibly related to changes in cognition or language). In turn, this led to the successful expansion of modern humans and now-modern behaviour beyond Africa, and the replacement of the remaining archaic populations. Thus morphological and behavioural evolution were decoupled, and 'morphological modernity' may have evolved before 'behavioural modernity'. This pattern is counterintuitive for those who argue that behavioural change lay behind the transformation of the archaic skeletal pattern into that of modern humans.

However, McBrearty & Brooks (2000) have argued that this view of events displays a Eurocentric bias and a failure to appreciate the depth and breadth of the African Middle Stone Age record that precedes the supposed 'human revolution' by at least 100,000 years. In their view, 'modern' features such as advanced technologies, increased geographic range, specialised hunting, aquatic resource exploitation, long-distance trade and the symbolic use of pigments, occur across a broad spectrum of Middle Stone Age industries. This suggests to them a gradual assembly of the package of modern human behaviours in Africa during the Late Middle–Early Late Pleistocene, and its later export to the rest of the world. Thus the origin of our species, behaviourally and morphologically, was linked with the appearance of Middle Stone Age technology, dated in many parts of Africa to more than 250,000 years ago.

CONCLUSIONS

It is still too early to determine definitively when and where 'modern' morphology and behaviour developed, especially when these concepts are apparently also so fluid. In my view, Africa was the ultimate source of the basic elements of both our anatomy and our behaviour. But it has also become evident that some supposedly unique attributes of modern human behaviours were present even in the Lower Palaeolithic outside of Africa (witness the evidence for systematic hunting of large mammals from sites such as Boxgrove and Schöningen). And the debate about Neanderthal, and specifically Châtelperronian, capabilities highlights the issue of potential versus performance. Were the Neanderthals developing complex behaviours independently of modern humans, or only because of contact with them (compare d'Errico *et al.*, 1998 with Mellars, 1999)? If we could bring up a Neanderthal child in a modern human society, could it achieve what we achieve, or would it be limited by its genetic and developmental endowment? Did behavioural innovations regularly and independently arise in different populations in human prehistory, but were often lost during population crises or extinctions, or did such innovations diffuse widely, even between distinct populations or species? Was the apparently unique role of Africa in modern human origins a result of a particular evolutionary pathway, or more a consequence of larger population size (Relethford & Jorde, 1999), with less bottlenecking, more continuity, and a greater potential both to make and to accumulate behavioural innovations?

While the (admittedly limited) evidence seems to point to a gradual assembly of modern human morphology and behaviour in Africa during the period from 300 to 100,000 years ago, rather than major punctuational events, genetic data continue to suggest that this may be too simple a story. A number of different genetic data sets suggest that there were major population bottlenecks during this period of time (Jorde *et al.*, 2000; Ingman *et al.*, 2000), with effective population size reduced to only a few thousand individuals. Such population crashes might indeed have produced saltational changes in morphology and behaviour within what must have been a very diverse *sapiens* clade. It will be exciting to see the evidence developing for or against these scenarios during this new century (for discussion of some new data see Balter, 2002).

DISCUSSION

Questioner: Why did they leave Africa?

Stringer: We don't know. Perhaps it was climatic and environmental changes, or population pressure, leading to range expansions. Global climate was highly variable during the later Pleistocene, over both long and short time-scales. Another interesting possibility is that the use of marine resources may be part of the reason, or at least provided new opportunities. There is growing evidence that humans (both *H. sapiens* and Neanderthals) were adapted or adapting to coastal life during the later Pleistocene. Early modern populations could have dispersed along littoral zones from North East Africa at times of Late Pleistocene low sea levels, taking them all the way to Indonesia. Australia would then have been within reach, as well as inland colonisations up river valleys.

Questioner: When did modern humans get to Australia?

Stringer: Probably at least 60,000 years ago, based on new dating for the Mungo 3 burial site and for archaeological sites in northern Australia. Sea-going craft would have been needed for repeated island hops, so boats must have existed by this time, despite the lack of direct evidence. The Mungo 3 burial is associated with the use of red ochre (possibly ceremonial). If the new datings are accurate, this would be the oldest known burial with red ochre, about double the age of examples from Europe. It is possible to argue that these early colonisers were not the ancestors of today's Aborigines, and that fully modern human behaviour arrived later, but in my view the earliest Australians were probably essentially 'modern', both anatomically and behaviourally.

Questioner: What is your view of evidence for the peopling of the island of Flores, 800,000 years ago?

Stringer: If the archaeological interpretations are accurate, primitive water craft were apparently used to get there, but the distances involved may have been much less than today. We know little of the exact palaeogeography at that time. I think the move to Australia would have been a much bigger jump, and I think that was only achieved by modern humans.

Questioner: Was *H. erectus* in Java?

Stringer: Yes, *H. erectus* was definitely in Java throughout much of the Pleistocene, and may have even overlapped with the arrival of *H. sapiens* in the region. There were regular land connections between South East Asia and Java during the Pleistocene, and numerous fossils of *erectus* have been discovered there.

References

Arsuaga, J.L., Bermudez de Castro, J.M. & Carbonell, E. (1997) The Sima de los Huesos hominid site. *Journal of Human Evolution*, **33**, 105–421 [whole volume].

Balter, M. (2002) What made humans modern? *Science*, **295**, 1219–25.

Bräuer, G., Yokoyama, Y., Falgueres, C. & Mbua, E. (1997) Modern human origins backdated. *Nature*, **386**, 337–8.

Duarte, C., Maurício, J., Pettitt, P.B., Souto, P., Trinkaus, E., van der Plicht, H. & Zilhão, J. (1999) The early Upper Paleolithic human skeleton from the Abrigo do Lagar Velho (Portugal) and modern human emergence in Iberia. *Proceedings of the National Academy of Sciences of the USA*, **96**, 7604–9.

d'Errico, F., Zilhão, J., Julien, M., Baffier, D. & Pelegrin, J. (1998) Neanderthal acculturation in western Europe? A critical review of the evidence and its interpretation. *Current Anthropology*, **39** (Supplement), S1–44.

Grün, R., Brink, J., Spooner, N., Taylor, L., Stringer, C., Franciscus, R. & Murray, A. (1996) Direct dating of Florisbad hominid. *Nature*, **382**, 500–1.

Grün, R., Spooner, N.A., Thorne, A., Mortimer, G., Simpson, J.J., McCulloch, M.T., Taylor, L. & Curnoe, D. (2000) Age of the Lake Mungo 3 skeleton, reply to Bowler and Magee and to Gillespie and Roberts. *Journal of Human Evolution*, **38**, 733–41.

Holden, C. (1998) No last word on language origins. *Science*, **28**, 1455–8.

Ingman, M., Kaessmann, H., Pääbo, S. & Gyllensten, U. (2000) Mitochondrial genome variation and the origin of modern humans. *Nature*, **408**, 708 –13.

Jorde, L., Watkins, W., Bamshad, M., Dixon, M., Ricker, C., Seielstad, M. & Batzer, M. (2000) The distribution of human genetic diversity: a comparison of mitochondrial, autosomal, and Y-chromosome data. *American Journal of Human Genetics*, **66**, 979–88.

Klein, R. (2000) Archeology and the evolution of human behavior. *Evolutionary Anthropology*, **9**, 17–36.

Krings, M., Capelli, C., Tschentscher, F., Geisert, H., Meyer, S., von Haeseler, A., Grossschmidt, K., Possnert, G., Paunovic, M. & Pääbo, S. (2000) A view of Neanderthal genetic diversity. *Nature Genetics*, **26**, 144–6.

Lahr, M. (1996) *The Evolution of Modern Human Diversity: A Study of Cranial Variation*. Cambridge: Cambridge University Press.

McBrearty, S. & Brooks, A. (2000) The revolution that wasn't: a new interpretation of the origin of modern human behavior. *Journal of Human Evolution*, **39**, 453–563.

McDermott, F., Stringer, C., Grün, R., Williams, C.T., Din, V. & Hawkesworth, C. (1996) New Late-Pleistocene uranium-thorium and ESR dates for the Singa hominid (Sudan). *Journal of Human Evolution*, **31**, 507–16.

Mellars, P. (1999) The Neanderthal problem continued. *Current Anthropology*, **40**, 341–50.

Relethford, J. & Jorde, L. (1999) Genetic evidence for larger African population size during recent human evolution. *American Journal of Physical Anthropology*, **108**, 251–60.

Stringer, C.B. (1992) Reconstructing recent human evolution. *Philosophical Transactions of the Royal Society of London (Biological)*, **337**, 217–24.

Stringer, C. (1994) Out of Africa – a personal history. In: *Origins of Anatomically Modern Humans* (eds M. Nitecki & D. Nitecki), pp. 149–72. New York: Plenum.

Stringer, C. (1998) Chronological and biogeographic perspectives on later human evolution. In: *Neanderthals and Modern Humans in Western Asia* (eds T. Akazawa, K. Aoki & O. Bar-Yosef), pp. 29–37. New York: Plenum.

Stringer, C. (2001a) Modern human origins: distinguishing the models. *African Archaeological Review*, **18**, 67–75.

Stringer, C. (2001b) Dating the origin of modern humans. In: *The Age of the Earth from 4004 BC to AD 2002* (ed. C. Lewis & S. Knell), pp. 265–74. London: Geological Society.

Thorne, A. & Wolpoff, M. (1992) The multiregional evolution of modern humans. *Scientific American*, **266**, 76–83.

Thorne, A., Grün, R., Mortimer, G., Spooner, N., Simpson, J., Mcculloch, M., Taylor, L. & Curnoe, D. (1999) Australia's oldest human remains: age of the Lake Mungo 3 skeleton. *Journal of Human Evolution*, **36**, 591–612.

Archaeology and the Origins of Modern Humans: European and African Perspectives

PAUL MELLARS

Summary. This chapter compares the archaeological evidence associated with the appearance of anatomically modern humans in Europe and Africa. In Europe there is a rapid appearance of new behavioural elements that are often seen to represent a 'revolution' in behavioural and perhaps cognitive terms, centred on *c*. 43–35,000 years before present (BP). In Africa, new behavioural elements seem to appear in a more gradual, mosaic, fashion but show many of the distinctive features of European Upper Palaeolithic culture by at least 70–80,000 BP, including seemingly explicit evidence for fully symbolic expression. The central problem remains that of assessing how far these well-documented changes in the archaeological record reflect not only major shifts in behavioural patterns, but also underlying shifts in the cognitive *capacities* for behaviour, including increasing complexity in the structure of language.

INTRODUCTION

THE AIM OF THIS chapter is to make two basic points about the patterns of human behavioural development associated with the emergence and spread of anatomically modern populations, which I believe can now be documented with some confidence in the associated archaeological records and which are potentially central to the theme of the present *Proceedings*. The first is that the appearance of anatomically modern populations over large areas of Europe appears to be associated with a wide range of changes that evidently reflect major shifts in the behaviour and possibly the cognition of the associated human groups, what archaeologists have generally referred to as the 'Middle to Upper Palaeolithic transition' or (more colourfully) the 'Upper Palaeolithic

Proceedings of the British Academy, **106**, 31–47, © The British Academy 2002.

revolution' (Gilman, 1984; Bar-Yosef, 1998b). The second is that the earliest well-documented occurrence of *many* of these distinctive new features can now be observed in the archaeological records from Africa and the immediately adjacent parts of south-west Asia at a very much earlier date than their appearance in European sites (McBrearty & Brooks, 2000; Deacon, 2000). Viewed in these terms, the archaeological evidence could be seen as at least consistent with the current Out of Africa scenarios of modern human origins, and perhaps as providing an independent line of support for this hypothesis. But as we shall see, this still leaves plenty of scope for debate over the precise significance of these patterns in terms of human behaviour and cognition.

THE EUROPEAN PERSPECTIVE

The basic features of the so-called Middle–Upper Palaeolithic transition in Europe have been analysed and debated at length (Mellars, 1973, 1982, 1989a, 1989b, 1996a; White, 1982; Kozlowski, 1990; Knecht *et al.*, 1993; Soffer, 1994; Gibson, 1996; Clark, 1997; Gamble, 1999; Klein, 1999; Hoffecker, 2002). In essence, we can now document major changes in at least a dozen different aspects of the archaeological records of the Middle–Upper Palaeolithic transition, all centred around the period 43–35,000 BP (in radiocarbon terms) and apparently reflecting shifts in many different dimensions of human culture and adaptation. Briefly, these can be summarised as follows (for a fuller discussion of these points see Mellars 1989b, 1996a).

1. A general shift from predominantly flake-based technologies to technologies based on the production of more regular and elongated blade forms, probably reflecting the introduction of indirect 'punch' techniques of flaking (Mellars, 1989b).
2. The introduction of new forms of stone tools (typical end-scrapers, various forms of burins, retouched bladelet forms, etc.), apparently implying shifts in several other related aspects of technology, such as skin working, hunting weaponry and more complex bone and antler technology (Mellars, 1989b; Gibson, 1996).
3. The appearance of a sharply increased element of 'imposed form' and morphological complexity in tool production, apparently reflecting an increase in the 'stylistic' as well as the functional component of the tools.
4. The appearance of complex and extensively shaped bone, antler and ivory tools, in a wide variety of forms. These again must reflect changes in other related aspects of technology (especially perhaps hunting projectiles) as well as new and more standardised norms of tool production (Knecht, 1993).

5 The effective explosion of explicitly decorative or ornamental items. These range from carefully perforated animal teeth and marine shells to a wide variety of laboriously shaped bead and pendant forms in a variety of raw materials (ivory, steatite, schist, etc.) (Figure 1). The majority of these almost certainly represent items of personal decoration, probably reflecting new ways of signifying the social role or status of individuals within the societies (White, 1993; Gamble, 1999; Kuhn *et al.*, 2001).

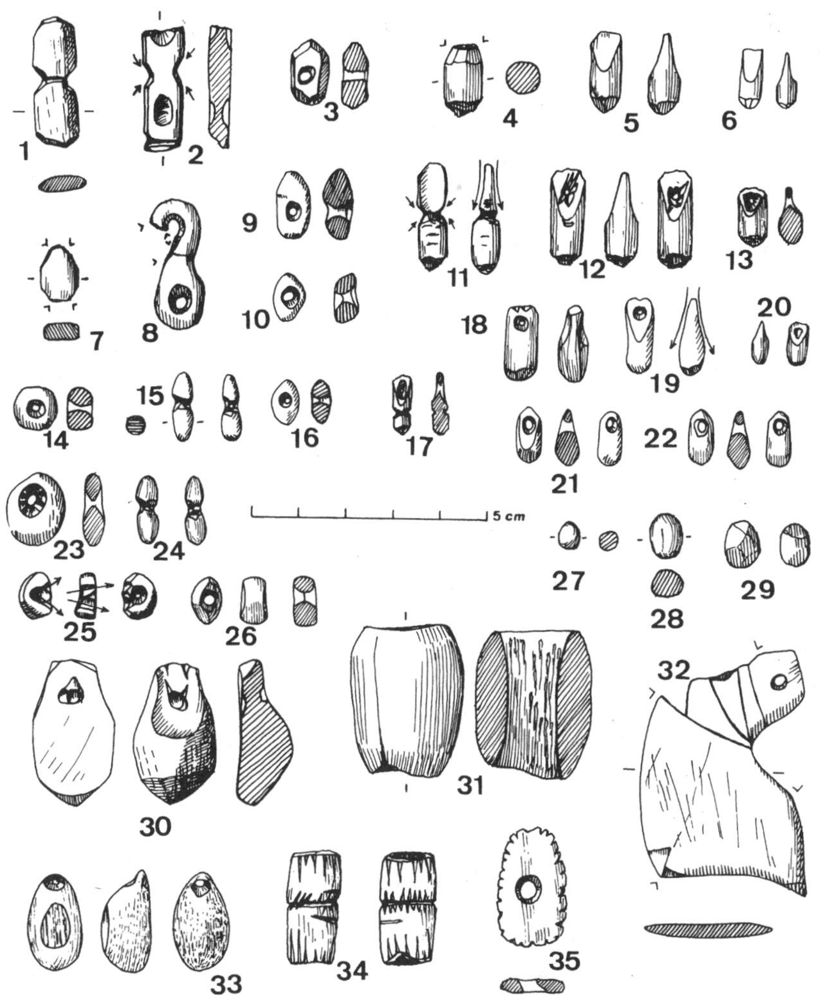

Figure 1. Perforated beads and other pendants manufactured from mammoth ivory and stone from the early Upper Palaeolithic levels of the Spy Cave, South Belgium.

6 The emergence of long-distance trading or exchange networks for the procurement of these ornamental items, extending in some cases, for marine shells and certain varieties of stone beads, up to distances of several hundred kilometres (White, 1993; Taborin, 1993; Gamble, 1999).

7 The appearance of extensively and regularly notched or incised bone artefacts, possibly representing in some cases systematic numerical or notation systems (Marshack, 1991).

8 The first emergence of varied and remarkably sophisticated forms of representational art. These range from representations of both male and female sex symbols, through elaborately carved statuettes of both animal and human figures to the recently discovered animal drawings in the Chauvet cave in south-eastern France (Hahn, 1993; Clottes, 2001).

9 The appearance of unambiguous musical instruments, best represented by the bird-bone flutes from the early Upper Palaeolithic levels of the Geissenklösterle cave in south Germany (Hahn & Münzel, 1995).

10 The emergence of more extensive, large-scale and apparently logistically organised systems for the procurement and distribution of lithic raw materials (Mellars, 1996a: 165–8; Gamble, 1999).

11 In at least some of the better documented areas of Europe (such as western France and the Czech Republic), a sharp increase in the numbers of occupied sites, apparently reflecting a marked increase in human population densities (Mellars, 1973, 1982, 2000: figures 1 and 2; Oliva, 1993; Svoboda, 1993; Demars, 1996).

12 More tentatively, the first evidence in Europe of explicitly 'ceremonial' human burials, reflected by the presence of clearly associated decorative or other grave goods, as documented apparently at Cueva Morín in northern Spain, Cro-Magnon in France and probably Mladeč in Moravia (Oliva, 1993).

These are, of course, the most archaeologically visible of the features marking the conventional Middle–Upper Palaeolithic transition, almost certainly just the tip of the iceberg of the overall spectrum of cultural and behavioural changes over this transition. In addition to the features listed, there are at least strong hints of changes in several other spheres, in the focus and range of subsistence strategies, in the maximum sizes of human residential groups, in the internal spatial organisation of occupation sites and, above all, in the spatial scale and complexity of social relationships (Mellars, 1973, 1982, 1996a; Soffer, 1994; Mithen, 1996; Gamble, 1999). Because these have generated more debate than some of the 'harder' aspects of the archaeological evidence, they may be better left out of the present discussion. Even without these more speculative features, however, the total spectrum of changes in the more well-documented aspects of the Middle–Upper Palaeolithic transition is sufficient to indicate a major transformation in human behavioural patterns, potentially no less significant in their character and consequences than those of the later Neolithic revolution (Mellars, 1989a, 1996a; Mithen, 1996; Sherratt, 1997; Bar-Yosef, 1998b, 2000).

If there is now general agreement as to at least the broad patterns of these changes in the archaeological records of Europe, debate continues to focus on two main issues. Exactly what significance do these changes have for the underlying cultural and cognitive parameters of the associated human populations? And how far can these changes be associated explicitly with the appearance of anatomically modern populations in Europe, and how far could they reflect some form of rapid and essentially independent evolution of the whole spectrum of Upper Palaeolithic technology and culture among the final Neanderthal populations themselves (D'Errico *et al.*, 1998; Zilhao & D'Errico, 1999; Mellars, 1999a; Gamble, 1999; Klein, 2000; Zilhao, 2001)? A full discussion of all these issues is beyond the scope of the present chapter. On the first issue, however, attention has focused mainly on the dramatic explosion of explicitly symbolic behaviour (however that term is defined) associated with the Middle–Upper Palaeolithic transition, as reflected not only in the emergence of complex, sophisticated, art and personal ornamentation, but also in the appearance of regularly notched and incised bones, bird-bone flutes and a sharply increased element of deliberately 'imposed form' in the shaping of both stone and (above all) bone, antler and ivory artefacts. To many workers this 'symbolic explosion' is exactly what one might anticipate from a major shift in the structure or complexity of language patterns, possibly associated with corresponding shifts in the neurological structure of the human brain (Mellars, 1991, 1998; Donald, 1991; Bickerton, 1995; Pinker, 1995; Noble & Davidson, 1996; Mithen, 1996; Deacon, 1997; Klein, 2000; Derek Bickerton's chapter in these *Proceedings*).

The second issue, of the potentially independent evolution of Upper Palaeolithic culture among the final Neanderthal populations of Europe, has generated some lively controversy in the literature over the past 2 or 3 years (D'Errico *et al.*, 1998; Zilhao & D'Errico, 1999; Mellars, 1999a, 2000; Zilhao, 2001). In broad terms there is no doubt that the majority of the most distinctive and dramatic features of the Middle–Upper Palaeolithic transition, especially those reflected in the symbolic explosion referred to above (i.e. art, personal ornamentation, long-distance transport of marine shells, bone flutes, etc.), coincide closely in chronological terms with the first well-documented appearance of anatomically modern populations within the different regions of Europe, associated in most cases with distinctively 'Aurignacian' or 'Proto-Aurignacian' technologies (Mellars, 1992, 1996a; Churchill & Smith, 2000; Kozlowski & Otte, 2000; Richards & Macaulay, 2000). The crux of the recent debate has focused on how far some of the other more 'technological' features of the Upper Palaeolithic 'package', such as increased levels of blade technology, the shaping of simple bone tools and the appearance of some new 'type-fossil' forms, may have emerged in certain areas shortly before the arrival of the first anatomically modern populations (D'Errico *et al.*, 1998; Zilhao & D'Errico, 1999; Mellars 1999a, 2000; Zilhao, 2001). This in turn would raise the question as to how far these developments should be seen as genuinely independent inventions on the part of the local Neanderthal populations, and how far they could be the result simply of what has sometimes been referred to as a 'bow-wave' effect of technological diffusion spreading some way in advance of the expanding anatomically and behaviourally modern populations. Ultimately, no doubt, agreement on this question will depend heavily on the associated dating evidence, which is of course notoriously problematic over the period in question, due partly to the massive problem of contamination effects of radiocarbon samples in this time range and the emerging pattern of major wiggles or plateaux in the radiocarbon time-scale between *c.* 45,000 and 30,000 BP (Aitken, 1990; Voelker *et al.*, 1998; Mellars, 1999b). In the meantime, most prehistorians find the close correlation between the rapid emergence of new behavioural features and the well-documented spread of new human populations across Europe difficult to visualise in purely coincidental terms (Mellars, 1992, 1999a, 2000; Bar-Yosef, 1998b, 2000; Klein, 1999, 2000; Kozlowski & Otte, 2000). But of course this still leaves plenty of scope for debate over the precise patterns of interaction between the incoming *sapiens* and local Neanderthal populations within the different regions of Europe, and the nature of any exchange or 'acculturation' of behavioural patterns between the two populations over this time range (Mellars, 1989a, 1999a; Harrold, 1989; Graves, 1991; Kozlowski & Otte, 2000). And this would not preclude the possibility of some major adaptations in behavioural patterns among the Neanderthal populations during their 250,000 years of development in Europe, especially in response to the rapid climatic oscillations of the later Middle and early Upper Pleistocene periods. Indeed, such adaptations would seem entirely predictable and inevitable in evolutionary terms (Mellars, 1996a: 348–52; Gamble, 1999). But to suggest that this led to the independent invention of the whole gamut of characteristically Upper Palaeolithic culture in several different areas of Europe, simultaneously and coincidentally with the spread of anatomically modern populations across the continent, would seem to imply an extraordinary and highly improbable evolutionary coincidence.

THE AFRICAN PERSPECTIVE

If we accept all the arguments advanced from the recent studies of both human skeletal remains and the associated studies of DNA and other genetic variations in present-day world populations, then by far the most likely point of origin of our species lies in Africa (Chris Stringer's chapter in these *Proceedings*; Stringer & McKie, 1996; Lahr & Foley, 1998; Relethford, 1998; Richards & Macaulay, 2000; Forster *et al.*, 2001; Templeton, 2002). The challenge to archaeologists is obvious. To what extent can we identify evidence that distinctively 'modern' patterns of technology and behaviour, broadly analogous to those reflected in the European Upper Palaeolithic, were indeed significantly earlier in Africa than in areas further to the north and west?

The interpretation of the intriguing but very patchy archaeological record from Africa certainly remains more controversial than that in Europe (Deacon, 1989, 2000; Thackeray, 1992; Clark, 1992; Yellen *et al.*, 1995; Henshilwood & Sealy, 1997; Ambrose, 1998; Deacon & Deacon, 1999; Klein, 1999, 2000; McBrearty & Brooks, 2000). Nevertheless, on one point there now seems to be a consensus: that at least in the southern parts of Africa there is evidence for the appearance of certain distinctively 'modern' features of both technology and apparently increased symbolic behaviour at a substantially earlier date than in any areas of Europe. The most explicit evidence at present comes from a number of sites close to the southern tip of Africa, most notably those of Border Cave, Klasies River Mouth, Boomplaas Cave and the newly excavated site of Blombos Cave, all in Cape Province. From the site of Klasies River Mouth, for example, we have evidence of stone tool industries that are in most respects remarkably similar to those from the classic Upper Palaeolithic sites in Europe, provisionally dated by several methods to around 70,000 BP (Figure 2). These so-called 'Howieson's Poort' industries combine high levels of blade technology manufactured from imported high-quality raw materials, with typical small end-scraper forms and a range of small, carefully shaped, triangular, trapeze and crescent forms that almost certainly represent the hafted inserts of either spears or conceivably arrows (Singer & Wymer, 1982; Deacon, 1989, 2000; Deacon & Deacon, 1999; McBrearty & Brooks, 2000). In northern Europe an almost identical range of forms is characteristic of the earliest stages of the Mesolithic (as at Star Carr), where they demonstrably served as the tips and barbs of wooden arrows (Clark, 1954). From the site of Blombos Cave we have a rather different industry, comprising large numbers of small, leaf-shaped, bifacially flaked projectile points, similar to those of the European Solutrian (Henshilwood & Sealy, 1997). The age of the latter site may be closer to 80,000 BP (Henshilwood *et al.*, 2002). Equally, if not more significant, is the presence at both Klasies River Mouth and Blombos Cave of a number of simple but extensively shaped bone tools, and the occurrence at both sites of large quantities of red ochre, including two recently discovered pieces from the Blombos Cave with carefully incised linear and criss-cross motifs along their edges (Figure 3). At present, these would appear to represent the earliest unambiguous examples of intentional 'design' motifs (or abstract art) recorded anywhere in the world (Henshilwood *et al.*, 2002).

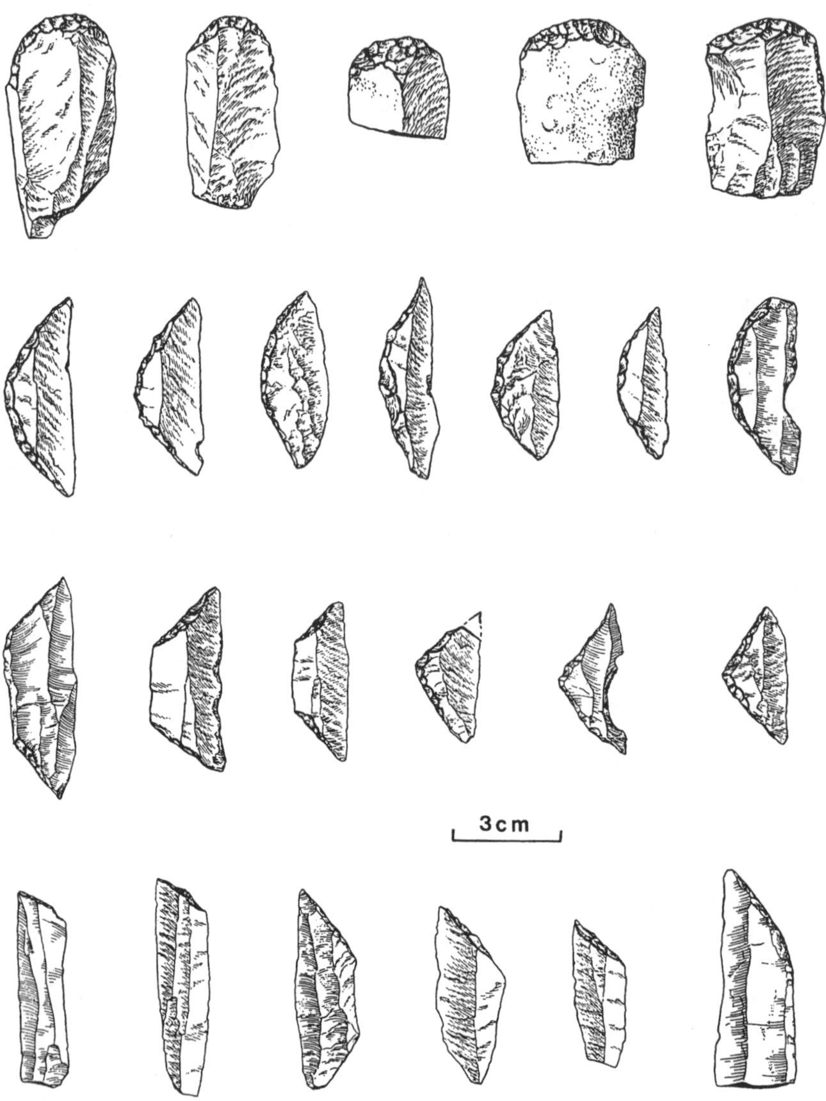

Figure 2. Stone tools from the Middle Stone Age Howieson's Poort levels at Klasies River Mouth, South Africa (*c.* 70,000 BP), showing typical end-scrapers and small geometric forms manufactured from blade segments, probably representing hafted inserts of arrows or spears.

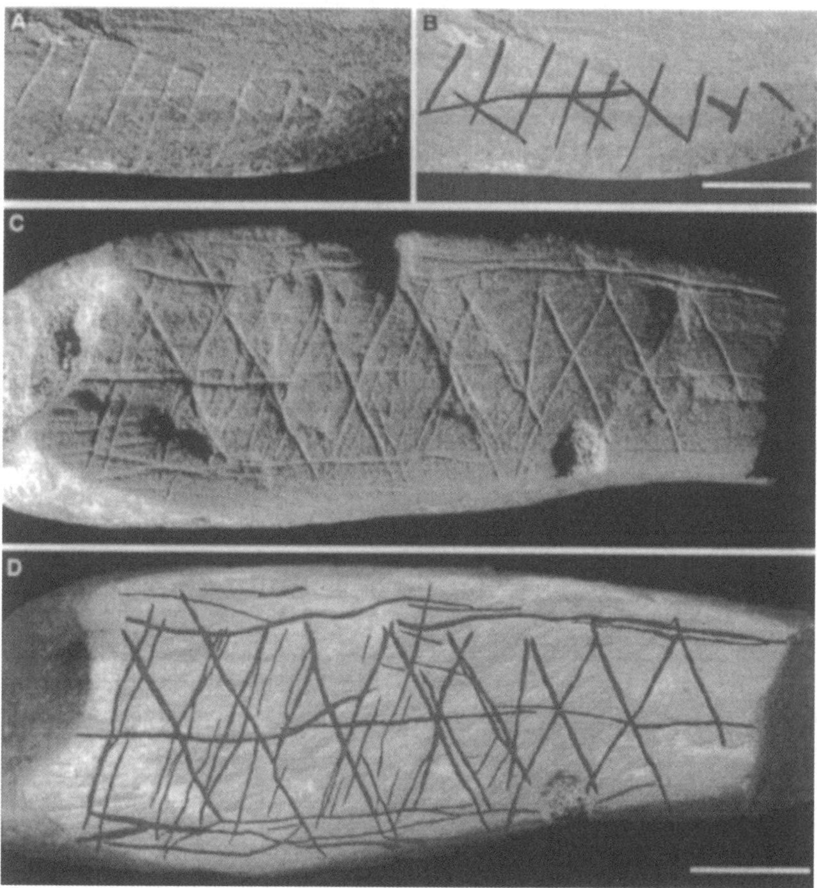

Figure 3. Engravings on red ochre from the Middle Stone Age levels of the Blombos Cave, South Africa, dated to *c.* 75–80,000 BP. Reprinted with permission from Henshilwood *et al.* (2002), *Science*, 295: 1278, figure 2. Copyright 2002 American Association for the Advancement of Science.

Further north in central and eastern Africa the archaeological records over the same time-range remain less well documented. From sites such as Mumba in Tanzania and Enkapune ya Muto in Kenya, however, there is evidence that broadly similar, essentially Upper Palaeolithic (or Later Stone Age), industries were being manufactured by at least 50,000 BP, and at the latter site there is evidence for the production of carefully shaped and perforated beads manufactured from segments of ostrich egg shell by around 40,000 BP (Ambrose, 1998). Similar beads from Mumba may date to 50,000 BP (McBrearty & Brooks, 2000). And from three separate sites at Katande (Zaire) there are rather more controversial claims for a series of remarkably modern-looking multiple-barbed bone spear points, dated to *c.* 90,000 BP (Yellen *et al.*, 1995; McBrearty & Brooks, 2000; Klein, 2000). Finally, immediately to the north of Africa, there is the crucially important evidence for what seem to be explicitly symbolic grave goods (including a large deer antler and a complete boar's jaw reportedly 'clasped in the arms' of one of the skeletons) associated with the burials of the distinctively anatomically modern humans at the two sites of Djebel Qafzeh (Figure 4) and Mugharet es Skhul in Israel, both securely dated to around 90–110,000 BP (Defleur, 1993; Bar-Yosef, 1998a, 2000; Valladas *et al.*, 1998). From the former site there is also evidence for the presence of a range of deliberately perforated sea shells in the same occupation levels (Inizan & Gaillard, 1978). If the latter items are indeed explicitly decorative or 'symbolic' objects, as all the evidence suggests, they precede by at least 50,000 years the widespread appearance of similar perforated sea shells, animal teeth and other forms of decorative pendants in the earliest Aurignacian levels in Europe, discussed above (Kuhn *et al.*, 2001).

CONCLUSIONS

Two features of the evidence discussed above seem to me especially significant. First, that many if not most of the distinctive *technological* features of the so-called 'package' of Upper Palaeolithic culture in Europe can now be documented clearly in certain African sites at least 20–30,000 years earlier than their occurrence in Europe. Here emphasis should be placed not so much on high levels of blade technology and associated transportation of high-quality raw materials (both of which have occasionally been documented in the European Mousterian; Mellars, 1996a) but more importantly on the occurrence of typical end-scrapers (apparently implying new skin-working technology), extensively shaped bone tools and small geometric forms clearly intended as inserts for composite, hafted artefacts, and possibly implying the emergence of archery (Deacon & Deacon, 1999; McBrearty & Brooks, 2000) (Figure 2). Secondly, and potentially far more significant, is the seemingly unambiguous evidence for the early emergence of explicit symbolic behaviour, reflected most clearly in the newly discovered geometrically incised ochre fragments from the Blombos cave (Figure 3; Henshilwood *et al.*, 2002), the regularly notched and incised bones from Klasies River Mouth (Singer & Wymer, 1982), the apparently ceremonial burials (of anatomically modern humans) from Skhul and Qafzeh (Figure 4), and the presence at the latter site of perforated sea-shell ornaments. None of these features can at present be convincingly identified at European sites earlier than *c.* 43–40,000 BP. As indicated earlier, the interpretation of this kind of explicitly symbolic behaviour in cultural and cognitive terms remains controversial, but many would see it as potentially indicative of

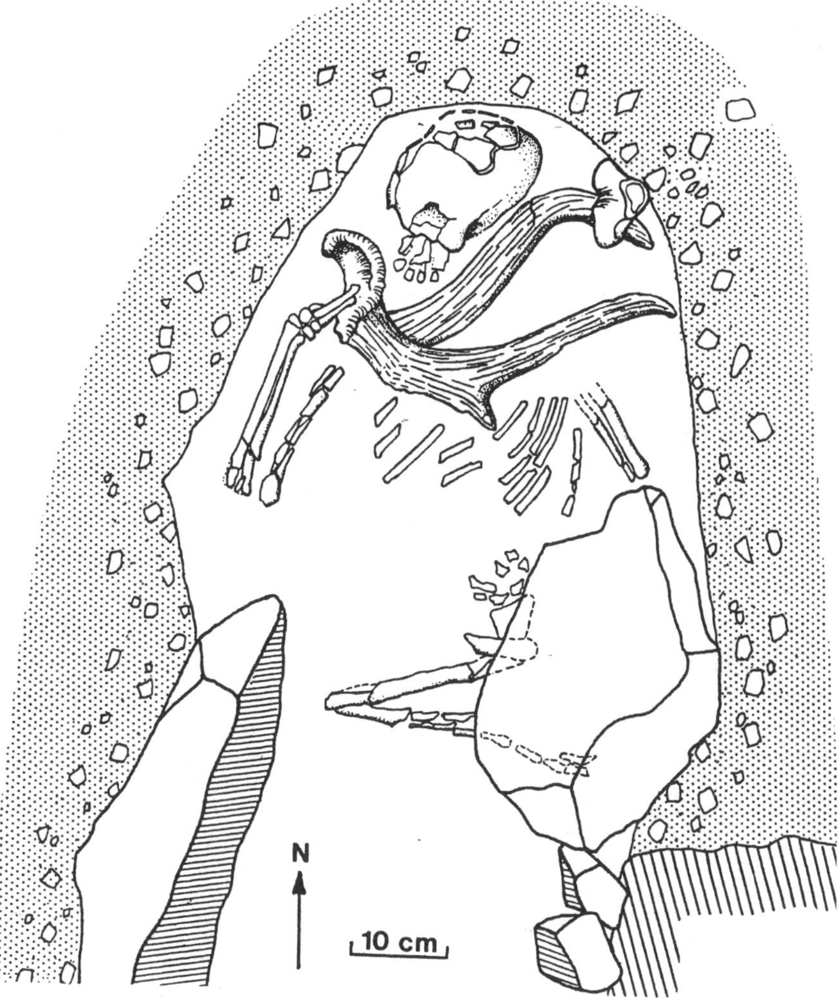

Figure 4. Burial of anatomically modern human skeleton associated with large fallow deer antlers from the Qafzeh Cave, Israel, dated to *c*. 90–100,000 BP.

the emergence of relatively complex language patterns, and quite possibly associated shifts in the neurological complexity of the brain (Donald, 1991; Mellars, 1991, 1998; Pinker, 1995; Noble & Davidson, 1996; Mithen, 1996; Bickerton, 1995; Derek Bickerton's chapter in these *Proceedings*).

Other aspects of the African evidence remain more controversial. McBrearty & Brooks (2000) and Deacon (2000) have made the obvious point that there is no reason to regard the emergence of all these features at the

African sites as a sudden or 'punctuational' event (i.e. as a 'revolution' in European terms) and have argued for a much more gradual, essentially mosaic, pattern of emergence of the distinctively 'modern' behavioural package over a long span of the African Middle Stone Age, extending back to *c.* 250–300,000 BP. This would place the evolution of these behavioural features broadly in parallel with the evolution of modern human anatomical features in Africa, as Chris Stringer (these *Proceedings*) has suggested. At the same time, as both Chris Stringer (these *Proceedings*) and Klein (2000) have pointed out, there is no way that we can exclude the possibility of relatively sudden, punctuational developments in human behaviour and mentality, potentially as a result of either major population bottlenecks or of genetic mutations influencing the structure of the brain (see also Derek Bickerton, these *Proceedings*).

One of the central questions remains exactly how and when these populations first dispersed beyond Africa. The initial appearance of anatomically modern populations in Israel at around 100,000 BP apparently did not extend into Europe, although it is conceivable that these populations expanded further to the south and east into south-east Asia and perhaps ultimately Australia (Lahr & Foley, 1998; Thorne *et al.*, 1999). In this context, the suggested redating of the clearly ceremonial red ochre human burial at Lake Mungo in Australia to around 60,000 BP (Thorne *et al.*, 1999) could be highly significant, reinforcing the evidence for early burial symbolism and ritual at the sites of Skhul and Qafzeh mentioned above. As at the two Israeli sites, the stone-tool technology associated with these early Australian remains seems to be essentially Middle Palaeolithic rather than Upper Palaeolithic in character, suggesting that the critical changes in human symbolic capacities may have occurred rather earlier in the evolution of anatomically modern humans than changes in associated lithic technology. The most likely scenario is that there was a second phase of dispersal of anatomically modern populations from eastern or northern Africa sometime before 45,000 BP, which brought certain basic elements of the distinctively Upper Palaeolithic-like package of both technology and associated symbolic and ceremonial activity initially into south-west Asia and subsequently into Europe (Ambrose, 1998; Klein, 1998, 2000; Bar-Yosef, 1998b, 2000). Needless to say there is no suggestion that fully Upper Palaeolithic culture (in the European sense) was transported wholesale into the sharply contrasting environments of last-glacial Europe, as both Deacon (2000) and McBrearty & Brooks (2000) have stressed. There would inevitably have been critical adaptations in both economy and technology, as well as in related demographic, social and symbolic patterns, associated with the dispersal of modern humans from the African savannas to the European tundras (for a fuller discussion of these points see Mellars, 1996b). But in any event it is clear that all of the present evidence puts the spotlight firmly onto southern Africa as the most likely cradle for the emergence of at least many features of the characteristically 'modern' behavioural package that we see in the European Upper Palaeolithic. To what extent this process involved a radical restructuring not only in the actual *patterns* of human behaviour but also in the underlying mental and cognitive *capacities* for new forms of behaviour (Renfrew, 1996), including language, remains one of the critical questions in modern human origins research.

References

Aitken, M.J. (1990) *Science-Based Dating in Archaeology*. London: Longmans.

Ambrose, S.H. (1998) Chronology of the Later Stone Age and food production in East Africa. *Journal of Archaeological Science*, **25**, 377–92.

Bar-Yosef, O. (1998a) The chronology of the Middle Paleolithic of the Levant. In: *Neanderthals and Modern Humans in Western Asia* (eds T. Akazawa, K. Aoki & O. Bar-Yosef), pp. 39–56. New York: Plenum Press.

Bar-Yosef, O. (1998b) On the nature of transitions: the Middle to Upper Palaeolithic transition and the Neolithic revolution. *Cambridge Archaeological Journal*, **8**, 141–63.

Bar-Yosef, O. (2000) The Middle and Early Upper Paleolithic in Southwest Asia and neighbouring regions. In: *The Geography of Neanderthals and Modern Humans in Europe and the Greater Mediterranean* (eds O. Bar-Yosef & D. Pilbeam), pp. 107–56. Bulletin No. 8. Cambridge: Peabody Museum of Harvard University.

Bickerton, D. (1995) *Language and Human Behavior*. Seattle: University of Washington Press.

Churchill, S.E. & Smith, F.H. (2000) Makers of the early Aurignacian of Europe. *Yearbook of Physical Anthropology*, **43**, 61–115.

Clark, J.G.D. (1954) *Excavations at Star Carr*. Cambridge: Cambridge University Press.

Clark, J.D. (1992) African and Asian perspectives on the origins of modern humans. In: *The Origin of Modern Humans and the Impact of Chronometric Dating* (eds M. J. Aitken, C. Stringer & P. Mellars), pp. 201–16. Philosophical Transactions of the Royal Society, Series B, 337, No. 1280. London: Royal Society.

Clark, G.A. (1997) The Middle–Upper Palaeolithic transition in Europe: an American perspective. *Norwegian Archaeological Review*, **30**, 25–53.

Clottes, J. (2001) *La Grotte Chauvet: L'Art Des Origines*. Paris: Seuil.

Deacon, H.J. (1989) Late Pleistocene palaeoecology and archaeology in the Southern Cape, South Africa. In: *The Human Revolution: Behavioural and Biological Perspectives on the Origins of Modern Humans* (eds P. Mellars & C. Stringer), pp. 547–64. Princeton: Princeton University Press.

Deacon, H.J. (2000) Modern human emergence: an African archaeological perspective. In: *Humanity from African Naissance to Coming Millennia—Colloquia in Human Biology and Palaeoanthropology* (eds P. V. Tobias, M. A. Raath, J. Maggi-Cecchi & G. A. Doyle), pp. 217–26. Florence: Florence University Press.

Deacon, H.J. & Deacon, J. (1999) *Human Beginnings in South Africa: Uncovering the Secrets of the Stone Age*. Cape Town: David Philip.

Deacon, T.W. (1997) *The Symbolic Species: The Co-Evolution of Language and the Human Brain*. Harmondsworth: Penguin Books.

Defleur, A. (1993) *Les Sépultures Moustériennes*. Paris: CNRS.

Demars, P.-Y. (1996) Demographie et occupation de l'espace au Paléolithique Supérieur et au Mésolithique en France. *Préhistoire Européenne*, **8**, 3–26.

D'Errico, F., Zilhao, J., Julien, M., Baffier, D. & Pelegrin, J. (1998) Neanderthal acculturation in western Europe? A critical review of the evidence and its interpretation. *Current Anthropology*, **39**, S1–S44.

Donald, M. (1991) *Origins of the Modern Mind: Three Stages in the Evolution of Culture and Cognition.* Cambridge: Harvard University Press.

Forster, P., Torroni, A., Renfrew, C. & Röhl, A. (2001) Phylogenetic star contraction applied to Asian and Papuan mtDNA evolution. *Molecular Biology and Evolution*, **18**, 1864–81.

Gamble, C. (1999) *The Palaeolithic Societies of Europe*. Cambridge: Cambridge University Press.

Gibson, K.R. (1996) The biocultural human brain, seasonal migrations, and the emergence of the Upper Palaeolithic. In: *Modelling the Early Human Mind* (eds P. Mellars & K. Gibson), pp. 33–46. Cambridge: McDonald Institute for Archaeological Research.

Gilman, A. (1984) Explaining the Upper Palaeolithic Revolution. In: *Marxist Perspectives in Archaeology* (ed. E. Springs), pp. 115–26. Cambridge: Cambridge University Press.

Graves, P. (1991) New models and metaphors for the Neanderthal debate. *Current Anthropology*, **32**, 513–41.

Hahn, J. (1993) Aurignacian art in Central Europe. In: *Before Lascaux: The Complex Record of the Early Upper Paleolithic* (eds H. Knecht, A. Pike-Tay & R. White), pp. 229–42. Boca Raton: CRC Press.

Hahn, J. & Münzel, S. (1995) Knochenflöten aus dem Aurignacien des Geissenklösterle bei Blaubeuren, Alb-Donau-Kreis. *Fundberichte Aus Baden-Württemberg*, **20**, 1–12.

Harrold, F.B. (1989) Mousterian, Chatelperronian and Aurignacian in Western Europe: continuity or discontinuity. In: *The Human Revolution: Behavioural and Biological Perspectives on the Origins of Modern Humans* (eds P. Mellars & C. Stringer), pp. 677–713. Edinburgh: Edinburgh University Press.

Henshilwood, C. & Sealy, J. (1997) Bone artefacts from the Middle Stone Age at Blombos Cave, Southern Cape, South Africa. *Current Anthropology*, **38**, 890–5.

Henshilwood, C.S., D'Errico, F., Yates, R., Jacobs, Z., Tribolo, C., Duller, G.A.T., Mercier, N., Sealy, J.C., Valladas, H., Watts, I. & Wintle, A.G. (2002) Emergence of modern human behavior: Middle Stone Age engravings from South Africa. *Science*, **295**, 1278–80.

Hoffecker, J.F. (2002) *Desolate Landscapes: Ice-Age Settlement in Eastern Europe*. New Brunswick: Rutgers University Press.

Inizan, M.-L. & Gaillard, J.M. (1978) Coquillages de Ksar-'Aqil: éléments de parure? *Paléorient*, **4**, 295–306.

Klein, R.G. (1998) Why anatomically modern people did not disperse from Africa 100,000 years ago. In: *Neanderthals and Modern Humans in Western Asia* (eds T. Akazawa, K. Aoki & O. Bar-Yosef), pp. 509–21. New York: Plenum Press.

Klein, R.G. (1999) *The Human Career*, 2nd edn. Chicago: University of Chicago Press.

Klein, R.G. (2000) Archaeology and the evolution of human behavior. *Evolutionary Anthropology*, **9**, 17–36.

Knecht, H. (1993) Splits and wedges: the techniques and technology of early Aurignacian antler working. In: *Before Lascaux: The Complex Record of the Early Upper Paleolithic* (eds H. Knecht, A. Pike-Tay & R. White), pp. 137–62. Boca Raton: CRC Press.

Knecht, H., Pike-Tay, A. & White, R. (1993) *Before Lascaux: The Complex Record of the Early Upper Paleolithic*. Boca Raton: CRC Press.

Kozlowski, J.K. (1990) A multi-aspectual approach to the origins of the Upper Palaeolithic in Europe. In: *The Emergence of Modern Humans: An Archaeological Perspective* (ed. P. A. Mellars), pp. 419–37. Edinburgh: Edinburgh University Press.

Kozlowski, J.K. & Otte, M. (2000) The formation of the Aurignacian in Europe. *Journal of Anthropological Research*, **56**, 513–34.

Kuhn, S.K., Stiner, M.C., Reese, D.S. & Güleç, E. (2001) Ornaments of the earliest Upper Paleolithic: new insights from the Levant. *Proceedings of the National Academy of Sciences of the USA*, **98**, 7641–6.

Lahr, M.M. & Foley, R. (1998) Towards a theory of modern human origins: geography, demography and diversity in recent human evolution. *Yearbook of Physical Anthropology*, **41**, 137–76.

McBrearty, S. & Brooks, A. (2000) The revolution that wasn't: a new interpretation of the origin of modern human behavior. *Journal of Human Evolution*, **39**, 453–563.

Marshack, A. (1991) *The Roots of Civilization*. Mount Kisco: Moyer Bell.

Mellars, P.A. (1973) The character of the Middle–Upper Palaeolithic transition in south-west France. In: *The Explanation of Culture Change: Models in Prehistory* (ed. C. Renfrew), pp. 255–76. London: Duckworth.

Mellars, P.A. (1982) On the Middle/Upper Paleolithic transition: a reply to White. *Current Anthropology*, **23**, 238–40.

Mellars, P.A. (1989a) Major issues in the emergence of modern humans. *Current Anthropology*, **30**, 349–85.

Mellars, P.A. (1989b) Technological changes across the Middle–Upper Palaeolithic transition: technological, social and cognitive perspectives. In: *The Human Revolution: Behavioural and Biological Perspectives on the Origins of Modern Humans* (eds P. Mellars & C. Stringer), pp. 338–65. Edinburgh: Edinburgh University Press.

Mellars, P.A. (1991) Cognitive changes and the emergence of modern humans in Europe. *Cambridge Archaeological Journal*, **1**, 63–76.

Mellars, P.A. (1992) Archaeology and the population-dispersal hypothesis of modern human origins in Europe. In: *The Origin of Modern Humans and the Impact of Chronometric Dating* (eds M. J. Aitken, C. B. Stringer & P. A. Mellars), pp. 225–34. Philosophical Transactions of the Royal Society, Series B, 337, No. 1280. London: Royal Society.

Mellars, P.A. (1996a) *The Neanderthal Legacy: An Archaeological Perspective from Western Europe*. Princeton: Princeton University Press.

Mellars, P.A. (1996b) The emergence of biologically modern populations in Europe: a social and cognitive 'revolution'? In: *Evolution of Social Behaviour Patterns in Primates and Man* (eds W. G. Runciman, J. Maynard Smith & R. I. M. Dunbar), pp. 179–202. Proceedings of the British Academy No. 88. London: The British Academy.

Mellars, P.A. (1998) Neanderthals, modern humans and the archaeological evidence for language. In: *The Origin and Diversification of Language* (eds N. G. Jablonski & L. C. Aiello), pp. 89–115. San Francisco: California Academy of Sciences.

Mellars, P.A. (1999a) The Neanderthal problem continued. *Current Anthropology*, **40**, 341–50.

Mellars, P.A. (1999b) Radiocarbon dating and the origins of anatomically modern populations in Europe. In: *Experiment and Design: Essays in Honour of John Coles* (ed. A. Harding), pp. 1–12. Oxford: Oxbow Books.

Mellars, P.A. (2000) The archaeological records of the Neanderthal–Modern human transition in France. In: *The Geography of Neanderthals and Modern Humans in Europe and the Greater Mediterranean* (eds O. Bar-Yosef & D. Pilbeam), pp. 35–47. Bulletin No. 8. Cambridge: Peabody Museum of Harvard University.

Mithen, S. (1996) *The Prehistory of the Mind: A Search for the Origins of Art, Religion and Science*. London: Thames & Hudson.

Noble, W. & Davidson, I. (1996) *Human Evolution, Language and Mind: A Psychological and Archaeological Inquiry*. Cambridge: Cambridge University Press.

Oliva, M. (1993) The Aurignacian in Moravia. In: *Before Lascaux: The Complex Record of the Early Upper Paleolithic* (eds H. Knecht, A. Pike-Tay & R. White), pp. 37–56. Boca Raton: CRC Press.

Pinker, S. (1995) *The Language Instinct*. Harmondsworth: Penguin books.

Relethford, J.H. (1998) Genetics of modern human origins and diversity. *Annual Review of Anthropology*, **27**, 1–23.

Renfrew, C. (1996) The sapient behaviour paradox: how to test for potential? In: *Modelling the Early Human Mind* (eds P. Mellars & K. Gibson), pp. 11–4. Cambridge: McDonald Institute for Archaeological Research.

Richards, M. & Macaulay, V. (2000) Genetic data and the colonization of Europe: genealogies and founders. In: *Archaeogenetics and the Population Prehistory of Europe* (eds C. Renfrew & K. Boyle), pp. 139–53. Cambridge: McDonald Institute Monographs.

Sherratt, A. (1997) Climatic cycles and behavioural revolutions: the emergence of modern humans and the beginning of farming. *Antiquity*, **71**, 271–87.

Singer, R. & Wymer, J. (1982) *The Middle Stone Age at Klasies River Mouth in South Africa*. Chicago: University of Chicago Press.

Soffer, O. (1994) Ancestral lifeways in Eurasia—the Middle and Upper Paleolithic records. In: *Origins of Anatomically Modern Humans* (eds M. H. Nitecki & D. V. Nitecki), pp. 101–9. New York: Plenum Press.

Stringer, C.B. & McKie, R. (1996) *African Exodus: The Origins of Modern Humanity*. London: Jonathan Cape.

Svoboda, J. (1993) The complex origins of the Upper Paleolithic in the Czech and Slovak republics. In: *Before Lascaux: The Complex Record of the Early Upper Paleolithic* (eds H. Knecht, A. Pike-Tay & R. White), pp. 23–36. Boca Raton: CRC Press.

Taborin, Y. (1993) Shells of the French Aurignacian and Gravettian. In: *Before Lascaux: The Complex Record of the Early Upper Paleolithic* (eds H. Knecht, A. Pike-Tay & R. White), pp. 211–28. Boca Raton: CRC Press.

Templeton, A.R. (2002) Out of Africa again and again. *Nature*, **416**, 45–51.

Thackeray, A.I. (1992) The Middle Stone Age south of the Limpopo River. *Journal of World Prehistory*, **6**, 385–440.

Thorne, A., Grün, R., Mortimer, G., Spooner, N.A., Simpson, J.J., McCulloch, M., Taylor, L. & Curnoe, D. (1999) Australia's oldest human remains: age of the Lake Mungo 3 skeleton. *Journal of Human Evolution*, **36**, 591–612.

Valladas, H., Mercier, N., Joron, J.-L. & Reyss, J.-L. (1998) GIF laboratory dates for Middle Paleolithic Levant. In: *Neanderthals and Modern Humans in Western Asia* (eds T. Akazawa, K. Aoki & O. Bar-Yosef), pp. 69–76. New York: Plenum Press.

Voelker, A.H.L., Sarnthein, M., Grootes, P.M., Erlenkeuser, H., Laj, C., Mazaud, A., Nadeau, M.-J. & Schleicher, M. (1998) Correlation of marine ^{14}C Ages from the Nordic seas with the GISP2 isotope record: implications for ^{14}C calibration beyond 25 ka BP. *Radiocarbon*, **40**, 517–34.

White, R. (1982) Rethinking the Middle/Upper Paleolithic transition. *Current Anthropology*, **23**, 169–92.

White, R. (1993) Technological and social dimensions of 'Aurignacian age' body ornaments across Europe. In: *Before Lascaux: The Complex Record of the Early Upper Paleolithic* (eds H. Knecht, A. Pike-Tay & R. White), pp. 277–300. Boca Raton: CRC Press.

Yellen, J.E., Brooks, A.S., Cornelissen, E., Mehlman, M.J. & Stewart, K. (1995) A Middle Stone Age worked bone industry from Katanda, upper Semliki valley, Zaire. *Science*, **268**, 553–6.

Zilhao, J. (2001) *Anatomically Archaic, Behaviorally Modern: The Last Neanderthals and Their Destiny (Kroon Lecture, 2001)*. Amsterdam: Amsterdam Archaeological Centre, University of Amsterdam.

Zilhao, J. & D'Errico, F. (1999) The chronology and taphonomy of the earliest Aurignacian and its implications for the understanding of Neanderthal extinction. *Journal of World Prehistory*, **13**, 1–68.

The Case for Saltational Events in Human Evolution

IAN TATTERSALL

Summary. Evolutionary saltation is a rather ill-defined Victorian concept that may contrast in various ways with gradual (linear) evolution. In comparing evolutionary modes it is thus more productive to bear in mind the difference between 'transformational' and 'taxic' mindsets, each of which sharply influences the way in which the evolutionary record is interpreted. It is also important to recognise the fact that, even under traditional notions of natural selection, we cannot think of the evolutionary process as one of fine-tuning the myriad systems that make up each individual and each taxon. Both individuals and species thrive or fail reproductively as the sums of their parts, not as the bearers of multiple features that can somehow be tracked independently of each other over vast spans of time. In this review I look at the human morphological and behavioural records in an attempt to discern pattern in human evolution, and particularly in the emergence of our own species, *Homo sapiens*, as narrowly defined. *Homo sapiens* is a variable species, although not notably more so than *Homo neanderthalensis*, which can itself be understood only in the context of its membership in a diverse clade of endemic European species. The record clearly indicates that both modern human morphology and modern human cognitive processes appeared rather suddenly, even saltationally, although not at the same time.

INTRODUCTION

I HAVE BEEN ASKED TO SPEAK at this Symposium about saltation and the evidence for it in the human evolutionary record. This I am happy to do; but I should point out straight away that this title is subject to quite extensive interpretation. Saltation is an essentially Victorian concept that, even if only implicitly, contrasts notions of slow, steady, intralineage change with competing patterns of

Proceedings of the British Academy, **106**, 49–59, © The British Academy 2002.

abrupt change among species. In other words, it pits the views of those such as Thomas Henry Huxley ('Nature does make jumps now and then') against those of Charles Darwin ('*Natura non facit saltum*') (Huxley, 1900). However, while the Darwinian notion of gradual evolution (accurate or otherwise) is a pretty clear-cut one, saltational change may in principle embrace not only short-term, within-lineage, innovations, wherein the acquisition of new features is more than simply incremental, but also processes of taxic diversification within clades. Notions of saltation, in other words, may invoke both linear, unidirectional, transformational change, and taxically divergent patterns of innovation: it is the scale and suddenness of such innovation that is at issue.

Given the fact that a century of debate has failed to produce an agreed-upon functional definition of saltation ('evolution by jerks' is perhaps as close as we have got so far), perhaps it would be better to adjust our terminology a little, and to recast the two principal models of evolutionary change currently on offer as transformational vs. taxic, as Niles Eldredge did 20 years ago (Eldredge, 1979; Tattersall, 1994). It is, after all, this division, above everything, that separates today's two major mindsets of evolutionary thought. Under the transformational model, we are obliged to look upon evolutionary change as a process of gradual, generation-by-generation, fine-tuning within lineages. The signal being sought is thus one of continuity in character evolution. The taxic approach, in contrast, looks for evidence of systematic diversification: a process of evolutionary experimentation that involves the generation of new species, each of which is free, subsequent to its origin, to compete with its close relatives (and with species of more remote taxic affiliation) in the ecological arena. In this latter view, evolutionary outcomes, hence larger patterns, are at least as much a product of ecological interactions as they are of processes of genetic innovation and winnowing within lineages.

The notion of fine-tuning of adaptations within lineages, and over the aeons, via the action of natural selection, has proven to be a highly seductive one, no less in palaeoanthropology than in other areas of palaeontology and systematics. Indeed, by the time that the grand 'Evolutionary Synthesis' had established its grip over evolutionary biology in the 1930s, it had become almost universally self-evident to evolutionary biologists that the action of natural selection (in the traditional, Darwinian, sense) was sufficient to explain virtually every evolutionary phenomenon. Evolution had become reduced to small, incremental, gene/gene frequency changes within lineages over long periods of time. However, while almost nobody would deny, even today, that natural selection must be a central process in the fixation of inherited novelties in local populations, a moment's reflection is sufficient to reveal that this cannot, even in principle, be the full story. When we speak of 'adaptation(s)' we rarely refer to entire organisms interacting within their environments. To most palaeoanthropologists, 'adaptations' are intellectually dissected-out characteristics of individual organisms and species. The formula seems to be: structure = adaptation. If a structure is there, it must be adaptive, for natural selection is ever vigilant, is it not? (Tattersall, 1994).

Well, as a matter of fact, no. Think about it for a moment. What is natural selection? It is any and all factors that result in a differential genetic contribution to the next generation by one individual relative to its contemporaries. And, crucially, every individual must succeed or fail in the inheritance stakes as the sum of its parts. Particular aspects of the individual genome cannot be differentially passed along to the next generation in isolation from others. Those specific attributes that we can, with sufficient ingenuity, characterise as 'adaptations,' are strongly, sometimes indissolubly, linked with numerous others in each individual; and each individual, in turn, consists of a huge number of such attributes. Individuals (and the populations to which they belong, for that matter) must necessarily thus compete in the genetic stakes as functional wholes, rather than as agglomerations of separate 'adaptations' that can be favoured or otherwise by natural selection. Certainly, new genetic innovations must arise as particularities *within* individuals; but their triage by natural selection has to take place inside the context of those individuals' overall reproductive success. Conversely, structures may survive and be passed along between generations not because they are notably advantageous, but simply because they do not significantly impede survival and reproduction. In a profound sense, then, we are mistaken to speak of such things as 'the evolution of bipedality' or the 'evolution of the brain,' as if these attributes were susceptible to independent tracking over time, separable from the larger packages of which they are part, or entirely free of random influences.

The upshot is that the history of the hominid family, like that of all other taxa, sums out not simply as the abstract aggregate of changes in a number of body systems that we may conveniently follow and examine independently of others. Rather, this history is one of the differential successes of hominid taxa that succeeded or failed as the totality of their parts. Heritable evolutionary innovations certainly arose inside those taxa, indeed within individuals, but in the long term larger evolutionary patterns have been determined by the overall success of those individuals, and of the taxa to which they belonged. It is thus critical that we be able to identify those taxa: essentially, species. And whether, at the current state of the palaeoanthropological art, we are in a position to do this is highly debatable. Clearly, this severely limits our options. Recently, for example, I attempted to characterise the pattern of hominid brain size evolution since the beginning of the Pleistocene (Tattersall, 1998a). On one level, of course, there is no problem. Clearly, if we compare average hominid brain sizes 2 million years ago with those of today, the trend is clear: today's brains are much larger. But if we wish to inquire beyond this, to address in more detail the matter of pattern in hominid brain evolution (even at the superficially simple level of asking whether brain size increase was gradual or episodic), the difficulties immediately mount. There are relatively few hominid fossils that are sufficiently well preserved to indicate reliable brain sizes. Even fewer can provide brain–body size relationships. And, among those fossils that can do this to a reasonable approximation, fewer still are dated with the precision which this exercise demands. Most importantly of all, however, we need to know the species to which the specimens involved belong; for, individually, brain sizes are notoriously variable, and if we are to extract useful information we have to understand what the average brain sizes and morphologies are for each species in our family tree.

And this points up a huge gap in our knowledge of hominid phylogeny. We have no agreed-upon idea at all of the species diversity represented in the large and still growing hominid fossil record. The tradition has, indeed, been to obscure pattern in that record by sweeping a huge variety of morphologies into such meaningless groupings as 'early *Homo*' or 'archaic *H. sapiens*'. The upshot is that, in the absence of anything even resembling a reasonable consensus on species diversity in the human clade, one is for the moment forced to give up on any useful attempt to discern pattern in human brain evolution. And, of course, until we have sorted out the basic systematics of the hominid record, we will be at a similar impasse in broaching most other questions, too.

SALTATION IN HUMAN EVOLUTION

Yet, on another level, it must still be possible to look at the record to see whether the dominant overall pattern is one of continuity or saltation. And, indeed, it is. However, given the ambiguity of the concept of saltation, I would strongly prefer to couch this search in slightly but significantly different terms. The question I would prefer to ask is: Does the evidence point towards uniformity, or in the direction of diversity? Is the morphological signal one of steady within-lineage change, or does it suggest the multiplication of species (which were then triaged among themselves)? The practical problem here for palaeontologists is, of course, that while morphological differentiation among local populations of widespread species is a routine phenomenon, and may, indeed, be the principal mechanism leading to the establishment of heritable novelties (Tattersall, 1994), it is independent of the event of speciation itself. There is thus no specifiable degree of morphological shift that can be taken as a firm indication that a reproductive barrier has intervened. Species populations can undergo major differentiation without the intervention of speciation (the creation of effectively discrete reproductive units); and speciation can take place in the absence of significant morphological innovation. In the first case, we risk overestimating species-level diversity on morphological criteria; and in the second we risk underestimation.

This is not a conundrum to which there is any clearly evident definitive solution. None the less, a survey of the living primate fauna makes it clear that it is the genus that is the *Gestalt* unit among Primates, and that species within genera do not typically contrast greatly in their hard-tissue characteristics. Thus, in any fossil assemblage in which two or more distinct 'morphs' can be discerned that display numerous hard-tissue distinctions, we may conservatively conclude that at least as many species are represented (Tattersall, 1986). This kind of non-quantifiable rule-of-thumb procedure may be less than perfectly satisfying intellectually; but in employing it we are much more likely to underestimate than to overestimate the number of species actually represented in our assemblage. This should mollify even those traditionalists who think it 'unbiological' not to regard virtually all variation as intraspecific. And, as I have argued elsewhere (Tattersall, 1986, 1992), to proceed in this manner will help ensure that we minimise any distortion of the actual pattern of evolution within the group.

However valid these musings might or might not be, however, some might find them a bit arid in the context of the subject of this Symposium, 'The Speciation of Modern *Homo sapiens*'. For, practicalities aside, they only obliquely address the two questions that ideally first need to be asked: 'How would we recognise a taxic event in the record of the origin of our own species?' and 'Who or what in the known record is our closest fossil relative?' (Asking who or what our ancestor is would be expecting a little too much for theoretical as well as practical reasons.) What's more, in our own case we face an additional difficulty beyond the routine one of morphological recognition: the behavioural innovations that characterise our species are at least as striking as the morphological ones (and have certainly been more fateful), and changes of the two kinds do not appear to have proceeded in tandem (Tattersall, 1998a, 1998b). With this in mind, what do the fossil and archaeological records themselves suggest about the origin of *Homo sapiens*?

THE HUMAN RECORD

There is abundant evidence for diversity in the hominid fossil record during the long period in which our family was confined to Africa. At about 1.8 million years ago in Kenya, for example, we have clear indications that at least four hominid species shared not only the same continent, but the same landscape. Indeed, it seems that we modern humans have been profoundly misled by our current status as the lone hominid in the world (Tattersall, 2000). It is natural enough for us to conclude that this is somehow the normal state of affairs, for this is what we are used to; but historically things have been otherwise and, in this regard at least, we should not permit the present to shape our expectations about the past. The dominant pattern we see throughout the human evolutionary record is, in fact, one of diversity rather than of linearity. It has also been one of highly episodic major innovation, so that overall our history fits the 'saltational' pattern, however defined. The 'archaic biped' body structure of the australopiths, for example, persisted for a very long time even as numerous species came and went, only to be superseded abruptly by the unanticipated appearance of *Homo ergaster*, the first hominid of more-or-less modern body form. Indeed, we have no inkling whatever as to the antecedents of this radically new kind of hominid; and even if we remove the species *rudolfensis* and *habilis* from the genus *Homo*, as Wood & Collard (1999) have recently and credibly suggested should be done, it is very difficult to find any evidence at all of the direct ancestry of *Homo* among the australopiths as currently represented.

What we do know, however, is that the subsequent fossil record of the genus *Homo* presents us with a picture of substantial morphological variety, although many traditionalists still apparently find it convenient to use the figleaf of 'archaic *H. sapiens*' to obscure this fact. And while it would be premature to claim that we understand the full nature of this diversity, there are powerful indications that during the later part of the Pleistocene a significant number of species was spun off within the hominid clade. The best regional example of this has come to light as the result of a reappraisal of the European and western Asian group to which the Neanderthals belong (Tattersall & Schwartz, 2000). It is as yet unclear what the exact affinities are of the 780 thousand-year-old species *Homo antecessor*, recently described on the basis of fairly sparse fossil evidence (Bermudez de Castro *et al.*, 1997). But it is becoming evident that there subsequently diversified in Europe a major hominid clade of which the Neanderthals are simply the best-documented component. Traditionally, effort in the analysis of this group of fossils was directed towards trying to force them into an essentially linear framework, under the guise of 'pre-*sapiens*,' 'protoneanderthals' and so forth; but predictably this has only served to foster confusion. Much of this confusion stems from the fact that (under the sway of linear thinking) time has often been considered paramount in the analysis of these fossils, as of others. Indeed, it is generally true that the ages of hominid fossils have often loomed far larger in exegeses of their phylogenetic relationships than have their morphologies. And while time is certainly a significant component of the fossil record, it has no necessary connection with phylogenetic affinity (Eldredge & Tattersall, 1975), although it is a factor available to be considered among others.

Morphological appraisal of this European record indicates the existence of a coherent clade that extends well beyond a single species. The best known member of this clade is the species *Homo neanderthalensis*, represented from something more than 200 thousand years ago to less than 30 thousand years. *Homo neanderthalensis* is highly distinctive, possessing numerous characteristics that separate it from all other hominids. And while some still try to cram it into an enlarged concept of *H. sapiens*, if *H. neanderthalensis* fails the test of specific distinctness, it is hard to conceive of any taxon that would pass such a test. We might as well give up on the practice of systematics and extend *H. sapiens* to include virtually all hominids of the last 2 million years (which, paleoanthropology being what it is, has actually been suggested: Wolpoff *et al.*, 1994).

Neanderthal characteristics include medial projections and posterior swellings within the nasal aperture; substantial maxillary sinuses that not only restrict the width of the nasal cavities but swell out the face below (and inside) the orbits; an unusually forwardly-positioned face and upper dentition, with which is correlated an anteriorly-shifted mandibular dentition and retromolar space; double-arched brow ridges that roll smoothly back and up into the frontal plane; a suprainiac depression that provides upper definition to occipital torus centrally; horizontal orientation of the lateral part of the occipital suture; and numerous characters of the post-cranial skeleton. There are also many cranial features that are generally distinctive of Neanderthals, but whose expression is sufficiently variable to exclude them from any hard-and-fast diagnosis; these include the projection of the occipitomastoid/juxtamastoid crest complex relative to the mastoid process, enlargement of the molar pulp cavities, and so forth. Indeed, the substantial Neanderthal sample is particularly valuable not only in presenting us with a morphology that is consistent over a large tract of time and space, but that shows intrapopulation variation analogous to what we see in our own species. Neanderthals varied among themselves no less than modern humans do (Schwartz & Tattersall, forthcoming).

When we look a bit farther afield, however, and particularly farther back in time, we encounter in the European record some specimens that possess some Neanderthal characteristics, but not others. The cranium from the German site of Steinheim, for example, has certain Neanderthal features in the front of the skull, notably Neanderthal-like brow ridges and a large nasal opening within which is a hint of a medial projection. However, its face is not inflated by the maxillary sinuses; and at the rear the skull is not notably Neanderthal-like, despite possessing a horizontal lateral occipital suture and a weakly developed occipital torus and suprainiac depression. What does this mean? Traditionally, this combination of features has been taken to indicate that the specimen represents an early stage in the lineage leading to Neanderthals; but it seems more likely that the Steinheim cranium represented a distinct member of a clade whose common ancestor actually possessed certain features, large nasal aperture, smoothly rolled brow ridges, and so forth, that were also retained by the Neanderthals. In other words, the Neanderthals were not as autapomorphic as even the staunchest advocates of placing them in *H. sapiens* normally admit they are. For, just like *H. sapiens*, they are a unique mixture of the primitive and the derived.

If Steinheim were a unique occurrence, of course, it might be possible to argue that Steinheim was simply a stage *en route* to Neanderthal status. But the picture is complicated by other specimens of not vastly different age. The best-preserved of these are from the Sima de los Huesos, in the Atapuerca Hills of Spain, and are about 300 thousand years old (Arsuaga *et al.*, 1997) While they also possess Neanderthal-like brow ridges, large nasal openings and horizontal occipital sutures, the Sima specimens are in other features clearly unlike both Neanderthals and Steinheim. The same can be said for other specimens from Europe, including the putatively ancestral specimens often referred nowadays to *Homo heidelbergensis*. Among the best cranial specimens thus classified are the 400 thousand-year-old Arago 21 and the beautifully preserved but uncertainly dated fossil from Petralona, in Greece. The species *H. heidelbergensis* has been considered ancestral to both *H. neanderthalensis* and *H. sapiens*; but the Petralona specimen, especially, displays intracranial (and most especially frontal) sinuses that are developed almost beyond belief. Since the primitive condition for the hominid clade is tiny cranial sinuses, and Neanderthal sinuses, while greatly developed compared with those of *H. sapiens*, are significantly smaller than those of *H. heidelbergensis*, it is hardly possible to regard the latter as a Neanderthal ancestor, as has been suggested from time to time. On the other hand, it could plausibly represent a distinct member of the larger Neanderthal-related clade that was particularly derived in this feature.

There is clearly much more to be done on the European Middle-to-Late Pleistocene hominid assemblage, but it already appears that we are unlikely ever to sort it out satisfactorily if we refuse to recognise taxic diversity within it. Whatever we have in Europe, it is not a steadily changing lineage, but an assortment of species whose interrelationships demand analysis. And this, in turn, points to a pattern of evolution that has involved repeated speciation. To put it another way, the dominant evolutionary mode in this case was clearly taxic rather than transformational. Whether it was saltational is, of course, another matter; in none of these cases (the listing of apomorphies aside) can we even hazard a reliable guess as to the historical nature of those speciations.

By about 200 thousand years ago, the Neanderthals seem to have achieved sole possession of Europe, a situation that persisted until their eviction by invading *H. sapiens* between about 40 and 30 thousand years ago. Exactly where these new hominids came from is unknown, although molecular studies and the sparse fossil evidence point to an ultimate origin in Africa (reviewed by Stringer & McKie, 1996). Analysis of the mtDNA extracted from the Feldhofer Neanderthal suggests that the *sapiens*/*neanderthalensis* lineages split well over 500 thousand years ago (Krings *et al.*, 1997), which seems reasonable;

and the very limited African fossil hominid record shows a substantial morphological variety over the past half-million years or so. About the most that can be said with certainty, however, is that this rather opaque record does not carry a clear signal of continuity. Again, the pattern was probably taxic rather than transformational.

THE ARRIVAL OF *HOMO SAPIENS*

At some time around 40 thousand years ago, anatomically and behaviourally modern *H. sapiens* entered Europe, with dates of comparable antiquity coming from both the western and the eastern extremities of the continent. Here we have straightforward evidence for a short-term event, although it cannot strictly be defined as saltational and might have had several components to it. The arriving *H. sapiens* (the 'Cro-Magnons') brought with them a distinctive new 'Upper Paleolithic' industry whose origin is as yet unknown. What we do know, however, is that, along with this stone-working technology they brought, or very soon developed, a host of symbolic expressions that included painting, carving, engraving, notation, music, elaborate body ornamentation, burial with grave goods, social stratification, and much else with which living humans can identify particularly closely. In all of these attributes the Cro-Magnons contrasted strongly not only with the Neanderthals, but with their own direct predecessors, the earliest humans of modern body form. Thus, as far as is known, in Africa all hominids of 100–200 thousand years ago, the period in which *H. sapiens* putatively emerged, possessed Middle Stone Age stone-working technology. This technology was functionally equivalent to the Mousterian toolkit that was made up to about 40 thousand years ago in the Levant, both by Neanderthals and by those anatomically modern humans (Qafzeh 9, for instance) who had made their appearance at not much under 100 thousand years ago (Valladas *et al.*, 1988). In Africa there are some hints in the sparse archaeological record of early (250–100 thousand years ago) behaviours that may presage behavioural modernity: flint mining, long-distance transport of materials, blade production, possible bone-working (Brooks, 2000), and even use of living space (Deacon, 1993). But hints are all we really have, as in the case of eastern Asia, where the Ngandong hominids, the last of the *H. erectus* clade, may have been evicted as little as 40 thousand years ago or less, and where the first Australians, who must have been formidable navigators, left no convincing evidence of themselves prior to about 60 thousand years ago.

Little as we can say with certainty, however, there seems to be no doubt that 'morphological *H. sapiens*' emerged from a background of considerable taxic variety among hominids Old World-wide at least 100 thousand years ago. This emergence thus took place substantially before we have any evidence that modern, symbol-mediated, behaviours formed part of the regular cultural repertoire of any hominid group. Of course, very occasional objects older than 40–50 thousand years ago have been interpreted as symbolic. But even if such interpretation is accurate, it only serves to underline that such expressions were not part of the larger cultural patterns of the societies that produced them. They were exceptions, not the rule. May we conclude, then, that the 'early moderns' (right back to the origin of *H. sapiens*) had possessed a potential for symbolic behaviours that remained fallow for at least 50 thousand years, until 'released' by some cultural stimulus? Certainly there is no evidence for any biological change (at least as expressed in the hard tissues) that could have had this effect; and in any event, perhaps we should not be surprised by this. For 'exaptation', or, if you prefer, 'pre-adaptation', has long been recognised as a key component of the evolutionary process. Birds, for example, possessed feathers for millions of years before adopting them as adjuncts to locomotion. The use of innovations is not necessarily confined to the contexts in which they arise.

It is possible to argue that all through the sporadic history of change and innovation in human prehistory, new species had mostly done what their predecessors had done, if perhaps a bit better, as the process of interspecific competition might determine. Only with the emergence of modern-behaving *H. sapiens* do we find this pattern disturbed. But it is disturbed with a vengeance. Modern-behaving *H. sapiens* is truly a new influence on the landscape, and is not simply an extrapolation of what went before. And if this is not an example of saltation, I can't imagine what might be.

DISCUSSION

Questioner: How is evolution of behaviour related to the anatomy?

Tattersall: Novelties, whether anatomical or behavioural, have to arise *within* species. Modern human cranial structure seems to have been established by at least 100 thousand years ago, and probably more, whereas evidence of significant hominid symbolic behaviours (hence cognition) is much younger than this. Therefore we must suppose that even if the potential for such behaviours was born with anatomically modern humanity, their expression had to await the invention of a cultural releasing factor (plausibly, language) some dozens of millennia later.

References

Arsuaga, J.L., Bermudez de Castro, J.M. & Carbonell, E. (1997) Special issue: the Sima de los Huesos hominid site. *Journal of Human Evolution*, **33**, 105–421.

Bermudez de Castro, J.M., Arsuaga, J.L., Carbonell, E., Rosas, A., Martinez, I. & Mosquera, M. (1997) A hominid from the lower Pleistocene of Atapuerca, Spain: possible ancestor to Neanderthals and moderns. *Science*, **276**, 1392–5.

Brooks, A. (2000) Modern human origins: archaeology and behavior. In: *Encyclopedia of Human Evolution and Prehistory* (eds E. Delson, I. Tattersall, J. Van Couvering & A. Brooks), pp. 434–42. New York: Garland Publishing.

Deacon, H.J. (1993) Southern Africa and modern human origins. In: *The Origin of Modern Humans and the Impact of Chronometric Dating* (eds M. Aitken, C. Stringer & P. Mellars), pp. 104–17. Princeton: Princeton University Press.

Eldredge, N. (1979) Alternative approaches to evolutionary theory. In: *Models and Methodologies in Evolutionary Theory* (eds J. Schwartz & H. Rollins), pp. 7–19. Bulletin 13. Carnegie: Carnegie Museum of Natural History.

Eldredge, N. & Tattersall, I. (1975) Evolutionary models, phylogenetic reconstruction, and another look at hominid phylogeny. In: *Approaches to Primate Biology* (ed. F.S. Szalay), pp. 218–42. Basel: Karger.

Huxley, L. (ed.) (1900) *Life and Letters of Thomas Henry Huxley*. London: Macmillan & Co.

Krings, M., Stone, A., Schmitz, R.W., Krainitzki, H., Stoneking, M. & Paabo, S. (1997) Neanderthal DNA sequences and the origin of modern humans. *Cell*, **30**, 19–30.

Schwartz, J.H. & Tattersall, I. (forthcoming) Morphology, variability and systematics: lessons from Krapina. *Proceedings of the Congress of Krapina.*

Stringer, C. & McKie, R. (1996) *African Exodus: The Origins of Modern Humanity*. London: Jonathan Cape.

Tattersall, I. (1986) Species recognition in human paleontology. *Journal of Human Evolution*, **15**, 165–75.

Tattersall, I. (1992) Species concepts and species identification in human evolution. *Journal of Human Evolution*, **22**, 341–9.

Tattersall, I. (1994) How does evolution work? *Evolutionary Anthropology*, **3**, 2–3.

Tattersall, I. (1998a) *The Origin of the Human Capacity*. New York: American Museum of Natural History [James Arthur Lectures on the Evolution of the Human Brain].

Tattersall, I. (1998b) *Becoming Human: Evolution and Human Uniqueness*. New York: Harcourt Brace.

Tattersall, I. (2000) Once we were not alone. *Scientific American*, **282**, 56–62.

Tattersall, I. & J.H.Schwartz. (2000) *Extinct Humans*. Colorado: Westview Press.

Valladas, H., Reyss, J.L., Joron, J.L., Valladas, G., Bar-Yosef, O. & Vandermeersch, B. (1988) Thermoluminescence dating of Mousterian 'proto-Cro-Magnon' remains from Israel and the origin of modern man. *Nature*, **331**, 614–6.

Wolpoff, M.H., Thorne, A.G., Jelinek, J. & Zhang, Y. (1994) The case for sinking *Homo erectus*: 100 years of Pithecanthropus is enough! *Courier Forsch-Inst. Senckenberg*, **171**, 341–61.

Wood, B. & Collard, M.(1999) The human genus. *Science*, **284**, 65–9.

Grades and Transitions in Human Evolution

MARK COLLARD

Summary. An assessment of the number of grades that have appeared in the course of human evolution is carried out in this chapter. Three grades are identified. The first is characterised by a species mean body mass of under 50 kg; a species mean stature of less than 150 cm; facultative bipedalism; relatively large teeth and jaws; a moderate size brain relative to body mass; and a relatively short period of maturation. The second grade is characterised by a species mean body mass of more than 50 kg; a species mean stature in excess of 150 cm; obligate bipedalism; relatively small teeth and jaws; a moderate size brain relative to body mass; and a relatively short period of maturation. The third grade is similar to the second in terms of body mass, stature, locomotor behaviour and masticatory system size; but exhibits a considerably higher level of encephalisation. It also exhibits delayed maturation. With varying degrees of certainty, *Ardipithecus ramidus*, *Australopithecus afarensis*, *Australopithecus africanus*, *Australopithecus anamensis*, *Australopithecus garhi*, *Homo habilis*, *Homo rudolfensis*, *Kenyanthropus platyops*, *Orrorin tugenensis*, *Paranthropus aethiopicus*, *Paranthropus boisei* and *Paranthropus robustus* can be assigned to the first grade, whereas *Homo antecessor*, *Homo ergaster*, *Homo erectus* and *Homo heidelbergensis* can be assigned to the second, and *Homo neanderthalensis* and *Homo sapiens* can be assigned to the third. The first grade appeared around 6 million years ago, probably in connection with the establishment of the human and chimpanzee lineages. The second grade probably emerged between 2.4 and 1.9 million years ago, and is associated with the appearance of *H. ergaster*. The third grade probably emerged between 500 and 242 thousand years ago.

Proceedings of the British Academy, **106**, 61–100, © The British Academy 2002.

INTRODUCTION

UNDERSTANDING THE evolution of any taxonomic group requires knowledge not only of genealogical issues such as species diversity and phylogeny, but also of adaptive trends, biogeographic patterns and other ecological issues (Huxley, 1958; Foley, 1984, 1999; Eldredge, 1985, 1986, 1989, 1990). However, in recent years hominid palaeontological research has focused primarily on the identification of species and the reconstruction of their phylogenetic relationships (Eldredge & Tattersall, 1975; Delson *et al.*, 1977; Tattersall & Eldredge, 1977; Corruccini & McHenry, 1980; Andrews, 1984; Stringer, 1984, 1987; Olson, 1985; Skelton *et al.*, 1986; Wood & Chamberlain, 1986; Tattersall, 1986, 1992; Chamberlain & Wood, 1987; Wood, 1988, 1989, 1991, 1992, 1993; Lieberman *et al.*, 1988, 1996; Groves, 1989; Skelton & McHenry, 1992; Kimbel & Martin, 1993; Rightmire, 1993, 1996, 1998, 2001; Corruccini, 1994; Strait *et al.*, 1997; Strait & Grine, 1999, 2001; Wolpoff *et al.*, 1994, 2001; Curnoe, 2001). Relatively few attempts have been made to elucidate patterns of hominid adaptation and biogeography and to link those patterns with potential causal processes (Oxnard, 1984; Foley, 1984, 1999; Wood & Collard, 1997, 1999a; Strait & Wood, 1999; Wolpoff, 1999; Collard & Wood, 1999; Eckhardt, 2000; McHenry & Coffing, 2000; Teaford & Ungar, 2000). With this imbalance in mind, in the present chapter I focus on the grade, a classificatory category that is based on adaptive equivalence (Huxley, 1958). My aim is to build on attempts that B. A. Wood and I have made to develop a grade classification for the fossil hominids (Wood & Collard, 1997; Collard & Wood, 1999). First, I discuss the concept of the grade, paying particular attention to its evolutionary basis. Secondly, I outline a taxonomy for the hominids, and describe their geographic and temporal distributions. Thirdly, I consider the means by which grades may be recognised in the hominid fossil record. Lastly, I review data pertaining to the adaptive strategies of the hominid species in order to determine the number of grades that have existed in human evolution.

GRADE CLASSIFICATION

Grade classification, as outlined by Huxley (1958), attempts to identify the adaptive types that have appeared in a morphological trend. An adaptive type is a taxon with a distinct phenotypic pattern or organisational plan that is seen in the fossil record to replace an older taxon with a less derived organisational plan. In some cases the replacement is straightforward, involving just two taxa. In others the replacement is more complex. The old organisational plan is first replaced by an array of new organisational plans. These taxa are then reduced in number by extinction, until only one is left. Regardless of the mode of replacement, the new taxon is called an 'adaptive type' because it must have been more efficient than the taxa it superseded. The rise and success of a new organisational plan is evidence that it was better adapted than the older organisational plan, and also better adapted than the organisational plan of any potential competitor. Like clades, grades are relative. They can only be delimited in relation to the trend being considered. Grades of all animals will be different from those of all vertebrates, which in turn will be different from the grades of all mammals. Likewise, the grades of all mammals will be different from the grades for separate trends of specialisation within the carnivores or the primates. Unlike clades, however, grades do not have to be monophyletic. They may also be polyphyletic, because convergent evolution can cause species from two or more distantly related lineages to arrive at the same adaptive type.

Huxley (1958) considers classifying by grades to be a palaeontological activity. However, Rosenzweig & McCord (1991) argue that the grade has a neontological equivalent: the 'fitness generating function' or 'G-function', which is an equation used to calculate the fitnesses of different phenotypes (Brown & Vincent, 1987; Rosenzweig *et al.*, 1987). A G-function takes into account the frequencies and densities of all the evolutionary factors affecting the success of an organism, and contains all the fitness trade-offs in terms of the costs and benefits an organism receives for living in a certain way in a particular time and place (Rosenzweig *et al.*, 1987; Rosenzweig & McCord, 1991). Because a G-function indicates which phenotypes are possible and shows the fitness reward an individual gets for emphasising any given trait, it implies the design rules that govern an organisational plan. An adaptive type is hence a G-function with a less severe fitness trade-off than the G-function, or G-functions, it replaces, and a grade is a G-function in a trade-off trend (Rosenzweig & McCord, 1991).

Rosenzweig *et al.* (1987) illustrate these concepts with a case of replacement in the evolution of the viper. Pit vipers have replaced true vipers in the Americas, and are currently replacing them throughout the Old World. The success of the pit vipers, Rosenzweig *et al.* (1987) suggest, is due to their ability to detect both infrared and visible light. Because the focal length of electromagnetic radiation varies with its wavelength, true vipers must trade-off sharpness of vision against the breadth of the spectrum they can see; they cannot focus sharply on both infrared and visible light. Pit vipers have overcome this limitation by dissociating the ability to sense infrared from the ability to detect visible light. They have developed what amounts to a second pair of eyes, their loreal pits, which unlike their true eyes are sensitive to infrared. By avoiding the compromise between wavelength and the sharpness of the image, the pit vipers have reduced the severity of their trade-off constraint relative to that of the true viper. They have become more efficient hunters, and are consequently in the process of forming another grade in the evolution the viper.

Rosenzweig & McCord (1991) highlight another illustrative example of a grade shift among the reptiles. The straight-necked turtles of the suborder Amphichelydia have been replaced several times by turtles that can flex their necks. In some instances this replacement was accomplished by turtles that flex their neck sideways (Pleurodira); in others it was carried out by turtles that flex their necks into an S-curve (Cryptodira). Rosenzweig & McCord (1991) argue that the crucial difference between straight-necked turtles and turtles that can flex their necks is the defensive capabilities of the latter. Unable to protect its head in its shell, Amphichelydia would have suffered from higher rates of predation than either Pleurodira or Cryptodira. Consequently it would have found it difficult to compete with them, especially for vacant niches. Rosenzweig & McCord (1991) contend that by evolving a flexible neck Pleurodira and Cryptodira improved their trade-off constraint to such an extent that they were able to replace Amphichelydia. In the process they became an adaptive type and a grade.

HOMINID TAXONOMY AND DISTRIBUTION

Opinions differ regarding the number of genera and species represented by the fossils assigned to Hominidae (e.g. Tattersall, 1986, 1992, 1996; Lieberman *et al.*, 1988, 1996; Groves, 1989; Wood, 1991, 1992, 1993; Wolpoff *et al.*, 1994; Rightmire, 1993, 1996, 1998, 2001; Wolpoff *et al.*, 1994, 2001; Wolpoff, 1999; Wood & Collard, 1999a, 1999b; Wood & Richmond, 2000; Asfaw *et al.*, 2002). Because there are both theoretical and practical reasons for erring on the side of too many rather than too few taxa (Tattersall, 1986, 1992, 2001; Lieberman *et al.*, 1996), a taxonomy that recognises six genera and 19 species is adopted here (Table 1).

The oldest genus, *Homo*, was established by Linnaeus in the mid-eighteenth century, along with the species to which modern humans are assigned, *H. sapiens* (Linnaeus, 1758). Seven fossil species are assigned to *Homo*. The name *H. neanderthalensis* was introduced in the mid-nineteenth century (e.g. King, 1864) for material recovered in the Neander Valley, Germany. However, the name has only recently been used widely (Tattersall, 1986, 1992; Stringer & Gamble, 1993; Wood & Richmond, 2000), as evidence demonstrating the morphological distinctiveness of the Neanderthals has accumulated (Hublin *et al.*, 1996; Schwartz & Tattersall, 1996; Ponce de León & Zollikofer, 2001; Lieberman *et al.*, 2002). Previously the fossils now assigned to *H. neanderthalensis* were included as a subspecies within *H. sapiens*. Material assigned to *H. neanderthalensis* has been found throughout Europe, as well as in central and south-west Asia (Stringer & Gamble, 1993). Current palaeontological evidence indicates that the Neanderthals emerged between 242 and 186

Table 1. Current hominid taxonomy, including formal taxonomic designations and approximate temporal and geographic ranges. Taxa are listed by date of initial publication. The symbol † before a taxon name indicates that the taxon is extinct. Parentheses around a citation indicate that the generic attribution of the taxon differs from the original attribution

Family Hominidae Gray 1825. Pliocene-present, world-wide
- Genus *Homo* Linnaeus 1758 [includes e.g. †*Pithecanthropus* Dubois 1894, †*Protanthropus* Haeckel 1895, †*Sinanthropus* Black 1927, †*Cyphanthropus* Pycraft 1928, †*Meganthropus* Weidenreich 1945, †*Atlanthropus* Arambourg 1954, †*Telanthropus* Broom & Robinson 1949]. Pliocene-present, world-wide
  - Species *Homo sapiens* Linnaeus 1758. Pleistocene-present, world-wide
  - Species †*Homo neanderthalensis* King 1864. Pleistocene, western Eurasia
  - Species †*Homo erectus* (Dubois 1892). Pleistocene, Africa and Eurasia
  - Species †*Homo heidelbergensis* Schoetensack 1908. Pleistocene, Africa and Eurasia
  - Species †*Homo habilis* L. S. B. Leakey *et al.* 1964. Pliocene-Pleistocene, Africa
  - Species †*Homo ergaster* Groves & Mazak 1975. Pleistocene, Africa and Eurasia
  - Species †*Homo rudolfensis* (Alexeev 1986). Pliocene-Pleistocene, East Africa
  - Species †*Homo antecessor* Bermudez de Castro *et al.* 1997. Pleistocene, western Eurasia
- Genus †*Australopithecus* Dart 1925 [includes †*Plesianthropus* Broom 1938]. Pliocene, Africa
  - Species †*Australopithecus africanus* Dart 1925. Pliocene, Africa
  - Species †*Australopithecus afarensis* Johanson *et al.* 1978. Pliocene, East Africa
  - Species †*Australopithecus anamensis* M. G. Leakey *et al.* 1995. Pliocene, East Africa
  - Species †*Australopithecus bahrelghazali* Brunet *et al.* 1996. Pliocene, East Africa
  - Species †*Australopithecus garhi* Asfaw *et al.* 1999. Pliocene, East Africa
- Genus †*Paranthropus* Broom 1938 [includes †*Zinjanthropus* L. S. B. Leakey 1959, †*Paraustralopithecus* Arambourg & Coppens 1967]. Pliocene-Pleistocene, Africa
  - Species †*Paranthropus robustus* Broom 1938. Pleistocene, southern Africa
  - Species †*Paranthropus boisei* (L. S. B. Leakey 1959). Pliocene-Pleistocene, East Africa
  - Species †*Paranthropus aethiopicus* (Arambourg & Coppens 1968). Pliocene, East Africa
- Genus †*Ardipithecus* White *et al.* 1995. Pliocene, East Africa
  - Species †*Ardipithecus ramidus* (White *et al.* 1994). Pliocene, East Africa
- Genus †*Kenyanthropus* M. G. Leakey *et al.* 2001. Pliocene, East Africa
  - Species †*Kenyanthropus platyops* Leakey *et al.* 2001. Pliocene, East Africa
- Genus †*Orrorin* Senut *et al.* 2001. Pliocene, East Africa
  - Species †*Orrorin tugenensis* Senut *et al.* 2001. Pliocene, East Africa

thousand years ago (Klein, 1999), although ancient DNA studies suggest that the Neanderthal lineage may have originated around 500 thousand years ago (Krings *et al.*, 1997, 1999). The last Neanderthal fossils date to around 30 thousand years ago (Smith *et al.*, 1999). The first evidence of *H. erectus* was recovered in Indonesia in the early 1890s (Dubois, 1892, 1894). Remains attributed to *H. erectus* have since been located elsewhere in Indonesia, as well as in mainland Eurasia and Africa (Ascenzi *et al.*, 2000; Wood & Richmond, 2000). The earliest *H. erectus* material may be from 1.9 million years ago, and the youngest reliably dated specimens are from around 200 thousand years ago (Wood & Richmond, 2000). The name *H. heidelbergensis* was introduced for the Mauer jaw in the early part of the last century (Schoetensack, 1908), but the taxon has only been widely used in the last couple of decades (Tattersall, 1986; Groves, 1989; Rightmire, 1996). Previously the Mauer specimen and related material were referred to as 'archaic *H. sapiens*'. *Homo heidelbergensis* is known from a number of African and European Middle Pleistocene sites (Rightmire, 1996, 2001; Wood & Richmond, 2000). Specimens assigned to *H. habilis* were first recovered at Olduvai Gorge in the early 1960s (Leakey *et al.*, 1964). Additional *H. habilis* fossils have since been discovered at a number of southern and eastern African localities, most notably Sterkfontein in South Africa (Hughes & Tobias, 1977; Grine *et al.*, 1993, 1996; Kimbel *et al.*, 1996; but see Kuman & Clarke, 2000) and Koobi Fora in Kenya (Wood, 1991, 1992). Current dating indicates that *H. habilis* appeared around 2.3 million years ago, and went extinct about 1.6 million years ago (Wood, 1991, 1992; Kimbel *et al.*, 1996). It has been suggested recently that the *habilis* hypodigm should be removed from *Homo* and placed in *Australopithecus* (Wolpoff, 1999; Wood & Collard, 1999a, 1999b; see also Kuman & Clarke, 2000) but this suggestion has not proved popular because it almost certainly makes *Australopithecus* paraphyletic (Strait & Grine, 2001; Tattersall, 2001). The species name *H. ergaster* was introduced in the mid-1970s (Groves & Mazak, 1975). However, it did not come into use until the early 1990s after several researchers argued that the specimens conventionally referred to as 'early African *H. erectus*' may be sufficiently distinct to be considered a different species (Andrews, 1984; Stringer, 1984; Wood, 1984, 1994). The validity of *H. ergaster* remains contested (e.g. Turner & Chamberlain, 1989; Brauer & Mbua, 1992; Rightmire, 1998; Asfaw *et al.*, 2002) and there is a pressing need for a comprehensive assessment of its taxonomic status. The best-preserved specimens assigned to *H. ergaster* come from the Lake Turkana region in Kenya and Dmanisi, Georgia (Wood, 1991; Walker & Leakey, 1993; Gabunia & Vekua, 1995; Gabunia *et al.*, 2001). Radiometric and faunal dating indicate that *H. ergaster* was extant between 1.9 million years ago and 1.5 million years ago. Originally proposed by Alexeev (1986), *H. rudolfensis* was not used until the 1990s, when it was suggested that part of the *H. habilis sensu lato* hypodigm should be recognised as a separate species (Groves, 1989; Wood, 1992). There is still some debate over the distinctiveness and composition of the hypodigm of *H. rudolfensis* (Wood, 1991, 1992; Rightmire, 1993) but most workers who recognise the taxon accept that it includes the cranium KNM-ER 1470. To date *H. rudolfensis* specimens have been found in deposits in Kenya and Malawi, and possibly Ethiopia, that date from 2.4 to 1.8 million years ago (Wood & Collard, 1999b). Recently, it has been argued that the *rudolfensis* hypodigm should be removed from *Homo* and assigned to either *Australopithecus* (Wolpoff, 1999; Wood & Collard, 1999a, 1999b) or *Kenyanthropus* (Leakey *et al.*, 2001; Lieberman, 2001). Bermudez de Castro *et al.* (1997) proposed the species *H. antecessor* on the basis of cranial and post-cranial fossils dated 0.7 million years ago from the site of Gran Dolina, Sierra de Atapuerca, Spain.

The second oldest hominid genus, *Australopithecus*, was established in the early part of the twentieth century (Dart, 1925). It has five fossil species assigned to it. The type species, *A. africanus*, was erected by Dart (1925) on the basis of an early hominid child's skull from Taung in southern Africa. Subsequent to the discovery of the Taung child, additional *A. africanus* fossils have been recovered at three South African sites: Makapansgat (Member 3), Gladysvale and, most notably, Member 4 at Sterkfontein. Currently *A. africanus* is dated from between 3.0 and 2.4 million years ago, although it is possible that it first appeared as far back as 3.5 million years ago (Clarke & Tobias, 1995; Clarke, 1998; Partridge *et al.*, 1999; but see McKee, 1996). Johanson *et al.* (1978) erected the species *A. afarensis* for material recovered from Laetoli, Tanzania, and Hadar, Ethiopia. *Australopithecus afarensis* is now also known from several other sites, including Maka, Belohdelie and Fejej in Ethiopia, and Koobi Fora in Kenya (Wood & Richmond, 2000). *Australopithecus afarensis* may be as old as 4.2 million years ago (Kappelman *et al.*, 1996), although most researchers currently consider its first appearance date to be 3.7 million years ago (Wood & Richmond, 2000). The last appearance date of *A. afarensis* is normally taken to be 3.0 million years ago (Wood & Richmond, 2000). Recently Strait *et al.* (1997) suggested that *A. afarensis* should be renamed *Praeanthropus africanus*, because their cladistic analyses indicated that its inclusion in *Australopithecus* made the latter paraphyletic. However, the International Commission of Zoological Nomenclature (1999) has suppressed the name *Praeanthropus africanus*, which means that if *A. afarensis* is to be removed from *Australopithecus* it should be called *Praeanthropus afarensis*. The third australopithecine species listed in Table 1, *A. anamensis*, was established in the mid-1990s for fossils from the sites of Kanapoi and Allia Bay, both of which are in Kenya (Leakey *et al.*, 1995). Recent work indicates that all of the fossils assigned to *A. anamensis* were deposited between *c.* 4.2 and 4.1 million years ago (Leakey *et al.*, 1998). The species name *A. bahrelghazali* was proposed on the basis of hominid fossils recovered in the Bahr el ghazal region of Chad, north-central Africa (Brunet *et al.*, 1995, 1996). Faunally dated to around 3.5 million years ago, these fossils greatly extended the known geographic range of *Australopithecus*, which had been restricted to eastern and southern Africa. Asfaw *et al.* (1999) established the last *Australopithecus* species listed in Table 1, *A. garhi*. Currently the *A. garhi* hypodigm comprises craniodental specimens that were recovered from the Hata beds of Ethiopia's Middle Awash region, and which date to around 2.5 million years ago. Post-cranial remains of comparable antiquity were also described by Asfaw *et al.* (1999) but, as they are not associated with diagnostic cranial remains, Asfaw *et al.* (1999) did not include them in the *A. garhi* hypodigm.

The genus *Paranthropus* was first recognised by Broom in the late 1930s (Broom, 1938). Three species are assigned to *Paranthropus* in the current taxonomy, the type species *P. robustus*, plus *P. boisei* and *P. aethiopicus*. Specimens assigned to *P. robustus* have been recovered from several South African cave sites, most notably Kromdraai, Swartkrans and Drimolen (Broom, 1938, 1949; Brain, 1993, 1994; Keyser *et al.*, 2000, Keyser, 2000). Current dating evidence suggests that *P. robustus* first appeared *c.* 1.9 million years ago and went extinct *c.* 1.5 million years ago. *Paranthropus boisei* was first recovered in the late 1950s at Olduvai Gorge, Tanzania (Leakey, 1958). It is now known from several other East African sites, including Koobi Fora in Kenya, Peninj in Tanzania, and Konso in Ethiopia (Leakey & Leakey, 1964; Tobias, 1965; Wood, 1991; Suwa *et al.*, 1997; Wood & Lieberman, 2001). Recently a partial maxilla was recovered at Melama in Malawi (Kullmer *et al.*, 1999). The oldest *P. boisei* specimens date to around 2.3 million years ago; the youngest date to around 1.3 million years ago (Wood *et al.*, 1994). *Paranthropus aethiopicus* fossils have been recovered at West Turkana, Kenya (Walker *et al.*, 1986), and from the Shungura Formation in Ethiopia's Omo Region (Arambourg & Coppens, 1968; Suwa, 1988; Wood *et al.*, 1994). *Paranthropus aethiopicus* is currently dated from between 2.5 million years ago and 2.3 million years ago (Wood *et al.*, 1994).

The remaining three genera, *Ardipithecus*, *Kenyanthropus* and *Orrorin*, have been established only recently. *Ardipithecus* was erected by White *et al.* (1995) for material that they had previously assigned to *Australopithecus* (White *et al.*, 1994). The material in question derives from deposits dated from *c.* 5.8 to 4.5 million years ago in the Middle Awash region of Ethiopia, and is assigned to the species *A. ramidus* (White *et al.*, 1994; Haile-Selassie, 2001). *Kenyanthropus* was established by Leakey *et al.* (2001) on the basis of fossils recovered from the Nachukui Formation, at Lomekwi, close to the western shore of Lake Turkana. The fossils, which date to *c.* 3.5 million years ago, have been assigned to the species *K. platyops* (Leakey *et al.*, 2001). As noted above, it has been suggested recently that the collection of fossils that are currently assigned to *H. rudolfensis* should be reassigned to *Kenyanthropus* as *K. rudolfensis* (Leakey *et al.*, 2001; Lieberman, 2001). If this suggestion is accepted, then the last appearance date of *Kenyanthropus* is 1.8 million years ago. *Orrorin* was erected by Senut *et al.* (2001) for material recovered from several localities in the Lukeino Formation in Kenya's Tugen Hills. The material dates to around 6 million years ago and has been assigned to the species *O. tugenensis* (Pickford & Senut, 2001; Senut *et al.*, 2001).

RECOGNISING HOMINID GRADES

Huxley (1958) suggested that for a taxon to be recognised as a grade it has to emerge and persist. In his view, emergence is proof of adaptive change, and persistence is evidence that the taxon is a successful adaptive type. However, these criteria are problematic for palaeoanthropologists. For taxa with long fossil records they work reasonably well, but persistence is a difficult criterion to apply to taxa with shorter evolutionary histories, such as *H. sapiens*, which probably arose only 200–150 thousand years ago. Accordingly, a different approach is adopted in this chapter, one that is not time-dependent and is applicable to both recently and more distantly evolved taxa (Wood & Collard, 1997; Collard & Wood, 1999).

For a mammalian taxon to emerge and persist, the individual animals that belong to it have to flourish in the face of the challenges posed by their environment to the extent that they can produce fertile offspring. To accomplish this they must meet three basic requirements: they must be able to maintain themselves in homeostasis despite fluctuations in the ambient levels of temperature and humidity, and in spite of any restrictions in the availability of water; they must acquire and process sufficient food to meet their minimum requirements for energy and for amino acids and trace elements; and they must be able convince a member of the opposite sex to accept them as a sexual partner. The ways in which a species meets these fundamental requirements is clearly dependent on its adaptive organisation. Thus, one method of assessing how many grades are represented in a sample of species is to look for major differences in the way in which they maintain homeostasis, acquire food and produce offspring. Many aspects of a primate's ontogeny and phenotype help it carry out these three tasks, but not all of them can be reconstructed reliably from the fossil record. Arguably, the most important of those that can be determined using palaeontological evidence are locomotor behaviour, body size, stature, sexual dimorphism, the relative size of the masticatory apparatus, relative brain size and the rate and pattern of development.

As a pervasive factor in the life of any motile organism, locomotion affects the maintenance of homeostasis, the acquisition of food, and the production of offspring. In primates body mass and stature affect many physiological, ecological and life-history variables, including thermoregulation, population density and home range (Wheeler, 1991, 1992; Ruff, 1991, 1993, 1994; Ruff & Walker, 1993; McHenry, 1994; Hens *et al.*, 2000). Sex differences in body mass have also been found to co-vary with important ecological and life-history variables in mammals, such as the intensity and frequency of male–male competition, and the operational sex ratio (Crook, 1972; Clutton-Brock *et al.*, 1977; Alexander *et al.*, 1979; Mitani *et al.*, 1996; Plavcan & van Schaik, 1997; Plavcan, 2001). The relative size of the masticatory apparatus of a species is linked to the effectiveness with which the food items consumed are rendered suitable for chemical digestion (Teaford & Ungar, 2000). For example, the relative size of the occlusal surface of the cheek teeth determines how efficiently a given quantity of food will be broken down. Likewise, the cross-sectional area of the mandibular body determines the amount of chewing-induced stress it can withstand, such that an individual with a large mandibular corpus can either break down tougher food items, or process larger quantities of less resistant food, more readily than one with a more slender mandibular body. Relative neocortex size in primates determines the principal social interactions that are involved in reproduction (Dunbar, 1992, 1995; Aiello & Dunbar, 1993). Primates with relatively large neocortices tend to live in large social groups, while those with relatively small neocortices usually live in small groups. This relationship most probably arises from the role of the neocortex in processing information about social relationships; a larger neocortex allows a greater number of relationships to be tracked and maintained, and hence a larger social group to be formed (Dunbar, 1992, 1995). Additionally, there is a positive correlation between relative neocortex size and behavioural flexibility (Reader & Laland, 2002). The length of the period of development is adaptively significant because it influences parental investment and the acquisition of learned behaviours (Beynon & Dean, 1988). Species with longer maturation periods are expected to exhibit greater parental investment and a larger number of learned behaviours than species with shorter periods of maturation (Beynon & Dean, 1988).

HOMINID ADAPTIVE TYPES

In this section, evidence pertaining to the key adaptive variables outlined above will be reviewed with a view to identifying groups among the hominids that may represent different grades.

Locomotor behaviour

The locomotor behaviour of *A. afarensis* is contested (Johanson & Coppens, 1976; Johanson & Taieb, 1976; Lovejoy, 1979, 1981, 1988; Johanson *et al.*, 1982; Stern & Susman, 1983; Susman *et al.*, 1984; Senut & Tardieu, 1985; Tague & Lovejoy, 1986; Latimer, 1991; Schmid, 1991; Hunt, 1994, 1996; Ohman *et al.*, 1997; Crompton *et al.*, 1998; Stern, 1999, 2000). Some characteristics are argued to indicate that *A. afarensis* employed modern human-like terrestrial bipedalism. Others are said to indicate that the bipedalism of *A. afarensis* involved less extension of the knee and hip than that of modern humans. Still other characteristics are posited as adaptations for climbing. On balance, a reasonable working hypothesis is that *A. afarensis* combined a form of terrestrial bipedalism with an ability to move about effectively in trees (Collard & Wood, 1999; McHenry & Coffing, 2000; Wood & Richmond, 2000). Recent analyses have indicated that the post-cranial skeletons of *A. africanus* and *A. anamensis* are similar to that of *A. afarensis* (McHenry, 1986, 1994; Abitbol, 1995; Clarke & Tobias, 1995; Leakey *et al.*, 1995; Lague & Jungers, 1996; McHenry & Berger, 1998; Ward *et al.*, 2001), which suggests that they too were facultative bipeds. The associated skeleton (BOU-VP-12/1) that may represent *A. garhi* differs from those of the other *Australopithecus* species in that it exhibits modern human-like elongation of the femur (Asfaw *et al.*, 1999). However, BOU-VP-12/1 also exhibits a forearm to upper arm ratio that is similar to *Pan* (Asfaw *et al.*, 1999), which suggests that it probably also combined bipedalism with climbing.

Few post-cranial fossils can definitely be attributed to *P. boisei*, but the available specimens suggest that, like *A. afarensis*, *A. africanus* and *A. anamensis*, *P. boisei* probably combined bipedal locomotion with proficient climbing (McHenry, 1973; Howell & Wood, 1974; Howell, 1978; Grausz *et al.*, 1988; Aiello & Dean, 1990). The post-cranial skeleton of *P. robustus* is also poorly known, and opinions differ over the functional interpretation of what material there is. For example, Susman (1988) suggests that it was more modern human-like in both its hands and its feet than *A. afarensis*, with the hand bones showing evidence of *Homo*-like manipulative abilities, while the foot bones indicate that it was more bipedal and less arboreal than *A. afarensis*. In contrast, a comparison of the distal humerus of the type specimen, TM 1517, with those of humans and apes indicates that the upper limbs of *P. robustus* were longer in relation to its lower limbs than is the case in modern humans (Aiello & Dean, 1990). Thus, it would appear that, even if *P. robustus* was not as arboreal as *A. afarensis*, *A. africanus* and *A. anamensis*, it is likely that its post-cranial morphology would have allowed it some arboreal capability.

The *H. habilis* hypodigm includes two fragmentary skeletons, OH 62 and KNM-ER 3735. The limb proportions of these specimens have been interpreted as evidence that *H. habilis* combined terrestrial bipedalism with climbing (Johanson *et al.*, 1987; Aiello & Dean, 1990; Hartwig-Scherer & Martin, 1991). Indeed, Hartwig-Scherer & Martin's (1991) study suggests that the intermembranal proportions, and therefore the mode of locomotion, of *H. habilis* were even less similar to those of modern humans than were those of *A. afarensis*. The mixed locomotor hypothesis is further supported by analyses of the hand bones associated with the type specimen OH 7 (Susman & Creel, 1979; Susman & Stern, 1979, 1982) and by analyses of the OH 8 foot (Kidd *et al.*, 1996).

The post-cranial evidence for *O. tugenensis* is limited, but the lower limb specimens that have been recovered suggest that it employed some form of bipedal locomotion (Senut *et al.*, 2001). The humeral and phalangeal remains, on the other hand, imply that *O. tugenensis* was a proficient climber (Senut *et al.*, 2001). Thus, like the australopithecines, paranthropines and *H. habilis*, *O. tugenensis* was most probably a facultative biped.

In contrast to the foregoing species, *H. ergaster* seems to have been an obligate terrestrial biped much like *H. sapiens*. Its lower limbs and pelvis indicate a commitment to bipedal locomotion that was equivalent to that seen in modern humans, and there is no evidence in the upper limbs for the sort of climbing abilities possessed by *Australopithecus*, *Paranthropus* and *H. habilis* (Walker & Leakey, 1993). Moreover, *H. ergaster* had a barrel-shaped thoracic cage and narrow waist, which implies that it may have been an efficient runner and/or able to travel long distances (Schmid, 1991; Aiello & Wheeler, 1995). The post-cranial skeleton of *H erectus* is relatively poorly known, with most of the relevant evidence consisting of pelves and femora. These bones differ from those of modern humans in some characters (for example greater robusticity, narrower medullary canal), but they are nonetheless consistent with modern human-like posture and gait (Wood & Richmond, 2000). The post-cranial remains of *H. antecessor*, *H heidelbergensis* and *H. neanderthalensis* are also consistent with modern human-like posture and gait (Stringer & Gamble, 1993; Roberts *et al.*, 1994; Arsuaga *et al.*, 1999; Carretero *et al.*, 1999).

Thus, on the basis of the locomotor inferences that can be made from their post-cranial morphology, the fossil hominids can be divided into two groups. The first group is composed of facultative bipeds. They combined a form of terrestrial bipedalism with an ability to climb proficiently. This group includes *A. afarensis*, *A. africanus*, *A. anamensis*, *A. garhi*, *O. tugenensis*, *P. robustus*, *P. boisei* and *H. habilis*. The second group comprises *H. antecessor*, *H. ergaster*, *H. erectus*, *H. heidelbergensis* and *H. neanderthalensis*, and is characterised by obligate terrestrial bipedalism. Currently little can be said about the locomotor repertoires of *A. ramidus*, *A. bahrelghazali*, *H. rudolfensis*, *K. platyops* and *P. aethiopicus*. No post-cranial fossils are reliably attributed to *A. bahrelghazali*, *K. platyops* or *P. aethiopicus*. Post-cranial fossils of *A. ramidus* have been found (White *et al.*, 1994, 1995) but no compelling evidence on its locomotor abilities is available at the moment. It has been claimed that the femora KNM-ER 1472 and KNM-ER 1481a and the pelvic bone KNM-ER 3228 represent *H. rudolfensis* (Wood, 1992; McHenry & Coffing, 2000). However, the attribution of these bones to *H. rudolfensis* is problematic, because the dates of the earliest *H. ergaster* specimens are within the *H. rudolfensis* time range (Wood, 1991; Wood & Collard, 1999a; Wood & Richmond, 2000). Also, it has been argued on morphological grounds that KNM-ER 1472 and KNM-ER 1481a represent *H. ergaster* (Kennedy, 1983; but see Trinkaus, 1984). As such, it is probably best to wait for evidence from associated skeletal evidence before assessing the locomotor habits of *H. rudolfensis* (Wood & Collard, 1999a, 1999b).

The hypothesised contrast between the locomotor repertoires of the two groups of hominids is supported by the work of Spoor *et al.* (1994, 1996). These authors used high-resolution computed tomography to examine the dimensions of the inner ear of a sample of extant primate species and modern humans. In line with the known relationship between the morphology of the inner ear, balance and locomotion, they found that the signature for the obligate terrestrial bipedalism of *H. sapiens* was different from the signature for the type of arboreally orientated locomotion of the great apes. Having established this predictive model, they then examined the inner ear morphology of specimens that have been assigned to *A. africanus*, *H. habilis*, *H. ergaster* and *H. erectus*. They found that the dimensions of the vestibular apparatus of the *Australopithecus* and *Paranthropus* specimens were similar to those of the great apes, while those of the *H. ergaster* and *H. erectus* specimens were similar to those of *H. sapiens*. This suggests, according to Spoor *et al.* (1994, 1996), that the former spent a substantial proportion of their time in an arboreal setting, while the latter was as much an obligate terrestrial biped as *H. sapiens*. Spoor *et al.* (1994, 1996) found that the vestibular dimensions of *H. habilis* were most similar to large terrestrial quadrupedal primates, which led them to conclude that *H. habilis* is unlikely to have been an obligate biped.

Body mass

Table 2 presents estimated mean body masses for *A. ramidus*, *A. afarensis*, *A. africanus*, *A. anamensis*, *H. ergaster*, *H. erectus*, *H. habilis*, *H. heidelbergensis*, *H. neanderthalensis*, *H. rudolfensis*, *P. aethiopicus*, *P. boisei* and *P. robustus*, together with anthropometrically recorded body masses for several *H. sapiens* groups. Two groups are evident in these data. One group comprises *A. ramidus*, *A. afarensis*, *A. africanus*, *H. habilis*, *H. rudolfensis*, *P. aethiopicus*, *P. boisei* and *P. robustus*. The other consists of *H. ergaster*, *H. erectus*, *H. heidelbergensis*, *H. neanderthalensis* and *H. sapiens*. The largest species in the former group, *P. boisei*, is estimated to have had a mean body mass of 41.3 kg, whereas the smallest species in the latter group, *H. ergaster*, is estimated to have had a mean body mass of 57.8 kg. *Australopithecus anamensis*, which has an estimated body mass of 51 kg, falls between these two groups, and therefore blurs the distinction between them. However, it is likely that 51 kg is not an accurate estimate of the species mean body mass of *A. anamensis* because it is derived from a single specimen that is thought to be male (Ward *et al.*, 2001). If it is assumed that *A. anamensis* displayed a level of sexual dimorphism similar to that seen in the other *Australopithecus* species (see below), then it is likely that its species mean body mass was less than 50 kg. Currently published body mass estimates are not available for *A. bahrelghazali*, *A. garhi*, *H. antecessor*, *K. platyops* and *O. tugenensis*. However, based on the size of the available post-cranial

Table 2. Hominid body mass. The body masses for the *H. sapiens* groups are from anthropometric studies. The fossil hominid body masses are derived from post-cranial data-based regression equations, except the estimates for *H. rudolfensis* and *P. aethiopicus*, which are based on cranial data. The taxa are listed in alphabetical order

Taxon		Body mass (kg)	Source of data
A. ramidus	Male	–	Data from Wood & Richmond (2000: 26)
	Female		
	Mean	40.0	
A. afarensis	Male	44.6	Data from McHenry (1994: table 1)
	Female	29.3	
	Mean	37.0	
A. africanus	Male	40.8	Data from McHenry (1994: table 1)
	Female	30.2	
	Mean	35.5	
A. anamensis	Male	51.0	Average of estimates based on promixal (55 kg) and distal (47 kg) dimensions of a single tibia, KNM-KP 29285, which is believed to have belonged to a male (Leakey *et al.*, 1995; Ward *et al.*, 2001)
	Female	–	
	Mean	51.0	
H. ergaster	Male	63.0	Data from McHenry (1994: table 2)
	Female	52.0	
	Mean	57.5	
H. erectus	Male	63.0	Data from McHenry (1994: table 2)
	Female	52.5	
	Mean	57.8	
H. habilis	Male	37.0	Data from McHenry (1994: table 2)
	Female	31.5	
	Mean	34.3	
H. heidelbergensis	Male	–	Average of five estimates. Four were computed using Hartwig-Scherer's (1993) *Homo* equation for tibial circumference/body mass, and values for tibial midshaft circumference for Boxgrove, Kabwe and two Atapuerca tibia given by Roberts *et al.* (1994). The fifth estimate is for Atapuerca Pelvis 1 and is taken from Arsuaga *et al.* (1999). The specimen estimates are 80.0 kg (Boxgrove), 66.5 kg (Kabwe), 49.5 kg (Atapuerca Tibia 1), 53.7 kg (Atapuerca Tibia 2), 94.0 kg (Atapuerca Pelvis 1)
	Female	–	
	Mean	68.7	
H. neanderthalensis	Male	73.7	Male value is the mean of estimates for Amud 1 (68.5 kg), La Chapelle (78.5 kg), La Ferrassie R (84.3 kg), Shanidar 4 (70.7 kg) and Shanidar 5 (66.6 kg) presented by Kappelman (1996: table 6). The female value is the estimate for Tabun C1 given by Kappelman (1996: table 6)
	Female	56.1	
	Mean	64.9	

Table 2. (Continued)

Taxon	Group		Body mass (kg)	Source of data
H. rudolfensis		Male	–	Orbital area based estimate for KNM-ER 1470 given by Kappelman (1996: table 4)
		Female	–	
		Mean	45.6	
H. sapiens	Aita	Male	60.9	Male and female data from Houghton (1996: tables 2.1 and 2.2)
		Female	54.1	
		Mean	57.5	
H. sapiens	Baining	Male	60.1	Male and female data from Houghton (1996: tables 2.1 and 2.2)
		Female	47.9	
		Mean	54.0	
H. sapiens	Bantu	Male	56.0	Male and female data from Wood (1995: table 29.2).
		Female	49.0	
		Mean	52.5	
H. sapiens	Karkar	Male	56.4	Male and female data from Houghton (1996: tables 2.1 and 2.2)
		Female	47.0	
		Mean	51.7	
H. sapiens	Manus	Male	60.2	Male and female data from Houghton (1996: tables 2.1 and 2.2)
		Female	48.2	
		Mean	54.2	
H. sapiens	Nagovisi	Male	58.6	Male and female data from Houghton (1996: tables 2.1 and 2.2)
		Female	49.1	
		Mean	53.9	
H. sapiens	Nasioi	Male	57.7	Male and female data from Houghton (1996: tables 2.1 and 2.2)
		Female	48.2	
		Mean	53.0	
H. sapiens	Ontong Java	Male	67.7	Male and female data from Houghton (1996: tables 2.1 and 2.2)
		Female	59.6	
		Mean	63.7	
H. sapiens	Pukapuka	Male	69.0	Male and female data from Houghton (1996: tables 2.1 and 2.2)
		Female	60.7	
		Mean	64.9	
H. sapiens	Samoa	Male	75.9	Male and female data from Houghton (1996: tables 2.1 and 2.2)
		Female	70.4	
		Mean	73.2	
H. sapiens	Tokelau	Male	69.7	Male and female data from Houghton (1996: tables 2.1 and 2.2)
		Female	70.6	
		Mean	70.2	
H. sapiens	Tolai	Male	60.6	Male and female data from Houghton (1996: tables 2.1 and 2.2)
		Female	55.1	
		Mean	57.9	
H. sapiens	Tonga (Foa)	Male	75.2	Male and female data from Houghton (1996: tables 2.1 and 2.2)
		Female	71.0	
		Mean	73.1	

Table 2. (Continued)

Taxon	Group		Body mass (kg)	Source of data
H. sapiens	Ulawa	Male	60.9	Male and female data from Houghton (1996: tables 2.1 and 2.2)
		Female	50.0	
		Mean	55.5	
H. sapiens	Mean	Male	63.5	Means of body masses of preceding 14 *H. sapiens* groups
		Female	52.3	
		Mean	59.7	
P. aethiopicus		Male	–	Orbital area based estimate for KNM-WT 17000 given by Kappelman (1996: table 4)
		Female	–	
		Mean	37.6	
P. boisei		Male	48.6	Data from McHenry (1994: table 1)
		Female	34.0	
		Mean	41.3	
P. robustus		Male	40.2	Data from McHenry (1994: table 1)
		Female	31.9	
		Mean	36.1	

evidence, it is reasonable to assume that the species mean body masses of *A. garhi* and *O. tugenensis* would have been relatively low, most probably less than 50 kg (Asfaw *et al.*, 1999; Senut *et al.*, 2001). It is also reasonable to assume, on the basis of the size of the available evidence, that the species mean body mass of *H. antecessor* was relatively high, most probably in excess of 50 kg (Bermudez de Castro *et al.*, 1997; Carretero *et al.*, 1999). Thus, the hominids fall into two groups in terms of species mean body mass. The first group comprises *A. ramidus*, *A. afarensis*, *A. africanus*, *A. anamensis*, *A. garhi*, *H. habilis*, *H. rudolfensis*, *O. tugenensis*, *P. aethiopicus*, *P. boisei* and *P. robustus*. These species have mean body masses lower than 50 kg. The second group comprises *H. antecessor*, *H. ergaster*, *H. erectus*, *H. heidelbergensis* and *H. neanderthalensis*. The body masses of these species exceed 50 kg.

Stature

Table 3 presents stature estimates for *A. afarensis*, *A. africanus*, *A. garhi*, *H. antecessor*, *H. ergaster*, *H. erectus*, *H. habilis*, *H. heidelbergensis*, *H. neanderthalensis*, *P. boisei* and *P. robustus*, as well as anthropometrically recorded statures for 13 groups of *H. sapiens*. These data clearly divide the hominids into two groups. One group consists of *A. afarensis*, *A. africanus*, *A. garhi*, *H. habilis*, *P. boisei* and *P. robustus*. These species have mean statures of less than 150 cm. The other group consists of *H. antecessor*, *H. ergaster*, *H. erectus*, *H. heidelbergensis*, *H. neanderthalensis* and *H. sapiens*. These

Table 3. Hominid stature. The statures for the *H. sapiens* groups are from anthropometric studies. The fossil hominid statures are derived from post-cranial data-based regression equations. The taxa are listed in alphabetical order

Taxon		Stature (cm)	Notes
A. afarensis	Male Female Mean	– – 128.0	Average of McHenry's (1991: table 2) estimates for AL 288-1ap (105 cm) and AL 333-3 (151 cm)
A. africanus	Male Female Mean	– – 124.4	Average of McHenry's (1991: table 2) estimates for Sts 14 (110 cm), Stw 25 (120 cm), Stw 99 (142 cm), Sts 392 (116 cm) and Stw 443 (134 cm)
A. garhi	Male Female Mean	– – 123.7	Average of Hens *et al.*'s (2000: table 8) five ape equation-based estimates for BOU-VP-35/1. The five estimates are 124.8 cm (Inverse), 123.8 cm (Classical), 124.4 cm (RMA), 121.4 cm (Ratio) and 124.0 cm (MA)
H. antecessor	Male Female Mean	– – 172.6	Average of the mean metatarsal estimate (170.9 cm) presented by Lorenzo *et al.* (1999), plus the mean radial (172.5 cm) and clavicular (174.5 cm) estimates reported by Carretero *et al.* (1999)
H. ergaster	Male Female Mean	185.0 – 185.0	Ruff & Walker's (1993) estimate of the adult stature of KNM-WT 15000, which is a near-complete skeleton of a male *H. ergaster* juvenile
H. erectus	Male Female Mean	– – 166.5	Average of McHenry's (1991: table 2) estimates for OH 34 (162 cm) and OH 28 (171 cm). The other *H. erectus* estimates provided by McHenry (1991) were not employed due to uncertainty regarding the taxonomic status of the specimens concerned (Wood & Collard, 1999a, 1999b)
H. habilis	Male Female Mean	– – 118.0	McHenry's (1991: table 2) estimate for OH 62Y. The other *H. habilis* estimates provided by McHenry (1991) were not employed due to uncertainty regarding the taxonomic status of the specimens concerned (Wood & Collard, 1999a, 1999b)
H. heidelbergensis	Male Female Mean	– – 176.0	Average of the estimates for Boxgrove 1 (175.3 cm), Berg Aukas 1 (181.6 cm) and Broken Hill E691 (174.0 cm) presented by Stringer *et al.* (1998) and the mean estimate (173.1 cm) derived from a humerus from Sima de los Huesos, Atapuerca, by Carretero *et al.* (1997)

Table 3. (Continued)

Taxon	Group		Stature (cm)	Notes
H. neanderthalensis		Male	–	Average of estimates for Spy 1 (167.0 cm) and Spy 2 (162.0 cm) given by Houghton (1996: table 3.14)
		Female	–	
		Mean	165.0	
H. sapiens	Aita	Male	159.6	Data from Houghton (1996: tables 2.1 and 2.2)
		Female	149.8	
		Mean	154.7	
H. sapiens	Baining	Male	157.7	Data from Houghton (1996: tables 2.1 and 2.2)
		Female	147.7	
		Mean	152.7	
H. sapiens	Karkar	Male	161.0	Data from Houghton (1996: tables 2.1 and 2.2)
		Female	151.1	
		Mean	156.1	
H. sapiens	Manus	Male	162.9	Data from Houghton (1996: tables 2.1 and 2.2)
		Female	151.0	
		Mean	157.0	
H. sapiens	Nagovisi	Male	160.5	Data from Houghton (1996: tables 2.1 and 2.2)
		Female	151.3	
		Mean	155.9	
H. sapiens	Nasioi	Male	163.2	Data from Houghton (1996: tables 2.1 and 2.2)
		Female	152.3	
		Mean	158.8	
H. sapiens	Ontong Java	Male	166.2	Data from Houghton (1996: tables 2.1 and 2.2)
		Female	156.0	
		Mean	161.1	
H. sapiens	Pukapuka	Male	168.8	Data from Houghton (1996: tables 2.1 and 2.2)
		Female	157.2	
		Mean	163.0	
H. sapiens	Samoa	Male	171.4	Data from Houghton (1996: tables 2.1 and 2.2)
		Female	159.2	
		Mean	165.3	
H. sapiens	Tokelau	Male	167.4	Data from Houghton (1996: tables 2.1 and 2.2)
		Female	161.0	
		Mean	164.2	
H. sapiens	Tolai	Male	163.6	Data from Houghton (1996: tables 2.1 and 2.2)
		Female	155.7	
		Mean	160.0	
H. sapiens	Tonga (Foa)	Male	171.3	Data from Houghton (1996: tables 2.1 and 2.2)
		Female	161.8	
		Mean	166.6	
H. sapiens	Ulawa	Male	162.9	Data from Houghton (1996: tables 2.1 and 2.2)
		Female	151.0	
		Mean	157.0	

Table 3. (Continued)

Taxon			Stature (cm)	Notes
H. sapiens	Mean	Male	164.3	Means of statures of preceding 13 *H. sapiens* groups
		Female	154.2	
		Mean	159.4	
P. boisei		Male	–	McHenry's (1991: table 2) estimate for KNM-ER 1500d, which he contends is a female. McHenry (1992) suggests that female *P. boisei* were 124 cm and male *P. boisei* were 137 cm, but it is not clear how these values were obtained. Thus, they were not used
		Female	–	
		Mean	115.0	
P. robustus		Male	–	Average of McHenry's (1991: table 2) estimates for SK 82 (126 cm), Sk 97 (137 cm) and SK 3155B (110 cm)
		Female	–	
		Mean	124.3	

species have mean statures in excess of 150 cm. As noted earlier there are no post-cranial remains that can be reliably attributed to *A. bahrelghazali*, *H. rudolfensis*, *K. platyops* and *P. aethiopicus*. Hence, it is not possible to estimate their species mean statures. Reliable stature estimates have yet to be published for *A. ramidus*, *A. anamensis* and *O. tugenensis*. However, the hypodigms of these species include post-cranial specimens, so rough estimates of stature are possible. Based on the size of the available post-cranial evidence, it is reasonable to assume that the species mean statures of *A. anamensis* and *O. tugenensis* would have been less than 150 cm (White *et al.*, 1994; Leakey *et al.*, 2001; Senut *et al.*, 2001). Thus, the hominids can be divided into two groups on the basis of stature. The first comprises *A. ramidus*, *A. afarensis*, *A. africanus*, *A. anamensis*, *A. garhi*, *H. habilis*, *O. tugenensis*, *P. boisei* and *P. robustus*, and is characterised by a species mean stature of less than 150 cm. The other group comprises *H. ergaster*, *H. erectus*, *H. antecessor*, *H. heidelbergensis*, *H. neanderthalensis* and *H. sapiens*, and is characterised by a species mean stature in excess of 150 cm.

Sexual dimorphism

Table 4 presents percentage body mass dimorphism values for *A. afarensis*, *A. africanus*, *H. ergaster*, *H. erectus*, *H. habilis*, *H. neanderthalensis*, *P. boisei* and *P. robustus*, plus several groups of *H. sapiens*. These data indicate that hominid species vary markedly in body mass sexual dimorphism. *Australopithecus afarensis* males were more than 50% larger than *A. afarensis* females, whereas in some modern human groups males and females are essentially the

Table 4. Hominid body size dimorphism. Male = male body mass. Female = female body mass. The sources of the body mass data are listed in Table 2. PBM = male body mass as a percentage of female body mass. The taxa are listed in alphabetical order

Taxon	Group	Male (kg)	Female (kg)	PBM (%)
A. afarensis		44.6	29.3	152
A. africanus		40.8	30.2	135
H. ergaster		63.0	52.0	121
H. erectus		63.0	52.5	120
H. habilis		37.0	31.5	117
H. neanderthalensis		73.7	56.1	131
H. sapiens	Aita	60.9	54.1	113
H. sapiens	Baining	60.1	47.9	125
H. sapiens	Bantu	56.0	49.0	114
H. sapiens	Karkar	56.4	47.0	120
H. sapiens	Nasioi	57.7	48.2	120
H. sapiens	Manus	60.2	48.2	125
H. sapiens	Nagovisi	58.6	49.1	119
H. sapiens	Ontong Java	67.7	59.6	114
H. sapiens	Pukapuka	69.0	60.7	114
H. sapiens	Samoa	75.9	70.4	108
H. sapiens	Tokelau	69.7	70.6	99
H. sapiens	Tolai	60.6	55.1	110
H. sapiens	Tonga (Foa)	75.2	71.0	106
H. sapiens	Ulawa	60.9	50.0	122
H. sapiens	Mean	63.5	52.3	121
P. boisei		48.6	34.0	143
P. robustus		40.2	31.9	126

same size (e.g. Tokelau and Tonga). Furthermore, the modern human sample indicates that within-species variation in sexual dimorphism can be considerable. In several *H. sapiens* groups males are 20% larger than females, while in others the sexes are more or less the same size. The extent of this intraspecific variability suggests that body mass sexual dimorphism estimates for fossil hominid groups should be interpreted cautiously. Overall, the data suggest that the species fall into two groups with regard to body mass sexual dimorphism data. The first comprises *A. afarensis*, *A. africanus*, *H. neanderthalensis*, *P. boisei* and *P. robustus*. The second comprises *H. ergaster*, *H. erectus*, *H. habilis* and *H. sapiens*. Body mass sexual dimorphism in the former group is high, ranging between 126% and 152%. In the latter group, body mass sexual dimorphism is moderate, ranging between 121% and 117%.

The position of *H. neanderthalensis* in the high body mass group does not accord with the results of studies that have examined dimorphism in skeletal features. Trinkaus (1980), for example, found that Neanderthal limb bones exhibit a similar level of sexual dimorphism to that seen in a large and geographically diverse sample of modern humans. Likewise, Smith's (1980) analysis of craniometric variables found that Neanderthal males were only between 2% and 10% larger than Neanderthal females. Most recently, Quinney & Collard (1997) found that Neanderthals display no more sexual dimorphism in their mandibles than Holocene humans. Thus, it is possible that the high body mass dimorphism value for *H. neanderthalensis* shown in Table 4 (131%) is misleading, and that the Neanderthals belong in the moderate dimorphism group with the other *Homo* species. It is also possible that the position of *H. ergaster* in the second group may need to be revised in the near future. Susman *et al.* (2001) have suggested recently that South African male *H. ergaster* may have averaged around 55 kg, while females of the species averaged about 30 kg. These estimates yield a percentage dimorphism of 183%, which is greater than any other hominid species.

At the moment it is not possible to estimate body mass dimorphism in *A. ramidus*, *A. anamensis*, *A. bahrelghazali*, *A. garhi*, *H. rudolfensis*, *H. antecessor*, *H. heidelbergensis*, *K. platyops*, *O. tugenensis* and *P. aethiopicus* using the same approach. However, the cranial and post-cranial remains of *A. anamensis* and *A. garhi* suggest that these species exhibited a similar level of sexual dimorphism to the other *Australopithecus* species (Asfaw *et al.*, 1999; Ward *et al.*, 2001). Additionally, analyses of body size variation in *H. heidelbergensis* indicate that this species had a level of body mass sexual dimorphism comparable to that of *H. sapiens* (Arsuaga *et al.*, 1997; Lorenzo *et al.*, 1998).

In sum, the hominids can be divided into two groups with regard to body mass sexual dimorphism. One group is characterised by high sexual dimorphism, the other by moderate sexual dimorphism. *Australopithecus afarensis*, *A. africanus*, *A. anamensis*, *A. garhi*, *P. boisei* and *P. robustus* can be relatively securely assigned to the first group. *Homo neanderthalensis* also appears to have exhibited high sexual dimorphism on the basis of post-cranial body mass estimates, but other evidence suggests that it may have displayed moderate body mass sexual dimorphism. *Homo erectus*, *H. habilis*, *H. heidelbergensis* and *H. sapiens* can be allocated to the second group with reasonable confidence. *Homo ergaster* can also be assigned to the moderate sexual dimorphism group on the basis of the body mass estimates presented in Table 4, but with less certainty.

Relative size of the masticatory apparatus

Table 5 gives species means for 11 variables from the lower posterior dentition and mandible for *A. africanus*, *H. ergaster*, *H. erectus*, *H. habilis*, *H. neanderthalensis*, *H. rudolfensis*, *H. sapiens*, *P. boisei* and *P. robustus*, together with mean body masses for the species. Figure 1 presents a dendrogram that was derived from the dental and mandibular data after they had been adjusted to counter the confounding effects of differential body mass. It is evident from the dendrogram that the species form two main groups in terms

Table 5. Hominid species means for 11 dental and mandibular measurements and body mass. The measurement codes follow Wood (1991). 141 = symphyseal height/mm; 142 = symphyseal breadth/mm; 150 = corpus height at M_1/mm; 151 = corpus width at M_1/mm; 271 = P_4 mesiodistal diameter/mm; 272 = P_4 buccolingual diameter/mm; 285 = M_1 mesiodistal diameter/mm; 286 = M_1 buccolingual diameter/mm; 313 = M_2 mesiodistal diameter/mm; 314 = M_2 buccolingual diameter/mm; 345 = square root of M_3 area/mm^2; BM = body mass/kg. The dental and mandibular data are taken from Wood & Collard (1999a). The body masses are from Table 2

Taxon	141	142	150	151	271	272	285	286	313	314	345	BM
A. africanus	41.0	20.0	33.0	23.0	9.3	11.0	13.2	12.9	14.9	14.1	14.8	35.5
H. ergaster	33.0	20.0	31.0	19.0	8.7	11.0	13.1	11.6	13.8	12.3	13.0	57.5
H. erectus	37.0	19.0	36.0	22.0	8.9	11.3	12.4	12.0	13.3	12.7	12.0	57.8
H. habilis	27.0	19.0	29.0	21.0	9.8	10.5	13.9	12.3	14.9	12.6	14.2	34.3
H. neanderthalensis	42.0	15.0	34.0	18.0	7.1	8.7	10.6	10.7	11.1	10.7	11.4	64.9
H. rudolfensis	36.0	23.0	36.0	23.0	10.5	12.0	14.0	13.2	16.4	13.7	15.8	45.9
H. sapiens	34.0	14.0	29.0	13.0	7.1	8.4	11.2	10.5	10.8	10.5	10.6	59.7
P. boisei	51.0	29.0	42.0	29.0	14.2	15.5	16.7	15.7	20.4	18.5	18.1	41.3
P. robustus	50.0	28.0	39.0	27.0	11.7	14.0	15.1	14.1	16.6	15.7	15.9	36.1

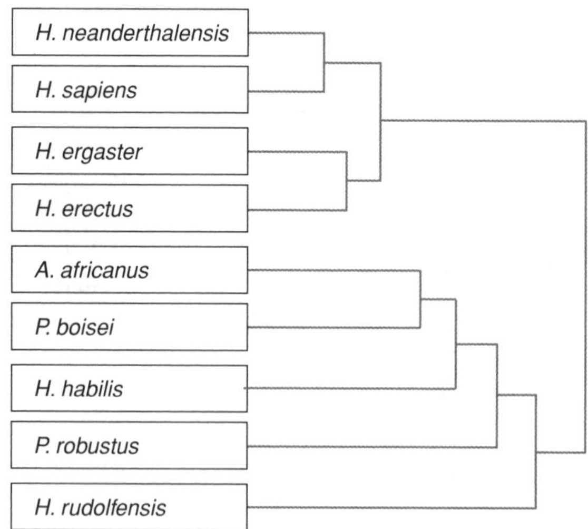

Figure 1. Dendrogram summarising similarities among hominid species in terms of the relative size of their masticatory apparatus. To obtain the dendrogram, the dental and mandibular species means presented in Table 5 were adjusted to counter the confounding effects of body size by dividing each of them by the cube root of the appropriate species mean body mass. Thereafter, the data were standardised and Euclidean distances among the taxa computed. Lastly, the Euclidean distances were used to construct a nearest neighbour dendrogram.

of the relative size of their teeth and mandibles. The first group comprises *A. africanus*, *H. habilis*, *H. rudolfensis*, *P. boisei* and *P. robustus*. The second comprises *H. ergaster*, *H. erectus*, *H. neanderthalensis* and *H. sapiens*. The species in the first group combine large teeth and jaws with a moderate body mass, whereas the species in the second group combine small teeth and jaws with a large body mass. While data for *A. ramidus*, *A. afarensis*, *A. anamensis*, *A. garhi* and *P. aethiopicus* were not included in the analysis, there are grounds to believe that, like *A. africanus*, *H. habilis*, *H. rudolfensis*, *P. boisei* and *P. robustus*, they were megadont (Walker *et al.*, 1986; Wood, 1991, 1995; Wood & Aiello, 1998; Asfaw *et al.*, 1999; Teaford & Ungar, 2000; Leakey *et al.*, 2001). Similarly, there is reason to think that the molars and mandibles of *H. antecessor* and *H. heidelbergensis* were small relative to their body mass, as is the case with *H. ergaster*, *H. erectus*, *H. neanderthalensis* and *H. sapiens* (Bermudez de Castro *et al.*, 1997; Wood & Richmond, 2000). Currently the relative size of the masticatory systems of *A. bahrelghazali* and *K. platyops* cannot be assessed. However, Leakey *et al.* (2001) note that the molars of KNM-WT 40000, the type specimen of *K. platyops*, are small, which may mean that *K. platyops* was not megadont. Overall, the available evidence suggests that the diets of *A. ramidus*, *A. afarensis*, *A. africanus*, *A. anamensis*, *A. garhi*, *H. habilis*, *H. rudolfensis*, *O. tugenensis*, *P. aethiopicus*, *P. boisei* and *P. robustus* required more bite force and processing than those of *H. antecessor*, *H. ergaster*, *H. erectus*, *H. heidelbergensis*, *H. neanderthalensis* and *H. sapiens*.

Relative brain size

As it is not possible to determine fossil hominid neocortex size with any certainty (Smith, 1996), the overall size of the brain is used as a proxy measure of neocortex size (Passingham & Ettlinger, 1974). Table 6 presents species mean estimates of absolute and relative brain size for *A. afarensis*, *A. africanus*, *H. ergaster*, *H. erectus*, *H. habilis*, *H. heidelbergensis*, *H. neanderthalensis*, *H. rudolfensis*, *H. sapiens*, *P. aethiopicus*, *P. boisei* and *P. robustus*. Relative brain size is in the form of the encephalisation quotient (EQ), which expresses brain size in relation to the estimated brain volume of a generalised placental mammal of the same body mass. The formula used here to calculate EQ is:

$$\text{EQ} = \text{observed endocranial volume}/0.0589(\text{body weight})^{0.76}$$

(Martin, 1981)

There are substantial differences in the mean absolute brain size of the australopithecines and paranthropines on the one hand, and the *Homo* species on the other. But most of these differences are almost certainly not meaningful when differences in body mass are taken into account. When this adjustment is made, the hominids cluster into two main groups (Figure 2). The first group

Table 6. Hominid absolute and relative brain size. CC = cranial capacity in cm^3; BM = body mass in kg; EQ = encephalisation quotient. The sources for the cranial capacity data are given in the fifth column of the table. The sources for the body mass data are given in Table 2. EQ was calculated using Martin's (1981) formula: EQ = observed endocranial volume/0.0589(body weight)$^{0.76}$. The taxa are listed in alphabetical order

Taxon	CC	BM	EQ	Source for CC
A. afarensis	404	37.0	2.3	Data from McHenry (1994: tables 1 and 3)
A. africanus	457	35.5	2.7	Data from Kappelman (1996: table 4)
H. ergaster	854	57.5	3.5	Computed from the values for KNM-Wt 15000 (909 cm^3), KNM-ER 3883 (804 cm^3) and KNM-ER 3733 (850 cm^3) given by Kappelman (1996: table 4)
H. erectus	1016	57.8	4.1	Computed from the values for Zhoukoudian XI (1015 cm^3), Zhoukoudian XII (1030 cm^3), Sangiran 17 (skull VIII) given by Kappelman (1996: table 4)
H. habilis	552	34.3	3.3	Data from Kappelman (1996: table 6)
H. heidelbergensis	1226	68.7	4.4	Average of values for Kabwe (1285 cm^3) and Steinheim (1110 cm^3) given by Kappelman (1996: table 4), and values for Atapuerca Skull 5 (1125 cm^3), Atapuerca Cranium 4 (1390 cm^3) and Atapuerca Cranium 6 (1220 cm^3) given by Arsuaga *et al.* (1997)
H. neanderthalensis	1512	64.9	5.7	Average of the values for Gibraltar 1 (1200 cm^3), Saccopastore (1245 cm^3), Le Moustier (1565 cm^3), La Chapelle (1625 cm^3), La Ferrassie (1689 cm^3), Amud 1 (1750 cm^3) given by Kappelman (1996)
H. rudolfensis	752	45.6	3.7	Data from Kappelman (1996: table 4)
H. sapiens	1355	59.7	5.4	Average of male and female values given by Kappelman (1996)
P. aethiopicus	410	37.6	2.3	Data from Kappelman (1996: table 4)
P. boisei	513	41.3	2.6	Average of the values for KNM-ER 732 (500 cm^3), KNM-ER 406 (510 cm^3) and OH 5 (530 cm^3) given by Kappelman (1996).
P. robustus	530	36.1	3.1	Data from McHenry (1994: table 3)

consists of *A. afarensis*, *A. africanus*, *H. ergaster*, *H. erectus, H. habilis*, *H. heidelbergensis*, *H. rudolfensis*, *P. aethiopicus*, *P. boisei* and *P. robustus*. Within the first group there are three subgroups. The first comprises

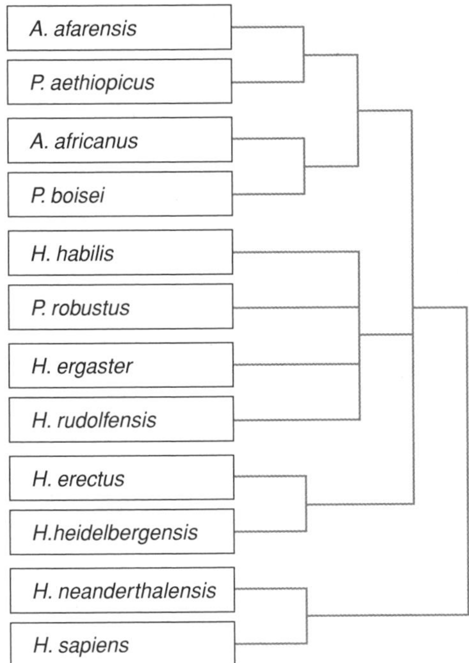

Figure 2. Dendrogram summarising similarities among hominid species in terms of the relative brain size.

A. afarensis, *A. africanus*, *P. aethiopicus* and *P. boisei*. The second comprises *H. ergaster*, *H. habilis*, *H. rudolfensis* and *P. robustus*. The third subgroup comprises *H. erectus* and *H. heidelbergensis*. The species in the first group are characterised by a brain that is moderate in size relative to body mass. Their EQs range between 2.3 and 4.4. The second group consists of *H. neanderthalensis* and *H. sapiens*. The species that form this group have large brains relative to their body masses. Their EQs are 5.4 and 5.7. Based on the available evidence it is reasonable to conclude that *A. garhi*, *K. platyops* and *H. antecessor* should also be assigned to the moderate size brain group (Bermudez de Castro *et al.*, 1997; Asfaw *et al.*, 1999; Leakey *et al.*, 2001). Currently it is not possible to estimate the relative brain size of *A. ramidus*, *A. anamensis*, *A. bahrelghazali* and *O. tugenensis*.

Development

Evidence pertaining to development is available for several species of *Australopithecus*, *Paranthropus* and *Homo* (Beynon & Dean, 1988; Smith, 1994; Dean, 1995, 2000; Tardieu, 1998; Clegg & Aiello, 1999; Moggi-Cecchi, 2000; Dean *et al.*, 2001). Analyses of dental and femoral development indicate that the developmental schedules of *A. afarensis*, *A. africanus*, *A. anamensis*, *H. habilis*, *H. rudolfensis*, *P. aethiopicus*, *P. boisei* and *P. robustus* were more similar to the developmental schedules of the African apes than to that of modern humans (Smith, 1994; Dean, 1995, 2000; Tardieu, 1998; Moggi-Cecchi, 2000; Dean *et al.*, 2001). Studies that have examined development in *H. ergaster* and *H. erectus* suggest that, while the pattern of development in these fossil species is similar to the pattern of development in *H. sapiens* (Beynon & Dean, 1988; Smith, 1994; Clegg & Aiello, 1999; Dean, 2000), the rate at which they developed was more ape-like than modern human-like (Dean, 2000; Dean *et al.*, 2001). Analyses of dental incremental markings indicate that the developmental schedule of *H. neanderthalensis* was comparable to that of *H. sapiens* (Dean *et al.*, 2001). Thus, the hominids for which evidence about development is available can be divided into two groups on the basis of their period of maturation. The first group comprises *A. afarensis*, *A. africanus*, *A. anamensis*, *P. aethiopicus*, *P. boisei*, *P. robustus*, *H. habilis*, *H. rudolfensis*, *H. ergaster* and *H. erectus*, and is characterised by a relatively short developmental period. The second group consists of *H. neanderthalensis* and *H. sapiens*. These species exhibit an extended period of development.

How many hominid grades?

Table 7 summarises the findings of the review. In the sample of hominids at least three grades can be recognised. The first of these is characterised by a species mean body mass less of than 50 kg; stature of less than 130 cm; facultative bipedalism; a relatively large masticatory system; a relatively small brain; and a rapid, ape-like developmental schedule. The second grade is characterised by a species mean body mass in excess of 50 kg; a stature in excess of 160 cm; obligate bipedalism; a relatively small masticatory system; an EQ of less than 4.5; and a short ape-like period of maturation. The third grade is similar to the second in terms of body mass, stature, locomotor behaviour and masticatory system size, but exhibits a considerably higher degree of encephalisation and delayed maturation. With varying degrees of certainty *A. ramidus*, *A. afarensis*, *A. africanus*, *A. anamensis*, *A. garhi*, *H. habilis*, *H. rudolfensis*, *K. platyops*, *O. tugenensis*, *P. aethiopicus*, *P. boisei* and *P. robustus* can be assigned to the first grade, whereas *H. antecessor*, *H. ergaster*, *H. erectus* and *H. heidelbergensis* can be assigned to the second, and *H. neanderthalensis* and *H. sapiens* to the third. Currently there is little evidence pertaining to the adaptive strategies of *A. bahrelghazali*.

It is noteworthy that sexual dimorphism is only partly concordant with the other adaptive variables. For example, most of the species that are allocated to the first grade on the basis of body mass, stature, locomotion, relative size

Table 7. Summary of adaptive characteristics of fossil hominid species and grade assignments. Parentheses indicate some uncertainty. ? = no data

Species	Type of bipedalism	Species mean body mass	Species mean stature	Degree of sexual dimorphism	Megadont?	Relative brain size	Long maturation period?	Grade
A. ramidus	?	Less than 50 kg	(Less than 150 cm)	?	(Yes)	?	?	1
A. afarensis	Facultative	Less than 50 kg	Less than 150 cm	High	(Yes)	Moderate	No	1
A. africanus	Facultative	Less than 50 kg	Less than 150 cm	High	Yes	Moderate	No	1
A. anamensis	Facultative	(Less than 50 kg)	(Less than 150 cm)	(High)	(Yes)	?	No	1
A. bahrelghazali	?	?	?	?	?	?	?	?
A. garhi	(Facultative)	(Less than 50 kg)	Less than 150 cm	(High)	(Yes)	(Moderate)	?	1
H. antecessor	Obligate	(More than 50 kg)	More than 150 cm	?	(No)	(Moderate)	?	2
H. ergaster	Obligate	More than 50 kg	More than 150 cm	(Moderate)	No	Moderate	No	2
H. erectus	Obligate	More than 50 kg	More than 150 cm	Moderate	No	Moderate	No	2
H. habilis	Facultative	Less than 50 kg	Less than 150 cm	Moderate	Yes	Moderate	No	1
H. heidelbergensis	Obligate	More than 50 kg	More than 150 cm	(Moderate)	(No)	Moderate	?	2
H. neanderthalensis	Obligate	More than 50 kg	More than 150 cm	(High)	No	Large	Yes	3
H. rudolfensis	?	Less than 50 kg	?	?	Yes	Moderate	No	1
H. sapiens	Obligate	More than 50 kg	More than 150 cm	Moderate	No	Large	Yes	3
K. platyops	?	?	?	?	?	(Moderate)	?	1
O. tugenensis	Facultative	(Less than 50 kg)	(Less than 150 cm)	?	(Yes)	?	?	1
P. aethiopicus	?	Less than 50 kg	?	?	(Yes)	Moderate	No	1
P. boisei	Facultative	Less than 50 kg	Less than 150 cm	High	Yes	Moderate	No	1
P. robustus	Facultative	Less than 50 kg	Less than 150 cm	High	Yes	Moderate	No	1

of the masticatory system, relative brain size and development, are strongly sexually dimorphic. However, one of the species, *H. habilis*, has the lowest percentage sexual dimorphism value of any fossil hominid species (117%). Similarly, *H. neanderthalensis*, which can be confidently assigned to the third grade on the basis of its body mass, stature, locomotion, relative size of the masticatory system, relative brain size and development, is considerably more sexually dimorphic than the species that are allocated to the second grade. The most probable explanation for this situation is that some of the fossil samples are biased in such a way that they under- or overestimate body mass sexual dimorphism. However, it is also possible that the evolution of body mass sexual dimorphism is decoupled from the evolution of the other adaptive variables, perhaps because it is influenced by sexual selection rather than natural selection (Eldredge, 1990).

With regard to timing, the oldest species in the first grade are *O. tugenensis* and *A. ramidus*. The former dates to around 6 million years ago. The oldest evidence for the latter is about 5.8–5.5 million years ago. The last species in the grade to go extinct is *P. boisei*, the most recent specimens of which date to around 1.4 million years ago (Wood *et al.*, 1994). The oldest species in the second grade is *H. ergaster*. The first appearance of this species is currently either 1.9 million years ago (the mandible, KNM-ER 1812, and the cranial fragment, KNM-ER 2598) or 1.85 million years ago (the cranial fragment, KNM-ER 1648) (Feibel *et al.*, 1989). However, given the nature of the stratigraphy at Koobi Fora (in excess of 500 thousand years are 'missing' in the sedimentary sequence prior to 1.9 million years ago) a date for the first appearance of *H. ergaster* of 1.85 or 1.9 million years ago is likely to be an underestimate (Collard & Wood, 1999). The last surviving species in the second grade is *H. heidelbergensis*. The youngest specimens that have been assigned to this species date to between 100 and 200 thousand years ago (Wood & Richmond, 2000). The oldest specimens allocated to the species that comprise the third grade, *H. neanderthalensis* and *H. sapiens*, date to between 242 and 186 thousand years ago (Klein, 1999). However, ancient DNA studies suggest that the lineages to which the species belong separated around 500 thousand years ago (Krings *et al.*, 1997, 1999). The third hominid grade has persisted into the present in the form of *H. sapiens*. Thus, in the course of human evolution there have been at least three grade shifts. The first occurred around 6 million years ago, probably in connection with the separation of the human and chimpanzee lineages. The second grade shift most probably took place between 2.4 and 1.9 million years ago, and is associated with the emergence of *H. ergaster*. The third grade shift, which involved the appearance of *H. neanderthalensis* and *H. sapiens*, probably occurred between 500 and 242 thousand years ago.

CONCLUSIONS

A review of the key adaptive characteristics of the hominids indicates that at least three grades have appeared in the course of human evolution. The first grade is characterised by a species mean body mass of less than 50 kg; stature of less than 130 cm; facultative bipedalism; a relatively large masticatory system; a relatively small brain; and a rapid, ape-like developmental schedule. The second grade is characterised by a species mean body mass in excess of 50 kg; a stature in excess of 160 cm; obligate bipedalism; a relatively small masticatory system; an EQ of less than 4.5; and a short, ape-like period of maturation. The third grade is similar to the second in terms of body mass, stature, locomotor behaviour and masticatory system size, but exhibits a considerably higher degree of encephalisation and delayed maturation. With varying degrees of certainty *A. ramidus*, *A. afarensis*, *A. africanus*, *A. anamensis*, *A. garhi*, *H. habilis*, *H. rudolfensis*, *K. platyops*, *O. tugenensis*, *P. aethiopicus*, *P. boisei* and *P. robustus* can be assigned to the first grade, whereas *H. antecessor*, *H. ergaster*, *H. erectus* and *H. heidelbergensis* can be assigned to the second, and *H. neanderthalensis* and *H. sapiens* to the third. Currently little can be inferred about the adaptive strategies of *A. bahrelghazali*. The first grade appeared around 6 million years ago, probably in connection with the establishment of the human and chimpanzee lineages. The second grade most probably emerged between 2.4 and 1.9 million years ago, and is associated with the emergence of *H. ergaster*. The third grade probably appeared between 500 and 242 thousand years ago.

ACKNOWLEDGEMENTS

I would like to acknowledge the support of the Arts and Humanities Research Board, the Natural Environment Research Council and the Wellcome Trust. I would like to thank Nicole Silverman; I would also like to thank Tim Crow, Eva Fairnell, James Rivington and Bernard Wood for their help in the preparation of this paper.

References

Abitbol, M.M. (1995) Reconstruction of the Sts 14 (*Australopithecus africanus*) pelvis. *American Journal of Physical Anthropology*, **96**, 143–58.

Aiello, L.C. & Dean, M.C. (1990) *An Introduction to Human Evolutionary Anatomy*. London: Academic Press.

Aiello, L.C. & Dunbar, R.I.M. (1993) Neocortex size, group size and the evolution of language. *Current Anthropology*, **34**, 184–93.

Aiello, L.C. & Wheeler, P.E. (1995) The expensive-tissue hypothesis: the brain and the digestive system in human and primate evolution. *Current Anthropology*, **36**, 199–221.

Alexander, R.D., Hoogland, J.L., Howard, R.D., Noonan, K.M. & Sherman, P.W. (1979) Sexual dimorphism and breeding systems in pinnipeds, ungulates, primates and humans. In: *Evolutionary Biology and Human Social Behavior* (eds N. Chagnon & W. Irons), pp. 402–35. Belmont: Duxbury.

Alexeev, V.P. (1986) *The Origin of the Human Race*. Moscow: Progress Publishers.

Andrews, P. (1984) An alternative interpretation of the characters used to define *Homo erectus*. *Courier Forschungsintitut Senckenberg*, **69**, 167–75.

Arambourg, C. & Coppens, Y. (1968) Découverte d'un australopithécin nouveau dans le gisements de l'Omo (Éthiopie). *South African Journal of Science*, **64**, 58–9.

Arsuaga, J.-L., Lorenzo, C., Carretero, J.-M., Gracia, A., Martínez, I., García, N., Bermúdez de Castro, J.-M. & Carbonell, E. (1999) A complete human pelvis from the Middle Pleistocene of Spain. *Nature*, **399**, 255–8.

Arsuaga, J.-L., Lorenzo, C., Carretero, J.-M., Lorenzo, C., Gracia, A., Martinez, I., Bermudez de Castro, J.M. & Carbonell, E. (1997) Size variation in Middle Pleistocene humans. *Science*, **277**, 1086–8.

Ascenzi, A., Mallegni, F., Manzi, G., Segre, A.G. & Segre Naldini, E. (2000) A re-appraisal of *Ceprano calvaria* affinities with *Homo erectus*, after the new reconstruction. *Journal of Human Evolution*, **39**, 443–50.

Asfaw, B., Gilbert, W.H., Beyene, Y., Hart, W.K., Renne, P.R., WoldeGabriel, G., Vrba, E.S. & White, T.D. (2002) Remains of *Homo erectus* from Bouri, Middle Awash, Ethiopia. *Nature*, **416**, 317–20.

Asfaw, B., White, T.D., Lovejoy, O., Latimer, B., Simpson, S. & Suwa, G. (1999) *Australopithecus garhi*: a new species of early hominid from Ethiopia. *Science*, **284**, 629–35.

Bermudez de Castro, J.M., Arsuaga, J.L., Carbonell, E., Rosas, A., Martinez, I. & Mosquera, M. (1997) A hominid from the Lower Pleistocene of Atapuerca, Spain: possible ancestor to Neanderthals and modern humans. *Science*, **276**, 1392–5.

Beynon, D. & Dean, M.C. (1988) Distinct dental development patterns in early fossil hominids. *Nature*, **335**, 509–14.

Brain, C.K. (1994) The Swartkrans Paleontological Research Project in perspective: results and conclusions. *South African Journal of Science*, **90**, 220–3.

Brain, C.K. (1993) *Swartkrans, A Cave's Chronicle of Early Man*. Pretoria: Transvaal Museum.

Brauer, G. & Mbua, E. (1992) *Homo erectus* features used in cladistics and their variability in Asian and African hominids. *Journal of Human Evolution*, **22**, 79–108.

Broom, R. (1938) The Pleistocene anthropoid apes of South Africa. *Nature*, **142**, 377–9.

Broom, R. (1949) Another new type of fossils ape-man (*Paranthropus crassidens*). *Nature*, **163**, 57.

Brown, J.S. & Vincent, T.L. (1987) A theory for the evolutionary game. *Theoretical Population Biology*, **31**, 140–66.

Brunet, M., Beauvilain, A., Coppens, Y., Heintz, E., Moutaye, A.H.E. & Pilbeam, D.E. (1995) The first australopithecine 2,500 kilometres west of the Rift Valley (Chad). *Nature*, **378**, 273–6.

Brunet, M., Beauvilain, A., Coppens, Y., Heintz, E., Moutaye, A.H.E. & Pilbeam, D.R. (1996) *Australopithecus bahrelghazali*, une nouvelle espèce d'Hominidé ancien de la région de Koro Toro (Tchad). *Comptes Rendus de l'Académie des Sciences*, **322**, 907–13.

Carretero, J.M., Arsuaga, J.L. & Lorenzo, C. (1997) Clavicles, scapulae and humeri from the Sima de los Huesos site (Sierra de Atapuerca, Spain). *Journal of Human Evolution*, **33**, 357–408.

Carretero, J.M., Lorenzo, C. & Arsuaga, J.L. (1999) Axial and appendicular skeleton of *Homo antecessor*. *Journal of Human Evolution*, **37**, 459–99.

Chamberlain, A.T. & Wood, B.A. (1987) Early hominid phylogeny. *Journal of Human Evolution*, **16**, 118–33.

Clarke, R. (1998) First ever discovery of a well-preserved skull and associated skeleton of *Australopithecus*. *South African Journal of Science*, **94**, 460–3.

Clarke, R. & Tobias, P.V. (1995) Sterkfontein Member 2 foot bones of the oldest South African hominid. *Science*, **269**, 521–4.

Clegg, M. & Aiello, L.C. (1999) A comparison of the Nariokotome *Homo erectus* with juveniles from a modern human population. *American Journal of Physical Anthropology*, **110**, 81–93.

Clutton-Brock, T.H., Harvey, P.H. & Rudder, B. (1977) Sexual dimorphism, socioeconomic sex ratio and body wight in primates. *Nature*, **269**, 797–800.

Collard, M. & Wood, B.A. (1999) Grades among the African early hominids. In: *African Biogeography, Climate Change and Early Hominid Evolution* (eds T.G. Bromage & F. Schrenk), pp. 316–27. Oxford: Oxford University Press.

Corruccini, R.S. (1994) How certain are hominid phylogenies? The role of confidence intervals in cladistics. In: *Integrative Paths to the Past: Palaeoanthropological Advances in Honour of F. Clark Howell* (eds R.S. Corruccini & R.L. Ciochon), pp. 167–83. New York: Prentice Hall.

Corruccini, R.S. & McHenry, H.M. (1980) Cladometric analysis of Pliocene hominids. *Journal of Human Evolution*, **9**, 209–21.

Crompton, R.H., Yu, L., Weijie, W., Gunther, M. & Savage, R. (1998) The mechanical effectiveness of erect and 'bent-hip, bent-knee' bipedal walking in *Australopithecus afarensis*. *Journal of Human Evolution*, **35**, 55–74.

Crook, J.H. (1972) Sexual selection, dimorphism and social organisation in the primates. In: *Sexual Selection and the Descent of Man, 1871–1971* (ed. B.G. Campbell), pp. 231–81. Chicago: Aldine.

Curnoe, D. (2001) Early *Homo* from Southern Africa: a cladistic perspective. *South African Journal of Science*, **97**, 186–90.

Dart, R.A. (1925) *Australopithecus africanus*: the man-ape of South Africa. *Nature*, **115**, 235–6.

Dean, M.C. (1995) The nature and periodicity of incremental lines in primate dentine and their relationship to periradicular bands in OH 16 (*Homo habilis*). In: *Aspects of Dental Biology; Paleontology, Anthropology and Evolution* (ed. J. Moggi-Cecchi), pp. 239–65. Florence: Angelo Pontecorboli.

Dean, M.C. (2000) Progress in understanding hominoid dental development. *Journal of Anatomy*, **197**, 77–101.

Dean, M.C., Leakey, M.G., Reid, D., Schrenk, F., Schwartz, G.T., Stringer, C.B. & Walker, A. (2001) Growth processes in teeth distinguish modern humans from *Homo erectus* and earlier hominids. *Nature*, **414**, 628–31.

Delson, E., Eldredge, N. & Tattersall, I. (1977) Reconstruction of hominid phylogeny: a testable framework based on cladistic analysis. *Journal of Human Evolution*, **6**, 263–78.

Dubois, E. (1892) Paleontologische and erzoekingen op Java. *Verslag Van Het Mijnwezen Batavia*, **3**, 10–4.

Dubois, E. (1894) *Pithecanthropus Erectus, Eine Menschenähnlich Ubergangsform Aus Java*. Batavia: Landsdruckerei.

Dunbar, R.I.M. (1992) Neocortex size as a constraint on group size in primates. *Journal of Human Evolution*, **22**, 469–93.

Dunbar, R.I.M. (1995) Neocortex size and group size in primates: a test of the hypothesis. *Journal of Human Evolution*, **28**, 287–96.

Eckhardt, R.B. (2000) *Human Paleobiology*. Cambridge: Cambridge University Press.

Eldredge, N. (1985) *Unfinished Synthesis*. New York: Oxford University Press.

Eldredge, N. (1986) Information, economics and evolution. *Annual Review of Ecological Systems*, **17**, 351–69.

Eldredge, N. (1989) *Macroevolutionary Dynamics: Species, Niches and Adaptve Peaks*. New York: McGraw-Hill.

Eldredge, N. (1990) Hierarchy and macroevolution. In: *Paleobiology* (eds D.E.G. Briggs & P.R. Crowther), pp. 124–9. Oxford: Blackwell.

Eldredge, N. & Tattersall, I. (1975) Evolutionary models, phylogenetic reconstruction and another look at hominid phylogeny. In: *Contributions to Primatology 5: Approaches to Primate Paleobiology* (ed. F.S. Szalay), pp. 218–42. Basel: Karger.

Feibel, C.S., Brown, F.H. & McDougall, I. (1989) Stratigraphic context of fossil hominids from the Omo Group deposits: Northern Turkana Basin, Kenya and Ethiopia. *American Journal of Physical Anthropology*, **78**, 595–622.

Foley, R.A. (1984) Early man and the Red Queen: tropical African community evolution and hominid adaptation. In: *Human Evolution and Community Ecology* (ed. R.A. Foley), pp. 85–110. London: Academic Press.

Foley, R.A. (1999) Evolutionary geography of Pliocene African hominids. In: *African Biogeography, Climate Change and Early Hominid Evolution* (eds T.G. Bromage & F. Schrenk), pp. 328–48. Oxford: Oxford University Press.

Gabunia, L. & Vekua, A. (1995) A Plio-Pleistocene hominid from Dmanisi, East Georgia. *Nature*, **373**, 509–12.

Gabunia, L., Antón, S.C., Lordkipanze, D., Vekua, A., Justus, A. & Swisher, C.C. III (2001) Dmanisi and dispersal. *Evolutionary Anthropology*, **10**, 158–70.

Grausz, H.M., Leakey, R.E., Walker, A.C. & Ward, C.V. (1988) Associated cranial and post-cranial bones of *Australopithecus boisei*. In: *Evolutionary History of the 'Robust' Australopithecines* (ed. F. E. Grine), pp. 127–32. New York: Aldine de Gruyter.

Grine, F.E., Demes, B., Jungers, W.L. & Cole, T.M. (1993) Taxonomic affinity of early *Homo* cranium from Swartkrans, South Africa. *American Journal of Physical Anthropology*, **92**, 411–26.

Grine, F.E., Jungers, W.L. & Schultz, J. (1996) Phenetic affinities among early *Homo* from East and South Africa. *Journal of Human Evolution*, **30**, 189–225.

Groves, C.P. (1989) *A Theory of Human and Primate Evolution*. Oxford: Oxford University Press.

Groves, C.P. & Mazak, V. (1975) An approach to the taxonomy of the hominidae: gracile Villafranchian hominids of Africa. *Casopis pro Mineralogii Geologii*, **20**, 225–47.

Haile-Selassie, J. (2001) Late Miocene hominids from the Middle Awash, Ethiopia. *Nature*, **412**, 178–81.

Hartwig-Scherer, S. (1993) Body weight prediction in early fossil hominids: towards a taxon-'independent' approach. *American Journal of Physical Anthropology*, **92**, 17–36.

Hartwig-Scherer, S. & Martin, R.D. (1991) Was 'Lucy' more human than her 'child'? Observations on early hominid post-cranial skeletons. *Journal of Human Evolution*, **21**, 439–49.

Hens, S.M., Konigsberg, L.W. & Jungers, W.L. (2000) Estimating stature in fossil hominids: which regression model and reference to use? *Journal of Human Evolution*, **38**, 767–84.

Houghton, P. (1996) *People of the Great Ocean: Aspects of Human Biology of the Early Pacific*. Cambridge: Cambridge University Press.

Howell, F.C. (1978) Hominidae. In: *Evolution of African Mammals* (eds V.J. Maglio & H.B.S. Cooke), pp. 154–248. Cambridge: Harvard University Press.

Howell, F.C. & Wood, B.A. (1974) Early hominid ulna from the Omo basin, Ethiopia. *Nature*, **249**, 174–6.

Hublin, J.-J., Spoor, F., Braun, M., Zonneveld, F. & Condemi, S. (1996) A late Neanderthal associated with Upper Palaeolithic artefacts. *Nature*, **381**, 224–6.

Hughes, A.R. & Tobias, P.V. (1977) A fossil skull probably of the genus *Homo* from Sterkfontein, Transvaal. *Nature*, **265**, 310.

Hunt, K.D. (1994) The evolution of human bipedality: ecology and functional morphology. *Journal of Human Evolution*, **26**, 183–202.

Hunt, K.D. (1996) The postural feeding hypothesis: an ecological model for the origin of bipedalism. *South African Journal of Science*, **9**, 77–90.

Huxley, J. (1958) Evolutionary processes and taxonomy with special reference to grades. *Uppsala Universitet Arssks*, **6**, 21–38.

International Commission of Zoological Nomenclature (1999) Opinion 1941. *Australopithecus afarensis* Johanson, 1978 (Mammalia, Primates): specific name conserved. *Bulletin of Zoological Nomenclature*, **56**, 223–4.

Johanson, D.C. & Coppens, Y. (1976) A preliminary anatomical diagnosis of the first Plio-Pleistocene hominid discoveries in the Central Afar, Ethiopia. *American Journal of Physical Anthropology*, **45**, 217–34.

Johanson, D.C. & Taieb, M. (1976) Plio-Pleistocene hominid discoveries in Hadar, Ethiopia. *Nature*, **260**, 293–7.

Johanson, D.C., Masao, F.T., Eck, G.G., White, T.D., Walter, R.C., Kimbel, W.H., Asfaw, B., Manega, P., Ndessokia, R. & Suwa, G. (1987) New partial skeleton of *Homo habilis* from Olduvai Gorge. *Nature*, **327**, 205–9.

Johanson, D.C., Taieb, M. & Coppens, Y. (1982) Pliocene hominids from the Hadar formation, Ethiopia (1973–77): stratigraphic, chronologic, and paleoenvironmental contexts, with notes on hominid morphology and systematics. *American Journal of Physical Anthropology*, **57**, 373–402.

Johanson, D.C., White, T.D. & Coppens, Y. (1978) A new species of the genus *Australopithecus* (Primates: Hominidae) from the Pliocene of eastern Africa. *Kirtlandia*, **28**, 1–14.

Kappelman, J. (1996) The evolution of body mass and relative brain size in fossil hominids. *Journal of Human Evolution*, **30**, 243–76.

Kappelman, J., Swisher, C.G. III, Fleagle, J.G., Yirga, S., Brown, T.M. & Feseha, M. (1996) Age of *Australopithecus afarenis* from Fejej, Ethiopia. *Journal of Human Evolution*, **30**, 139–46.

Kennedy, G. (1983) Some aspects of femoral morphology in *Homo erectus*. *Journal of Human Evolution*, **12**, 587–616.

Keyser, A. (2000) The Drimolen skull: the most complete australopithecine cranium and mandible to date. *South African Journal of Science*, **96**, 189–92.

Keyser, A., Menter, C.G., Moggi-Cecchi, J., Rayne Pickering, T. & Berger, L.R. (2000) Drimolen: a new hominid-bearing site in Gauteng, South Africa. *South African Journal of Science*, **96**, 193–7.

Kidd, R.S., O'Higgins, P. & Oxnard, C.E. (1996) The OH 8 foot: a reappraisal of the functional morphology of the hindfoot using a multivariate analysis. *Journal of Human Evolution*, **31**, 269–91.

Kimbel, W.H. & Martin, L.B. (1993) *Species, Species Concepts and Primate Evolution*. New York: Plenum Press.

Kimbel, W.H., Johanson, D.C. & Rak, Y. (1996) Systematic assessment of a maxilla of *Homo* from Hadar, Ethiopia. *American Journal of Physical Anthropology*, **103**, 235–62.

King, W. (1864) The reputed fossil man of the Neanderthal. *Quaternary Journal of Science*, **1**, 88–97.

Klein, R. G. (1999) *The Human Career*. Chicago: Chicago University Press.

Krings, M., Geisert, H., Schmitz, R.W., Krainitzki, H. & Pääbo, S. (1999) DNA sequence of the mitochondrial hypervariable region II from the Neanderthal type specimen. *Proceedings of the National Academy of Sciences of the USA*, **96**, 5581–5.

Krings, M., Stone, A., Schmitz, R.W., Krainitzki, H., Stoneking, M. & Pääbo, S. (1997) Neanderthal DNA sequences and the origin of modern humans. *Cell*, **90**, 19–30.

Kullmer, O., Sandrock, O., Abel, R., Schrenk, F., Bromage, T.G. & Juwayeyi, Y.M. (1999) The first *Paranthropus* from the Malawi Rift. *Journal of Human Evolution*, **37**, 121–7.

Kuman, K. & Clarke, R.J. (2000) Stratigraphy, artefact industries and hominid associations for Sterkfontein, Member 5. *Journal of Human Evolution*, **38**, 827–47.

Lague, M.R. & Jungers, W.L. (1996) Morphometric variation in Plio-Plesitocene hominid distal humeri. *American Journal of Physical Anthropology*, **101**, 401–27.

Latimer, B. (1991) Locomotor adaptations in *Australopithecus afarensis*: the issue of arboreality. In: *Origine(s) de la Bipédie Chez les Hominidés* (eds Y. Coppens & B. Senut), pp. 169–76. Paris: Cahiers de Paléoanthropologie, Editions du CNRS.

Leakey, L.S.B. (1958) Recent discoveries at Olduvai Gorge. *Nature*, **188**, 1050–52.

Leakey, L.S.B. & Leakey, M.D. (1964) Recent discoveries of fossil hominids in Tanganyika: at Olduvai and near Lake Natron. *Nature*, **202**, 5–7.

Leakey, L.S.B., Tobias, P.V. & Napier, J.R. (1964) A new species of the genus *Homo* from Olduvai Gorge, Tanzania. *Nature*, **202**, 7–9.

Leakey, M., Feibel, C.S., McDougall, I. & Walker, A. (1995) New four million-year-old species from Kanapoi and Allia Bay, Kenya. *Nature*, **376**, 565–71.

Leakey, M., Feibel, C.S., McDougall, I., Ward, C. & Walker, A. (1998) New specimens and confirmation of an early age for *Australopithecus anamensis*. *Nature*, **393**, 62–6.

Leakey, M., Spoor, F., Brown, F.H., Gathogo, P.N., Kiarie, C., Leakey, L. & McDougall, I. (2001) New hominid genus from eastern Africa shows diverse middle Pliocene lineages. *Nature*, **410**, 433–40.

Lieberman, D.E. (2001) Another face in our family tree. *Nature*, **410**, 419–20.

Lieberman, D.E., McBratney, B.M. & Krovitz, G. (2002) The evolution and development of cranial form in *Homo sapiens*. *Proceedings of the National Academy of Sciences of the USA*, **99**, 1134–9.

Lieberman, D.E., Pilbeam, D.R. & Wood, B.A. (1988) A probablistic approach to the problem of sexual dimorphism in *Homo habilis*: a comparison of KNM-ER 1470 and KNM-ER 1813. *Journal of Human Evolution*, **17**, 503–11.

Lieberman, D.E., Wood, B.A. & Pilbeam, D.R. (1996) Homoplasy and early *Homo*: an analysis of the evolutionary relationships of *H. habilis sensu stricto* and *H. rudolfensis*. *Journal of Human Evolution*, **30**, 97–120.

Linnaeus, C. (1758) *Systema Naturae*. Stockholm: Laurentii Salvii.

Lorenzo, C., Arsuaga, J.L. & Carretero, J.M. (1999) Hand and foot remains from the Gran Dolina Early Pleistocene site (Sierra de Atapuerca, Spain). *Journal of Human Evolution*, **37**, 501–22.

Lorenzo, C., Carretero, J.M., Arsuaga, J.L., Gracia, A. & Martinez, I. (1998) Intrapopulational body size variation and cranial capacity in Middle Pleistocene humans: the Sima de los Huesos sample (Sierra de Atapuerca, Spain). *American Journal of Physical Anthropology*, **106**, 19–33.

Lovejoy, C.O. (1979) A reconstruction of the pelvis of AL-288 (Hadar Formation, Ethiopia) (Abstract). *American Journal of Physical Anthropology*, **50**, 460.

Lovejoy, C.O. (1981) The origin of man. *Science*, **211**, 311–50.

Lovejoy, C.O. (1988) The evolution of human walking. *Science America*, **259**, 118–25.

McHenry, H.M. (1973) Early hominid humerus from East Rudolf, Kenya. *Science*, **180**, 739–41.

McHenry, H.M. (1986) The first bipeds: a comparison of the *A. afarensis* and *A. africanus* post-cranium and implications for the evolution of bipedalism. *Journal of Human Evolution*, **15**, 177–91.

McHenry, H.M. (1991) Femoral length and stature in Plio-Pleistocene hominids. *American Journal of Physical Anthropology*, **85**, 149–58.

McHenry, H.M. (1992) Body size and proportions in early hominids. *American Journal of Physical Anthropology*, **87**, 407–31.

McHenry, H.M. (1994) Early hominid postcrania: phylogeny and function. In: *Integrative Paths to the Past: Palaeoanthropological Advances in Honor of F. Clark Howell* (eds R.S. Corruccini & R.L. Ciochon), pp. 251–68. Englewood Cliffs: Prentice Hall.

McHenry, H.M. & Berger, L.R. (1998) Body proportions in *Australopithecus afarensis* and *A. africanus* and the origin of genus *Homo*. *Journal of Human Evolution*, **35**, 1–22.

McHenry, H.M. & Coffing, K. (2000) *Australopithecus* to *Homo*: transformations in body and mind. *Annual Review of Anthropology*, **29**, 125–46.

McKee, J.K. (1996) Faunal evidence and Sterkfontein Member 2 foot bones of early hominid. *Science*, **271**, 1301.

Martin, R.D. (1981) Relative brain size and basal metabolic rate in terrestrial vertebrates. *Nature*, **293**, 57–60.

Mitani, J., Gros-Louis, J. & Richards, A.F. (1996) Sexual dimorphism, the operational sex ratio, and the intensity of male competition in polygynous primates. *American Naturalist*, **147**, 966–80.

Moggi-Cecchi, J. (2000) Fossil children: what can they tell us about the origin of the genus *Homo*? In: *The Origin of Humankind* (eds M. Aloisi, B. Battaglia, E. Carafoli & G.A. Danieli), pp. 35–50. Amsterdam: IOS Press.

Ohman, J.C., Krochta, T.J., Lovejoy, C.O., Mensforth, R.P. & Latimer, B. (1997) Cortical bone distribution in the femoral neck of hominoids: implications for the locomotion of *Australopithecus afarensis*. *American Journal of Physical Anthropology*, **104**, 117–31.

Olson, T.D. (1985) Cranial morphology and systematics of the Hadar Formation hominids and '*Australopithecus*' *africanus*. In: *Ancestors: The Hard Evidence* (ed. E. Delson), pp. 102–19. New York: Liss.

Oxnard, C.E. (1984) *The Order of Man: A Biomathematical Anatomy of the Primates*. New Haven: Yale University Press.

Partridge, T.C., Shaw, J., Heslop, D. & Clarke, R.J. (1999) A new hominid skeleton from Sterkfontein, South Africa: age and preliminary assessment. *Journal of Quaternary Science*, **14**, 293–8.

Passingham, R.E. & Ettlinger, G. (1974) A comparison of cortical functions in man and the other primates. *International Review of Neurobiology*, **16**, 233–99.

Pickford, M. & Senut, B. (2001) The geological and faunal context of Late Miocene hominid remains from Lukeino, Kenya. *Comptes Rendus de l'Académie des Sciences, Série IIa*, **332**, 145–52.

Plavcan, M.J. (2001) Sexual dimorphism in primate evolution. *Yearbook of Physical Anthropology*, **44**, 25–53.

Plavcan, J.M. & van Schaik, C.P. (1997) Intrasexual competition and body weight dimorphism in anthropoid primates. *American Journal of Physical Anthropology*, **103**, 37–68.

Ponce de León, M.S. & Zollikofer, C.P.E. (2001) Neanderthal cranial ontogeny and its implications for late hominid diversity. *Nature*, **412**, 534–8.

Quinney, P.S. & Collard, M. (1997) Sexual dimorphism in the mandible of *Homo neanderthalensis and Homo sapiens*: morphological patterns and behavioural implications. In: *Archaeological Sciences 1995* (eds A. Sinclair, E. Slater & J.A. Gowlett), pp. 420–5. Oxford: Oxbow.

Reader, S.M. & Laland, K.N. (2002) Social intelligence, innovation, and enhanced brain size in primates. *Proceedings of the National Academy of Sciences of the USA*, **99**, 4436–41.

Rightmire, G.P. (1993) Variation among early *Homo* crania from Olduvai Gorge and the Koobi Fora Region. *American Journal of Physical Anthropology*, **90**, 1–33.

Rightmire, G.P. (1996) The human cranium from Bodo, Ethiopia: evidence for speciation in the Middle Pleistocene? *Journal of Human Evolution*, **31**, 21–39.

Rightmire, G.P. (1998) Evidence from facial morphology for similarity of Asian and African representatives of *Homo erectus*. *American Journal of Physical Anthropology*, **106**, 61–85.

Rightmire, G.P. (2001) Patterns of hominid evolution and dispersal in the Middle Pleistocene. *Quaternary International*, **75**, 77–84.

Roberts, M.B., Stringer, C.B. & Parfitt, S.A. (1994) A hominid tibia from Middle Pleistocene sediments at Boxgrove, UK. *Nature*, **369**, 311–3.

Rosenzweig, M.L. & McCord, R.D. (1991) Incumbent replacement: evidence for long-term evolutionary progress. *Paleobiology*, **17**, 202–13.

Rosenzweig, M.L., Brown, J.S. & Vincent, T.L. (1987) Red Queen and ESS: the coevolution of evolutionary rates. *Evological Ecology*, **1**, 59–94.

Ruff, C.B. (1991) Climate and body shape in hominid evolution. *Journal of Human Evolution*, **21**, 81–105.

Ruff, C.B. (1993) Climatic adaptation and hominid evolution: the thermoregulatory imperative. *Evolutionary Anthropology*, **2**, 53–60.

Ruff, C.B. (1994) Morphological adaptation to climate in modern and fossil hominids. *Yearbook of Physical Anthropology*, **37**, 65–107.

Ruff, C.B. & Walker, A. (1993) Body size and body shape. In: *The Nariokotome* Homo ergaster *Skeleton* (eds A. Walker & R. E. Leakey), pp. 234–65. Berlin: Springer Verlag.

Schmid, P. (1991) The trunk of the australopithecines. In: *Origine(s) de la Bipedie Chez les Hominides* (eds Y. Coppens & B. Senut), pp. 225–34. Paris: Editions du CNRS.

Schoetensack, O. (1908) *Der Unterkiefer Des* Homo Heidelbergensis *Aus Den Sanden Von Mauer bei Heidelberg*. Leipzig: Engelman.

Schwartz, J.H. & Tattersall, I. (1996) Significance of some previously unrecognized apomorphies in the nasal region of *Homo neanderthalensis*. *Proceedings of the National Academy of Sciences of the USA*, **93**, 10852–4.

Senut, B. & Tardieu, C. (1985) Functional aspects of Plio-Pleistocene hominid limb bones: implications for taxonomy and phylogeny. In: *Ancestors: The Hard Evidence* (ed. E. Delson), pp. 193–201. New York: Alan R. Liss.

Senut, B., Pickford, M., Gommery, D., Mein, P., Cheboi, K. & Coppens, Y. (2001) First hominid from the Miocene (Lukeino Formation, Kenya). *Comptes Rendus de l'Académie des Sciences, Série IIa*, **332**, 137–44.

Skelton, R.R. & McHenry, H.M. (1992) Evolutionary relationships among early hominids. *Journal of Human Evolution*, **23**, 309–49.

Skelton, R.R., McHenry, H.M. & Drawhorn, G.M. (1986) Phylogenetic analysis of early hominids. *Current Anthropology*, **27**, 21–43.

Smith, B.H. (1994) Patterns of dental development in *Homo*, *Australopithecus*, *Pan*, and *Gorilla*. *American Journal of Physical Anthropology*, **94**, 307–25.

Smith, F.H. (1980) Sexual differences in European Neanderthal crania with special reference to the Krapina remains. *Journal of Human Evolution*, **9**, 359–75.

Smith, F.H., Trinkaus, E., Pettitt, P.B., Karavanic, I. & Paunovic, M. (1999) Direct radiocarbon dates for Vindija G1 and Velika Pecina Late Pleistocene hominid remains. *Proceedings of the National Academy of Sciences of the USA*, **96**, 12281–6.

Smith, R.J. (1996) Biology and body size in human evolution: statistical inference misapplied. *Current Anthropology*, **37**, 309–49.

Spoor, F., Wood, B.A. & Zonnefeld, F. (1994) Implications of early hominid labyrinthine morphology for evolution of human bipedal locomotion. *Nature*, **369**, 645–8.

Spoor, F., Wood, B.A. & Zonnefeld, F. (1996) Evidence for a link between human semicircular canal size and bipedal behaviour. *Journal of Human Evolution*, **30**, 183–7.

Stern, J.T. (1999) The cost of bent-knee, bent-hip bipedal gait. A reply to Crompton *et al. Journal of Human Evolution*, **36**, 567–70.

Stern, J.T. (2000) Climbing to the top: a personal memoir of *Australopithecus afarensis*. *Evolutionary Anthropology*, **9**, 113–33.

Stern, J.T. & Susman, R.L. (1983) The locomotor anatomy of *Australopithecus afarensis*. *American Journal of Physical Anthropology*, **60**, 279–317.

Strait, D. & Grine, F.E. (1999) Cladistics and early hominid phylogeny. *Science*, **285**, 1210.

Strait, D. & Grine, F.E. (2001) The systematics of *Australopithecus garhi*. *Ludus Vitalis*, **9**, 109–35.

Strait, D. & Wood, B.A. (1999) Early hominid biogeography. *Proceedings of the National Academy of Sciences of the USA*, **96**, 9196–200.

Strait, D.S., Grine, F.E. & Moniz, M.A. (1997) A reappraisal of early hominid phylogeny. *Journal of Human Evolution*, **32**, 17–82.

Stringer, C.B. (1984) The definition of *Homo erectus* and the existence of the species in Africa and Europe. *Courier Forschungsinstitut Senckenberg*, **69**, 131–43.

Stringer, C.B. (1987) A numerical cladistic analysis for the genus *Homo*. *Journal of Human Evolution*, **16**, 135–46.

Stringer, C.B. & Gamble, C. (1993) *In Search of the Neanderthals*. London: Thames and Hudson.

Stringer, C.B., Trinkaus, E., Roberts, M.B., Parfitt, S.A. & MacPhail, R.I. (1998) The Middle Pleistocene human tibia from Boxgrove. *Journal of Human Evolution*, **34**, 509–47.

Susman, R.L. (1988) Hand of *Paranthropus robustus* from Member 1, Swartkrans: fossil evidence for tool behavior. *Science*, **240**, 781–4.

Susman, R.L. & Creel, N. (1979) Functional and morphological affinities of the subadult hand (OH 7) from Olduvai Gorge. *American Journal of Physical Anthropology*, **51**, 311–31.

Susman, R.L. & Stern, J.T. (1979) Telemetered electromyography of flexor digitorum profundis and flexor digitorum superficialis in *Pan troglodytes* and implications for interpretation of the OH 7 hand. *American Journal of Physical Anthropology*, **50**, 565–74.

Susman, R.L. & Stern, J.T. (1982) Functional morphology of *Homo habilis*. *Science*, **217**, 931–4.

Susman, R.L., de Ruiter, D. & Brain, C.K. (2001) Recently identified postcranial remains of *Paranthropus* and early *Homo* from Swartkrans Cave, South Africa. *Journal of Human Evolution*, **41**, 607–29.

Susman, R.L., Stern, J.T. & Jungers, W.L. (1984) Aboreality and bipedality in the Hadar hominids. *Folia Primatologica*, **43**, 113–56.

Suwa, G. (1988) Evolution of the 'robust' australopithecines in the Omo succession: epistemology and fossil evidence. In: *Evolutionary History of the 'Robust' Australopithecines* (ed. F.E. Grine), pp. 199–222. New York: Aldine de Gruyter.

Suwa, G., Asfaw, B., Beyene, Y., White, T.D., Katoh, K. & Nagaọka, S. (1997) The first skull of *Australopithecus boisei*. *Nature*, **389**, 489–92.

Tague, R.G. & Lovejoy, C.O. (1986) The obstetrics of AL 288-1 (Lucy). *Journal of Human Evolution*, **15**, 237–55.

Tardieu, C. (1998) Short adolescence in early hominids: infantile and adolescent growth of the human femur. *American Journal of Physical Anthropology*, **107**, 163–78.

Tattersall, I. (1986) Species recognition in human paleontology. *Journal of Human Evolution*, **15**, 165–76.

Tattersall, I. (1992) Species concepts and species identification in human evolution. *Journal of Human Evolution*, **22**, 341–9.

Tattersall, I. (1996) *The Fossil Trail*. New York: Oxford University Press.

Tattersall, I. (2001) Classification and phylogeny in human evolution. *Lud Vit*, **9**, 137–42.

Tattersall, I. & Eldredge, N. (1977) Fact, theory, and fantasy in human paleontology. *American Science*, **65**, 204–11.

Teaford, M. & Ungar, P. (2000) Diet and the evolution of the earliest human ancestors. *Proceedings of the National Academy of Sciences of the USA*, **97**, 13506–11.

Tobias, P.V. (1965) The early *Australopithecus* and *Homo* from Tanzania. *Anthropologie Prague*, **3**, 43–8.

Trinkaus, E. (1980) Sexual differences in Neanderthal limb bones. *Journal of Human Evolution*, **9**, 377–97.

Trinkaus, E. (1984) Does KNM-ER 1481A establish *Homo erectus* at 2.0 myr BP? *American Journal of Physical Anthropology*, **64**, 137–9.

Turner, A. & Chamberlain, A.T. (1989) Speciation, morphological change and the status of African *Homo erectus*. *Journal of Human Evolution*, **18**, 115–30.

Walker, A. & Leakey, R. (1993) *The Nariokotome* Homo ergaster *Skeleton*. Berlin: Springer-Verlag.

Walker, A.C., Leakey, R.E., Harris, J.M. & Brown, F.H. (1986) 2.5-Myr *Australopithecus boisei* from west of Lake Turkana, Kenya. *Nature*, **322**, 517–22.

Ward, C.V., Leakey, M.G. & Walker, A. (2001) Morphology of *Australopithecus anamensis* from Kanapoi and Allia Bay, Kenya. *Journal of Human Evolution*, **41**, 255–368.

Wheeler, P.E. (1991) The influence of bipedalism on the energy and water budgets of early hominids. *Journal of Human Evolution*, **21**, 117–36.

Wheeler, P.E. (1992) The thermoregulatory advantages of large body size for hominid foraging in savannah environments. *Journal of Human Evolution*, **23**, 351–62.

White, T.D., Suwa, G. & Asfaw, B. (1994) *Australopithecus ramidus* a new species of early hominid from Aramis, Ethiopia. *Nature*, **371**, 306–12.

White, T.D., Suwa, G. & Asfaw, B. (1995) Corrigendum: *Australopithecus ramidus* a new species of early hominid from Aramis, Ethiopia. *Nature*, **375**, 88.

Wolpoff, M. (1999) *Paleoanthropology*. New York: McGraw-Hill.

Wolpoff, M.H., Hawks, J., Frayer, D.W. & Hunley, K. (2001) Modern human ancestry at the peripheries: a test of the replacement theory. *Science*, **291**, 293–7.

Wolpoff, M., Thorne, A.G., Jelenik, J. & Zhang, Y. (1994) The case for sinking *Homo erectus*: 100 years of *Pithecanthropus* is enough! *Courier Forschungsinstitut Senckenberg*, **171**, 341–61.

Wood, B.A. (1984) The origin of *Homo erectus*. *Courier Forschungsinstitut Senckenberg*, **69**, 99–111.

Wood, B.A. (1988) Are 'robust' australopithecines a monophyletic group? *Evolutionary History of the 'Robust' Australopithecines* (ed. F.E. Grine), pp. 269–84. New York: Aldine de Gruyter.

Wood, B.A. (1989) Hominid relationships: a cladistic perspective. In: *The Growing Scope of Human Biology* (eds L.H. Schmitt, L. Freeman & N.W. Bruce), pp. 83–102. Perth: University of Western Australia Press.

Wood, B.A. (1991) *Koobi Fora Research Project. Vol. 4. Hominid Cranial Remains.* Oxford: Clarendon Press.

Wood, B.A. (1992) Origin and evolution of the genus *Homo*. *Nature*, **355**, 783–90.

Wood, B.A. (1993) Early *Homo*: how many species? In: *Species, Species Concepts, and Primate Evolution* (eds W.H. Kimbel & L.B. Martin), pp. 485–522. New York: Plenum Press.

Wood, B.A. (1994) Taxonomy and evolutionary relationships of *Homo erectus*. *Courier Forschungsinstitut Senckenberg*, **171**, 159–65.

Wood, B.A. (1995) Evolution of the early hominid masticatory system: mechanisms, events and triggers. In: *Paleoclimate and Evolution with Emphasis on Human Origins* (eds E.S. Vrba, G.H. Denton, T.C. Partridge & L.H. Burckle), pp. 438–48. New Haven: Yale University Press.

Wood, B.A. & Aiello, L.C. (1998) Taxonomic and functional implications of mandibular scaling in early hominids. *American Journal of Physical Anthropology*, **105**, 523.

Wood, B.A. & Chamberlain, A.T. (1986) *Australopithecus*: grade or clade? In: *Major Topics in Primate and Human Evolution* (eds B.A. Wood, L. Martin & P. Andrews), pp. 248–70. Cambridge: Cambridge University Press.

Wood, B.A. & Collard, M. (1997) Grades and the evolutionary history of early African hominids. In: *Archaeological Sciences 1995* (eds A. Sinclair, E. Slater & J.A. Gowlett), pp. 445–8. Oxford: Oxbow Books.

Wood, B.A. & Collard, M. (1999a) The human genus. *Science*, **284**, 65–71.

Wood, B.A. & Collard, M. (1999b) The changing face of genus *Homo*. *Evolutionary Anthropology*, **8**, 195–207.

Wood, B.A. & Lieberman, D.E. (2001) Craniodental variation in *Paranthropus boisei*: a developmental and functional perspective. *American Journal of Physical Anthropology*, **116**, 13–25.

Wood, B.A. & Richmond, B.G. (2000) Human evolution: taxonomy and paleobiology. *Journal of Anatomy*, **196**, 19–60.

Wood, B.A., Wood, C.G. & Konigsburg, L.W. (1994) *Paranthropus boisei*: an example of evolutionary stasis? *American Journal of Physical Anthropology*, **95**, 117–36.

II

LANGUAGE AND THE EVOLUTION OF THE BRAIN

Asymmetry as the defining characteristic of the human brain

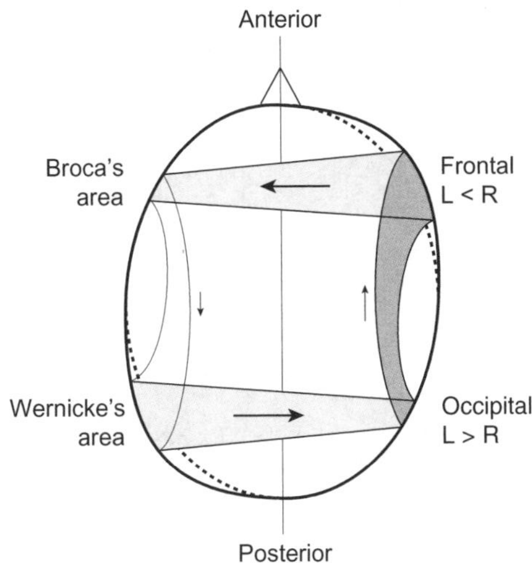

The right-frontal to left-occipital 'torque' of the human brain and its influence on inter- and intra-hemispheric connectivity of areas of heteromodal association cortex (from Crow TJ. Nuclear schizophrenic symptoms as a window on the relationship between thought and speech. Brit J Psychiat 1998; 173: 303-309).

From Protolanguage to Language

DEREK BICKERTON

Summary. Although evolution is normally conceived of as a gradual process, it can produce an appearance of catastrophism where functions change ('pre-adaptation') or where gradual changes in two or more components impinge on one another. The fossil and archaeological records argue strongly for some such development in the case of human language. It is argued that language as we know it requires the conjunction of three things: (1) an event structure derived from reciprocal altruism; (2) the capacity to use unstructured symbolic units (protolanguage); and (3) sufficient 'spare' neurones (i.e. neurones uncommitted to any single function) to maintain the coherence of internally generated messages in brains designed by evolution to attend primarily to the environment. These developments co-occurred only in the human species, accounting for the uniqueness of human language.

INTRODUCTION

GRADUALISM is to be expected in evolution. Adaptations that arise from within the existing pool of variation do not, almost by definition, stray far from current norms. Mutations also occur but, broadly speaking, the more extreme the mutation the more likely it is to be dysfunctional: the two-headed calves and similar 'hopeless monsters' usually do not even reproduce.

Accordingly, it is normal to assume that language and cognition also evolved gradually, that there was a steady and progressive increase in these faculties within the hominid line. Many (for example Hauser, 1996) still assume, despite the surely unanswerable (and certainly unanswered!) arguments of Bickerton (1990, 1995), that language is somehow continuous with other animal systems of communication, and evolved seamlessly out of them. Even those who accept discontinuity between language and other systems

Proceedings of the British Academy, **106**, 103–120, © The British Academy 2002.

(Pinker & Bloom, 1990; Newmeyer, 1991; Pinker, 1994; Bloom, 2000) argue that language must have developed over a series of stages, each more complex than the last.

In taking this stance, writers on the evolution of language are simply taking their cue from those who have studied the evolutionary record. For instance, according to Falk (1993: 226):

> The fossil and archaeological record picks up around two million years ago in East Africa. And what a record it is! Brain size 'took off' and subsequently doubled ... Recorded tool production also accelerated in Homo, spanning from initial clunky stone tools to contemporary computer, space and biological engineering.

Tobias (1971) states that:

> Long continued increase in size and complexity of the brain was paralleled for probably a couple of million years by *long-continued elaboration and 'complexification'* ... of the culture. The feedback relationship between the two sets of events is *as indubitable as it was prolonged in time.* [emphasis added]

According to Deacon (1997):

> [*Homo erectus*]'s brains and their symbolic communication were undoubtedly co-evolving together, *even if the tools they were using were not progressing at a comparable pace*. [emphasis added]

One might, if ignorant of prehistory, assume from Falk's (1993) account that the harpoon was invented about 1 million years ago, bridges and boats about 500 thousand years ago, and the wheel perhaps 200 thousand years ago; even such a scenario would leave a fearsome amount to be invented in the last 50 thousand or so years. Nothing in this account suggests that in the first 1.95 million years of Falk's astounding record almost nothing happened: the clunky stone tools became less clunky and slightly more diversified stone tools, and everything beyond that, from bone tools to supercomputers, happened in the last one-fortieth of the period in question. Similarly, for the 'long-continued elaboration and complexification of the culture' envisaged by Tobias (1971), and for the 'co-evolution of brains and symbolic communication' envisaged by Deacon (1997), there is simply not one scintilla of evidence: simply a blind faith that, if evolution is gradual, and we are where we are, we must have got here, far as it may seem, in a series of incremental steps.

At least some writers are prepared to give us the actual facts. According to Jellinek (1977), reviewing the Lower Palaeolithic, the tools that 'were not progressing at a comparable pace', comparable, that is, to the mysterious and wholly unevidenced 'culture' and 'symbolic communication', serve as a perfectly reliable index of our ancestors' capacities:

> We invariably make assumptions relating to the cognitive abilities of the hominids who produced the artefacts. Some of these assumptions can be based on the uniformity of the industries over periods of hundreds of thousands of years. The absence of evidence of innovation and differentiation in the tool forms that can be observed over these prolonged intervals *can be taken as evidence against the presence of the conceptual abilities relating to abstraction and synthesis that characterize modern Homo sapiens.* (Jellinek, 1977: 15; emphasis added)

It is, of course, possible to claim that humans had these abilities, or were gradually developing them, but had not yet put them into practice. But there is absolutely no evidence for such a claim, which goes against common sense: if a species has capacities that would increase its adaptive fitness, then the norm is that it utilises such capacities from the beginning.

There is an argument from the few surviving groups of hunter-gatherers that I would like to defuse here. Because these are undoubtedly modern humans, but have a technology that has not changed appreciably over the last few tens of thousands of years, it is argued that our ancestors could have similarly gone for hundreds of thousands of years without expressing their cognitive capacities in their culture. Any such argument ignores some crucial facts. First, modern humans who lack an elaborate material culture are in a small minority. Secondly, they are mainly people who have been forced into less productive regions by more complex societies; indeed, in the opinion of some anthropologists, they may not be the preservers of some antediluvian 'state of nature', but former agriculturists who have regressed, or been regressed, to a marginal condition that may reflect very inaccurately true pre-agricultural societies. Thirdly, we should consider not what some marginal minority does, but what a species as a whole does. One might be prepared to accept that some pre-*sapiens* hominids may have failed to heed the imperious call to progress from an enhanced cognition, but surely not all of them!

Yet it is clear that the refusal to implement these alleged gains from gradual cognitive evolution was universal in pre-*sapiens Homo*:

> The overriding impression of the technological evidence in the archaeological record is one of almost unimaginable monotony. Perhaps the most overwhelming example of this is Acheulian of Olduvai Gorge, where for approximately a million years no significant innovation is discernible ... We are talking about tens of thousands of hominids maintaining patterns of technological traditions without discernible change. (Jellinek, 1977: 28)

Moreover, cultural diffusion was clearly at work: Acheulian hand axes have recently been discovered as far east as Japan. Is it possible to think of any *sapiens* innovation that has travelled for the best part of 10 thousand miles without undergoing the slightest change?

The mismatch between the fossil and archaeological records forms an acute embarrassment for those who believe that human cognitive capacities, including language, developed gradually. Certainly brain size increase continued throughout the period and, although its pace may have varied, there are no obvious discontinuities in the record. In culture, the reverse is the case. There is an almost level plateau for almost 2 million years, then a rapid and dizzying ascent to our present state, an ascent which seems far from over. When a scientific puzzle of this magnitude relates so closely to ourselves, one would expect an enormous concentration on solving it. Far from it, most writers in the field have gone into denial, exaggerating pre-*sapiens* achievements, minimising our own, attempting always to explain away rather than to explain. One is irresistibly reminded of the reactions of a Victorian family when one of its members conceived an illegitimate child.

However, many are agreed that the emergence of language has contributed greatly to the increased intelligence and creativity of modern humans. Also, as Jellinek pointed out in the excerpts above, one normally expects increments in creativity and intelligence to be somehow reflected in the traces hominids invariably left behind them. The result is a paradox, the paradox of gradualness. On the one hand, language must have developed gradually because every complex adaptation in evolution (the eye, for example) has developed gradually. On the other, language cannot have developed gradually, because the combined record of fossilised remains and hominid and human artefacts is wholly inconsistent with gradual development. The question then simply is, is it possible to get from protolanguage to language without invoking some kind of saltation, or without denying one or other of the two halves of the paradox?

PRE-ADAPTATION

There is a rather obvious way in which apparently sudden evolutionary developments can be accounted for. That is pre-adaptation, to give it its traditional term, although 'exaptation' (Gould & Vrba, 1982) may well be preferred, avoiding as it does the faint flavour of teleology that the traditional term drags with it. If some prior (and of course gradual) development could be co-opted to serve the ends of language, then we can dispense with gradual development (which, in so far as it involves language, has more arguments going against it than there is space for here; for a summary see Bickerton, 1998). As has been set forth elsewhere (Bickerton, 1998; Calvin & Bickerton, 2000), a crucial development in social intelligence could have supplied the infrastructure that would, in a single step, have changed structureless protolanguage into a highly structured (although morphology-poor) human language. This development is briefly summarised below.

Reciprocal altruism (Trivers, 1971) is widespread among primates (de Waal, 1982; Strum, 1987; Smuts, 1987). Reciprocal altruism generally takes the form of dyadic relationships in which the partners cement their alliance by the exchange of favours (mutual grooming, food sharing, support in conflict situations, etc.) on the principle of 'I'll scratch your back if you scratch mine'. But reciprocal altruism is subject to exploitation by cheaters (Cosmides & Tooby, 1992). If reciprocal altruism is not to collapse, individuals involved in alliances must have some way of telling whether they are being cheated or not. In other words, they have to be able to keep score, approximately, not necessarily with absolute mathematical accuracy, in order to ensure that they do not waste their energy by giving more than they receive. For clearly, the animal or person that does not waste effort on a cheater is a fitter person or animal, and therefore more likely to reproduce his or her genes than are those who allow themselves to be cheated.

But in order to 'keep score', the individual has to set up a rough calculus to show that that individual has been the agent in doing favours not significantly oftener than the other, and has been the recipient of favours at least as often as the other. Also, as some actions involve a third entity, food, for example, in cases of food sharing, three abstract roles have to be considered: agent, who performs the action; recipient or goal, who is the object of it; and sometimes the third thing, which we may call the theme or entity affected by the action. It has been suggested that other things beside reciprocal altruism may have been implicated in setting up these roles. Indeed they may, but to concede this is not to deny the importance of the role of reciprocal altruism, or even to diminish it. For in no context other than that of reciprocal altruism is it necessary to set up abstract roles like agent, goal, theme; roles that must at one time be occupied by animal X, at another by animal Y. It has also been pointed out that reciprocal altruism is not limited to primates; vampire bats engage in reciprocal exchanges of blood ('so why don't vampire bats have language?' is sometimes the implied, or even explicit, follow-up). Vampire bats do not have language because, in the first instance, they only exchange blood; because they do not (unlike primates) exchange other things, there is no need for them to set up the abstract category 'action' (covering things as diverse as grooming, fighting and food exchanging) to join the abstract role categories into what will turn out to be a template for the most basic language structure. In the second instance, reciprocal altruism only provides syntax, not language *per se*, and in order to make any use of syntax you have to have at least words (or signs) that can be mapped into syntactic structures. As vampires have no protolanguage, cheater detection remains for them (as it has done for non-hominid primates) cheater detection and nothing more; its potentialities for giving structure to language must lie latent until there is some sort of language to give structure to.

Thus, probably even before hominids began (and certainly by the time they began), there was a system developed for keeping score that involved casting the memories of any given event into a single mould, a mould that resembled a small play with a single action and one, two or three participants (agents, goals and themes). Lo and behold, a unique set of circumstances provided one brand of primate with symbolic utterances (for an account of how this came about see Bickerton, 2002), some of which described actions (verbs) and some of which described the kind of entities that might be agents, goals or themes of these actions (nouns). All that remained was to take each utterance and cast it in the mould already formed for it by episodic memory. Instead of a purely pragmatic mode of mapping thought to utterance, there would be a rigid framework that enabled automatic recovery of meaning on the part of the hearer and made possible the construction of ever longer and more complex sentences. There would be syntax, where there was none before.

But at this point one may well ask, why was there none before? Why, if cheater detection was in place in a variety of antecedent species, if events were already being cast into the agent–theme–goal mode, did not protolanguage become language as soon as it appeared, 2 million years ago?

WHAT BRAIN SIZE DOES

One might, of course, claim that language did appear, much as we know it, 2 million years ago. But that merely makes more enormous the delay in implementing (in terms of concrete artefacts) the cognitive advantages that a true language made available. Moreover, it leaves the history of our species as unexplained as do most other scenarios. Any adequate account of how our species came into existence has to explain how it was that the brain grew to at least its present size without changing the hominid way of life in any significant manner, and then, without further increase, made possible the stunning explosion of creativity that characterised our species. (The complaint, quite often heard, that our species spent more than half its lifetime without changing much is, again, another way of trying to hide the illegitimate baby: there are a number of factors that would have inevitably yielded a slowish start, small numbers, lack of competitive pressure, need to fine-tune and explore the new medium, and so on, quite apart from the fact that earliest dates for novel artefacts are continually being pushed back as investigations spread out of Europe into areas that we occupied at an earlier date. See the chapter in these *Proceedings* by Paul Mellars.) So far, no convincing explanation of the brain–culture mismatch has been produced.

One way to deal with this is to suppose that brain size increased for reasons that have nothing to do with intelligence: for example, the 'radiator theory' of Falk (1990), according to which the brain's venous drainage system, enlarging in parallel with overall brain growth, prevented the brain from overheating during long chases across the savanna in hot sunlight. The theory remains controversial, but if correct provides a neat explanation of how the brain could have vastly increased in size without, apparently, adding very much to cognitive capacity.

Another proposal suggests how increase in brain size for specific purposes might have little to do with general cognitive processes. Calvin (1983) has pointed out the problems presented by aimed throwing over distance, which probably played a significant part in the foraging activities of early hominids. Throwing encounters a problem in exact timing: to determine accurately the launch window for a throw of 8 metres, firing of the neurones involved has to be accurate to within a millisecond. Such accuracy is impossible unless timing is averaged out over a large number of neurones. But extra neurones added for such a purpose do not in and of themselves guarantee any cognitive increment.

We might also propose that there was some degree of cognitive improvement, but not of a kind that would generate the creativity that characterises modern humans. For instance, Wynn (1999: 275) considers that our ancestors of 300 thousand years ago may have had Piagetian 'operational intelligence' but that the nature of their artefacts, unchanged over hundreds of thousands of years and virtually identical from Britain to Japan, suggests that 'whatever ... makes modern symbolic systems volatile ... was missing ... '.

None of these approaches are, of course, mutually exclusive. Their combination would indeed have yielded bigger brains, but need not, in and of itself, have made us the kind of animal we are today.

Still, only half the equation is solved by such developments. They suggest how our brains might have grown radically without correspondingly radical cognitive gains; they are silent on the issue of how, once brains had reached a certain size, immense changes in behaviour became possible. Nor have we dealt with the problem that closed the last section: why, if some primitive form of language has been around for 2 million years or more, and if the cheater-detection template had been around even longer, did language not become syntactic almost immediately. For if language as we know it began so early, we have to explain two further things: how it failed, for nearly 2 million years, to enhance significantly our cognition, and what additional factor, entering on the scene only with modern humans, caused us to start behaving so differently from other animals.

In order to understand what is happening, we have to consider what brains were built to do. Brains were built to serve the purposes of the animals that contained them; to preserve their lives, help them find food and mates, spread their genes. As such, they were (and for other creatures, still are) primarily reactive mechanisms. Animals observe their environment, detect within it features that are associated with dangers and opportunities, and react appropriately to these. Those reactions range from automatic reflexes to quite varied and sophisticated responses. But while they can draw generalisations from environmental features, learn, in fact, to a greater or lesser extent, nothing obliges us to believe that they are capable of reflecting on them. There is no evidence that they consciously compare newer with older experiences, recast their strategies in accord with those comparisons, or perform any of the many other reflective activities that are a constant commonplace of human life.

The difference between what brains normally do and what ours can be made to do lies simply in this: while the mental activities of other species are driven largely and perhaps almost exclusively by external stimuli (dreaming in some mammals is an obvious exception, but this is unguided and uncontrolled activity), ours may in addition be driven by internal stimuli. The idea of school shootings may suddenly come uppermost in my mind (there is no need for it to have been explicitly triggered by any particular newscast or op-ed piece) and I can thereupon elaborate a policy, even a whole series of policies, for combating them; I may not hit on a very effective policy, but that is not the point. The point is that I can do it. And should it be argued that I can do such things merely as a consequence of being immersed in a rich culture, then I can ask, how is it that I, but not chimpanzees, dolphins and hominids, have such a rich culture to be immersed in?

Whether the capacity to have internally generated thoughts preceded protolanguage, or arose as a result of it, I do not propose to argue here. Suffice it to say that once one has the most minimal protolinguistic ability, the capacity must be there. Linguistic acts may be, but do not have to be, the result of external stimuli. If I tell you that your fly zip is open, your fly zip does not have to be open, I may merely want to embarrass you or throw you off balance while I pursue some social agenda of my own. I have internally generated the sequence, 'I say, don't look now [he will of course] but your fly is open'.

In order for me to say this, I must have dispatched a sequence containing three clauses, 10 words and at least (depending on how you analyse it) 26 phonemes to the organs of speech. I only have to garble that message slightly, 'don't lick now, but your flaw is open', in order to make complete nonsense of it. But to send that message, I have somehow (we still do not really have more than a skeletal idea how) merged and blended dozens of neural signals from different parts of the brain and yet maintained them as a series of coherent signals so that the motor organs of speech could execute them precisely and accurately.

All of this would be wonderful enough if the rest of the brain, all the bits and pieces engaged in other tasks, not to mention all the bits and pieces that want to talk but would like to say something different, would just shut down and listen up as soon as the Imperial 'I' announced His Imperial Intention of saying something. But of course there is no 'Imperial I', no 'executive suite' (Dennett, 1990), there is no-one in there but neurones, each doing its own thing, and trained by hundreds of millions of years of brain evolution to attend to a world outside, a world laden with life-threatening and life-enhancing phenomena, rather than to its own internal natterings. Given this situation, the fact that I can get from a set of firing neurones to a comprehensible utterance is nothing short of miraculous.

In short, what I am saying is that the emergence of protolanguage imposed on the brain tasks of a kind that no brain had ever been required to perform before, tasks beside which the accurate timing of launch windows becomes trivial. And assuredly, when the brain began to perform such tasks, it was not very good at them. At the beginning, this would hardly have mattered, because the earliest meaningful utterance probably consisted (like those of young children) of single symbols, so few in number that they could be uttered with a good deal of variation and still, usually, be understood. The problem arose when longer messages had to be assembled.

If the already-existing template was to be imposed on unstructured, protolinguistic, output, this meant that a coherent neural signal had to be maintained throughout the merging of the neural signals that coded for each unit (word) of the message. But maintaining a coherent signal through a series of merges was something that no previous brain had been built to do. (Note that other systems of animal communication are marked by their inability to combine units.) The problems involved in timing a launch window were added to the problems involved in maintaining a coherent internally generated message over what, for the brain, were protracted lengths of time (hundreds, if not thousands, of milliseconds).

And why was internal generation a problem? Because the brain was built to attend to stuff coming in from outside, not to stuff coming out from inside, certainly not to stuff coming from wherever in the neocortex words started to be stored, by a devious route that remains to be traced, to the organs that controlled vocal articulation. Before the complex problems arising from linguistic communication could be solved, two things had to be achieved:

1 enough spare (recruitable) units to be able to maintain a coherent and complex message over time against all the competing neural activity in the brain;
2 enough of the right kind of connections to allow such a message to travel unhindered and uncorrupted to its destination.

Let us look more closely at each of these in turn.

By 'recruitable' units I mean individual neurones or assemblies of neurones that are not irrevocably committed to some existing function (sight or hearing, for example) but that, while they might at various times perform various other functions, could, when required, be co-opted (Calvin, 1996) to serve linguistic functions. Such units would in fact be in Darwinian competition with one another (Calvin, 1998). Rather than a model in which 'I' decide what 'I' want to say, then say it, we must accustom ourselves to a model in which, at any given moment, a number of different sentences are trying to say themselves, one captures sufficient neurones to put it together, and afterwards 'I' convince 'myself' that that was what 'I' intended to say all along. Under such conditions, a quite substantial 'critical mass' of recruited neurones must be assembled for any one potential sentence to win out against the competition.

The connection problem cannot be solved until enough neurones are in place. Neuronal connections are to a very large extent epigenetic; their formation (and certainly their strengthening, when they exist) is driven by the amount of information they are required to carry. Fortunately everywhere in the brain is (remotely, somehow) connected to everywhere else, otherwise a novelty like language could never have got off the ground. But constant use is necessary to maintain and strengthen these connections, so a large enough quantity of spare neurones must be in place before the connections can be perfected (not to say that there has not been, for some time, a beneficial spiral of advancement between added neurones and improved connections). But what do we mean by 'a large enough quantity'?

In terms of absolute numbers, we have no way (at present) of answering this question. All that can be done is to show the neuronal increment required by adding an additional unit to a message. That way, we can get at least a rough idea of the brain size required if complex sentences are to be produced.

Let us assume that to merge two lexical units into a single message (as distinct from sending lexical item one, uttering it, then sending lexical item two and uttering it) doubles the time required for sending the message and also doubles the number of spare neurones required to keep the message coherent (Figure 1). So doubling brain size (or rather, doubling that part of brain size devoted to spare neurones) gets you very little; there is no point in creating a unitary message, you might just as well keep sending words one at a time.

Suppose you want to add a further unit. Assuming that the number of additional neurones required for a coherent signal incorporating an additional unit

Figure 1. One merge. W = word, sign or other symbolic unit.

is a constant; call that number x. Then adding a further unit will require going from x to $2x$. But this will not simply make possible three-unit utterances. The brain is not a linear computer; it can conduct any number of operations simultaneously, in parallel. Accordingly, it can carry out two merges simultaneously, then merge the result, thereby obtaining the capacity to produce strings of up to four units (Figure 2).

But let us consider a moment what is at issue here. We are now faced with competition between two possible ways of sending a message. The first is the existing protolinguistic way, in which words are dispatched singly to the organs of speech. The second is the new linguistic way, into which the neural signals representing the units of a message are pre-assembled into a single complex signal before being despatched to the organs of speech. The first way is unstructured, therefore there is no principled way of determining what relationship the words have to one another (pragmatic knowledge must suffice). The second way is structured. The longer the messages sent by the first method, the greater the danger of misinterpretation as possible ambiguous readings pile up. The second method can transmit much longer messages than the first with an almost negligible risk of ambiguity.

However, if the message is short, the first method has a lot going for it. Provided the message is firmly rooted in the here and now, messages of up to four or five units will be reasonably unambiguous in context. Moreover, as each word is separately executed, signals will not have to be maintained for long, therefore there is a good chance that each signal will be faithfully executed. However, as soon as signal time is prolonged, there is the risk (which may have been quite high until the second method became fully established) that part of

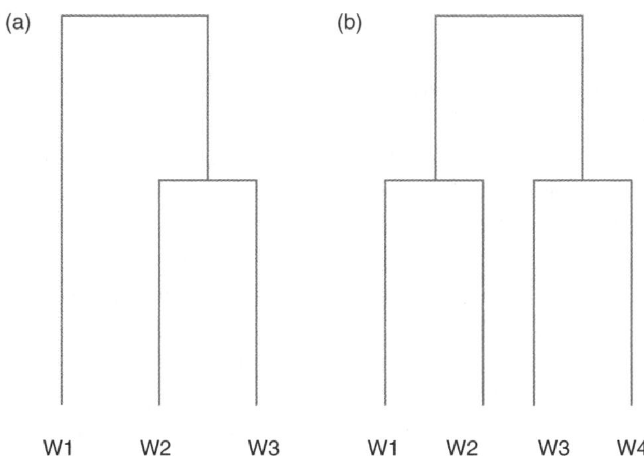

Figure 2. Two merges.

the message may be corrupted and, instead of incongruity, blank incomprehension will result.

In other words, for a species that cannot perform more than two merges, it is probably safer to rely on protolanguage than to try casting utterances into the linguistic mould. The significance of this fact is that a species could be capable of syntactic structures sharply limited in length and complexity and yet continue to employ protolanguage until a threshold, the threshold that allows for creation of utterances with five or more units, has been reached and passed.

That threshold comes with the capacity to perform three successful merges (Figure 3). As Figure 3 shows, that capacity, thanks to the brain's parallel processing, allows utterances of five to eight words to be assembled and produced. Capacity to produce utterances of up to 16 words results from only one additional merge (Figure 4). At this point we have, to all intents and purposes, reached the level of normal human discourse. Seldom, if ever, do we utter sentences of more than 16 words; if we do, they almost certainly consist of detachable portions that can be assembled separately and dispatched separately, while written sentences can, needless to say, be constructed in a much more leisurely and disconnected manner.

Now let us consider what this involves in terms of incremental brain growth, and see if there is any way in which, even very tentatively, we can tie that growth to the fossil record. Elsewhere (Bickerton, 1990; Calvin & Bickerton, 2000) I have suggested that protolanguage probably originated with *Homo erectus*. We assume that, to have this capacity, there must have been a portion of the brain potentially available for it (and probably for other tasks as well); a portion, in other words, that was not irrevocably committed to monitoring digestion or blood flow, recording input from peripheral cells, controlling motor activity, and so forth. In the beginning, if its value is x, that will have to be augmented to $4x$ to make syntax worthwhile, but only to $5x$ to yield full human capacity: a very steep threshold. Or, to put it another way, the brain would require a substantial increase to bring it even within reach of syntax, but only a small additional increment to make full human syntax potentially available.

What follows cannot, for the moment, go beyond pure speculation, yet it may be worthwhile in that it at least provides a target that can be shot at, in a landscape that still remains entirely unexplored. Let us arbitrarily suppose that at the beginning of the process, roughly 10% of the brain consists of recruitable neurones. The figure is arbitrary, but it cannot be massively less than this (otherwise no cognitive processes would be available) nor massively more (otherwise not enough neurones would be left to perform basic functions).

The brain size of *H. erectus* was 950 cubic centimetres (cm^3) (Tobias, 1987). Assuming that approximately one-tenth of this volume was recruitable, the value of x is 95. Therefore $4x$ is 380. Then 380 plus 950 is 1330. The average brain size of our species is 1350 cm^3.

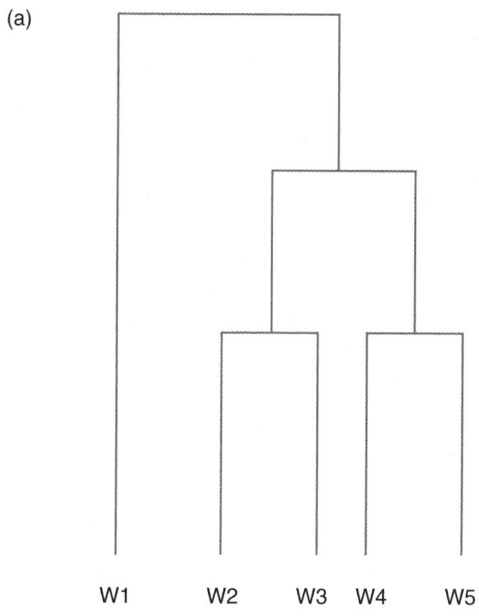

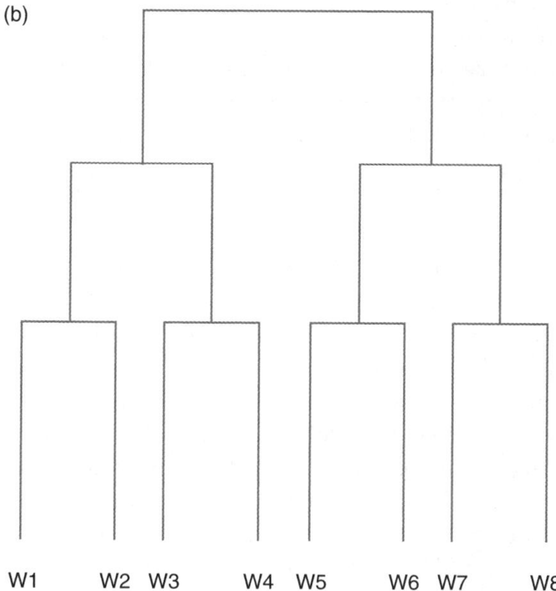

Figure 3. Three merges.

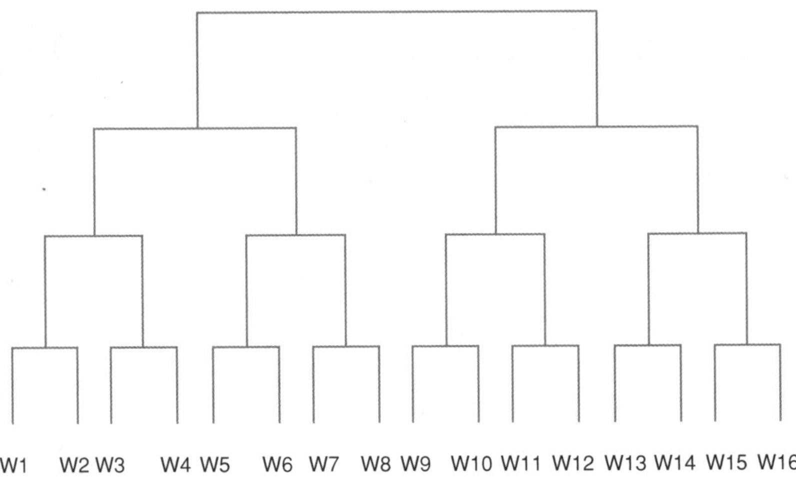

Figure 4. Four merges.

A wild guess? A numerical trick? Obviously such a result should be treated with the greatest caution, if not outright scepticism. Its value is not as a precise calculation: it is heuristic. It shows one way in which it might be possible to solve the problem with which this chapter began: that our brain expanded enormously with very little to show for it, then without expanding any more it suddenly became capable of (almost!) anything.

Until we can resolve this problem, we shall understand next to nothing about how we evolved.

THE REST OF THE STORY

There is one obvious objection to the story I have outlined. This is, that the brains of Neanderthals were larger than those of our immediate ancestors. If the explanation I have given is the correct one, how is it that we are here now instead of the Neanderthals? Why did Neanderthals not achieve language before we did, and all the cognitive benefits that came with it, leading inevitably to their victory in the interspecies competition in glacial Europe?

There are several matters that such a response ought to take into account. First is the fact that total brain size is not at issue here. It is the amount of brain that contains recruitable neurones. This amount may well have been less in Neanderthals than in modern humans.

Supporting this is the robust–gracile distinction. Neanderthals were much more robust than Cro-Magnons, and therefore could be expected to devote more of their larger brain to housekeeping matters. Therefore it is not unreasonable to suppose that, although in gross size their brains may have exceeded ours, they were at least a merge or two behind us, and were therefore limited to protolanguage, while our ancestors enjoyed the benefits of full human language.

However, there is no need even to invoke a difference in merger capacity. As noted in the previous section, number of disposable neurones does not alone determine the issue. Once those neurones are available, the connections between them have to be direct and robust enough to carry messages with the required speed and accuracy. Those connections would have taken time to forge. It follows that Neanderthals and modern humans could have been on a par merger-wise, yet the latter could have had a more developed system of connections. This could have occurred if there had merely been a difference of a few tens of thousands of years in reaching the merge threshold.

This argument acquires greater force when one considers that improved connections were not necessarily limited to the pathway between cortex and motor organs. The motor organs themselves had to undergo a good deal of refinement. There would be little point in achieving the capacity to produce sentences of up to 16 units if the execution of such sentences had been muddled, blurred and difficult to process by hearers. The achievement of syntax must have exercised a very strong selective pressure towards improved vocal abilities; again, a relatively short time-lag would have given modern humans an edge in intergroup communication, if not necessarily in thinking. That edge, however slight, might have been sufficient to extinguish the competition.

There is yet a further possibility of reconciling the present account with human–Neanderthal comparisons. Although language has long been a favourite among explanations of human victory, it is far from the only contender. It could be that humans and Neanderthals were linguistic equals; the structure of their vocal organs, their artefacts, and their behaviour do not encourage this view, but it cannot be ruled out. Some factor of cultural organisation, or introduced diseases, could then have made the crucial difference.

Thus there is nothing in human–Neanderthal relations that excludes the scenario presented here. On the positive side, it presents the only explanation I know of for the curious trajectory of our ancestors. Thus, despite the unavoidable elements of speculation it contains, it merits serious consideration as a rough sketch, rather than a sophisticated roadmap, of a way into the mysteries of our past.

DISCUSSION

Questioner: About a threshold of neurones in the brain, Neanderthals have large brains, but you always said that Neanderthal's lacked speech ...

Bickerton: I'm not sure about that anymore. The Neanderthal is a problem. It's not enough to just have a big brain. It's not so much a threshold of neurones as it is developing long massive pathways. Once you have the brain size, then you have to develop the pathways from generation to generation. We don't really need anything as critical as language versus protolanguage to explain how we won out and the Neanderthal didn't. All that is basically needed are other things that could tip the brain toward language, that is some speculation about social life, Neanderthals lived very different isolated social lives.

Questioner: Take the use of the harpoon and the symmetrical hand axe? Wouldn't everything you said apply to these as well?

Bickerton: I don't think so: the harpoon changed rapidly and the hand axe didn't change.

Questioner: Apes have tools. If you will make a hand axe, you need to have a plan. Is there any difference between the skills of apes and humans? For example, no ape can thread a needle.

Bickerton: You may be right. But my guess is that no hominids could do that as well.

Comment: Well then that's the essence of being human.

Bickerton: When you're making a tool, no one else is making it. Language is different: sometimes you're the agent, while other times he's the agent.

Questioner: Why did you upgrade the Neanderthal?

Bickerton: I can't see why if their brain size is big, they couldn't stumble through a few sentences. Think of it this way: there could be two parts of the brain not talking to each other and then lo and behold there is a connection formed between them. Is it a mutation? Epigenetic? Clearly there is different parcelation but these things have to interact and to specify where they go. It's not enough to have parcelation.

Questioner: Words are processed in 200 milliseconds, it shows that many parallel processes decipher the words to form communications; if you speak on and on how does this cascade work?

Bickerton: It seems to me that what is actually happening, once you get the brain in its original mode, is that it can't say sentences all the time. I don't see how this has a bearing on what I am saying.

Questioner: There is a threshold brain size, by which we have humans. Why are you not saying there is a threshold brain size for language? Then Neanderthals might have language.

Bickerton: Then why, if they had language, didn't they do more with it? When you have the brain capacity, you can manipulate the environment to suit yourself. If you want your genes to reproduce you are going to have to use your capacity to make things easier for yourself. We don't see any evidence of serious valuable creative activity in the Neanderthals or early on.

References

Bickerton, D. (1990) *Language and Species*. Chicago: Chicago University Press.

Bickerton, D. (1995) *Language and Human Behavior*. Seattle: University of Washington Press.

Bickerton, D. (1998) Catastrophic evolution: the case for a single step from protolanguage to full human language. In: *Approaches to the Evolution of Language* (eds J. R. Hurford, M. Studdert-Kennedy & C. Knight), pp. 341–58. Cambridge: Cambridge University Press.

Bickerton, D. (2002) *Foraging Versus Social Intelligence in the Evolution of Protolanguage.* In: *The Transition to Language* (ed. A. Wray), pp. 207–25. Oxford: Oxford University Press.

Bloom, P. (2000) *Review of Lingua ex Machina. New York Times Review of Books.* New York: New York Times.

Calvin, W.H. (1983) A stone's throw and its launch window: timing precision and its implications for language and hominid brains. *Journal of Theoretical Biology*, **104**, 121–35.

Calvin, W.H. (1996) *The Cerebral Code: Thinking a Thought in the Mosaic of the Mind.* Massachusetts: MIT Press.

Calvin, W.H. (1998) Competing for consciousness: a Darwinian mechanism at an appropriate level of explanation. *Journal of Consciousness Studies*, **5**, 389–404.

Calvin, W.H. & Bickerton, D. (2000) *Lingua Ex Machina: Reconciling Darwin and Chomsky with the Human Brain*. Massachusetts: MIT Press.

Cosmides, L. & Tooby, J. (1992) Cognitive adaptations for social exchange. In: *The Adapted Mind* (eds J. H. Barkow, L. Cosmides & J. Tooby), pp. 163–228. Oxford: Oxford University Press.

Deacon, T. (1997) *The Symbolic Species: The Co-evolution of Language and the Brain.* New York: W.W. Norton.

Dennett, D. (1990) *Consciousness Explained.* New York: Little, Brown.

Falk, D. (1990) The radiator hypothesis. *Behavioral and Brain Sciences*, **13**, 333–81.

Falk, D. (1993) Sex differences in visuospatial skills. In: *Tools, Language and Cognition in Human Evolution* (eds K. R. Gibson & T. Ingold), pp. 216–29. Cambridge: Cambridge University Press.

Gould, S.J. & Vrba, E.S. (1982) Exaptation – a missing term in the science of form. *Paleobiology*, **8**, 4–15.

Hauser, M. (1996) *The Evolution of Communication*. Massachusetts: MIT Press.

Jellinek, A. (1977) The lower Paleolithic. *Annual Review of Anthropology*, **6**, 1–32.

Newmeyer, F.J. (1991) Functional explanations in linguistics and the origins of language. *Language and Communication*, **11**, 1–28.

Pinker, S. (1994) *The Language Instinct*. New York: Morrow.

Pinker, S. & Bloom, P. (1990) Natural language and natural selection. *Behavioral and Brain Sciences*, **13**, 707–84.

Smuts, B. (1987) *Primate Societies*. Chicago: University of Chicago Press.

Strum, S.C. (1987) *Almost Human: A Journey into the World of Baboons.* New York: Norton.

Tobias, P.V. (1971) *The Brain in Hominid Evolution*. New York: Columbia University Press.

Tobias, P.V. (1987) The brain of *Homo habilis*: a new level of organization in cerebral evolution. *Journal of Human Evolution*, **16**, 741–61.

Trivers, R.L. (1971) The evolution of reciprocal altruism. *Quarterly Review of Biology*, **46**, 35–57.

de Waal, F.B.M. (1982) *Chimpanzee Politics: Power and Sex among Apes*. London: Cape.

Wynn, T.G. (1999) The evolution of tools and symbolic behavior. In: *Handbook of Human Symbolic Evolution* (eds. A. Lock & C. R. Peters), pp. 261–87. Oxford: Blackwell Science.

Is the Neural Basis of Vocalisation Different in Non-Human Primates and *Homo sapiens*?

DETLEV PLOOG

Summary. From an evolutionary perspective the voice was a prerequisite for the emergence of speech. Speech, the most advanced mode of vocal communication, became possible only after gradual transformations of the sound-producing system and its central nervous control, in co-evolution with the transformations of the auditory system, had taken place. There are two systems in the brain that produce and control vocal behaviour. The first is very old phylogenetically. In non-human primates and humans it comprises limbic structures, all of which funnel into the peri-aqueductal grey of the midbrain. If this matrix is destroyed all land-living vertebrates become mute. The second system, the neocortical voice pathway as part of the pyramidal tract, emerged in non-human primates and developed in substance from monkeys to humans; it is indispensable for the voluntary control of the voice. The destruction of this system has no influence on the monkey's spontaneous vocal behaviour; in humans, however, it has disastrous consequences and makes speech impossible. The hypothesis is advanced that the last step in the evolution of the phonatory system in the brain was the outgrowing and augmenting of the fine fibre portion of the pyramidal tract synapsing directly with the motor nuclei for the vocal cords and the tongue, so that the direct and voluntary control of vocal behaviour became possible. The question raised in my title, 'Is the neural basis of vocalisation different in primates and *Homo sapiens*', must, of course, be answered with 'yes'. The neural basis is in fact quite different. Explaining this difference and its consequences for the evolution of language and speech is the purpose of this chapter.

Proceedings of the British Academy, **106**, 121–135, © The British Academy 2002.

INTRODUCTION

Vocal behaviour is a prerequisite for speech behaviour. It has a very long evolutionary history in vertebrates, whilst speech behaviour has a short one. If we follow the evolutionary history of vocal behaviour from toads and frogs to reptiles and mammals, not to mention birds, we can observe profound changes within the sound-producing apparatus, from the tripartite to the quadripartite larynx, including the well-known descent of the human larynx during the first months of ontogenesis. That this transformation is necessary for the development of intelligible speech is demonstrated by children with Down's syndrome, where this descent is incomplete. The genetically determined malfunction impairs speech considerably.

Changes in the internal laryngeal voice-producing muscles and other evolutionary transformations of the larynx should also be considered here. Only in the human species does the vocal muscle (m. thyreoarytaenoideus lateralis) send fine fibres into the medial part of the vocal cords, which allows extremely fine tuning of the cords, a prerequisite for the human faculty of singing.

Each little step in the evolution of the voice-producing system led to a higher complexity of vocal behaviour, resulting in different species-specific vocal repertoires, which are used almost exclusively for sexual selection and social communication. Even genetically determined slight differences in the vocal expressions of a given species, so-called 'dialects', have selective consequences (Ploog *et al.*, 1975). I am convinced that vocal gestures are at the roots of the evolution of language, and not body gestures, as Michael Corballis and others believe. Audio-vocal behaviour as opposed to gestural behaviour is advantageous in the dark, and also while hunting, harvesting, cooking, making tools and performing many other daily routines during which communication takes place while the hands are busy. The chief argument, however, is the co-evolution of the vocal and auditory systems, which is rarely mentioned in this debate. Therefore, before I come to the neural basis of vocalisation in non-human primates and humans I will briefly comment on the auditory part of the communication system (Ploog, 1981, 1988, 1990).

AUDIO-VOCAL SIGNALLING

Very early in the evolution of the vertebrates audio-vocal signals function in sexual selection. Female frogs, for example, distinguish between the mating croaks of mature and immature male frogs, and they prefer the voices of their own population to those of neighbouring populations. Moreover, the evolution of the ear from reptiles to mammals, and especially primates, is striking. While reptiles have only one bone for sound transmission in the middle ear, mammals have three, which required substantial bony transformations of the lower jaw. Why is it that the sound-producing system and the sound-decoding system have co-evolved over 200 or more million years? The answer is that audio-vocal behaviour was the most successful type of behaviour in the increasingly complex sexual selection processes and social communication. And finally, why is it that the new-born human baby is capable of distinguishing certain phonemes from others that are all universal in the languages of *Homo sapiens*? In my opinion, the co-evolution of auditory and vocal behaviour, including the prominent evolution of the brain structures involved, forms the basis of the evolution of language.

TWO SYSTEMS FOR VOCAL BEHAVIOUR

The remainder of my chapter deals with the brain structures involved in vocalisation. My colleagues and I investigated the central nervous organisation of phonation in the squirrel monkey, a small South American primate that is endowed with a rich vocal repertoire. We found that the great variety of calls predominantly regulates the complex social behaviour of these monkeys. Only a few calls refer to external events, especially and differentially to aerial and terrestrial predators. All call types are only slightly modifiable fixed-action patterns, which means that they are innate (Winter *et al.* 1966, 1973; Herzog & Hopf, 1984). There are at least two subspecies with slightly different vocal repertoires that are genetically transmitted (Ploog, 1986, 1995). Our findings in squirrel monkeys are also basically relevant for macaques and other primate species.

Most, if not all, natural vocal expressions can be elicited reliably and repeatedly by electrical stimulation of specific brain sites in the awake animal (Jürgens & Ploog, 1970). In addition, chemical stimulation and pharmacological blockade of specific brain sites have been used to explore further functional properties of the vocal system in the brain (Jürgens & Lu, 1993).

Figure 1 gives an overview of those primate brain structures from which species-specific vocalisations can be elicited electrically (indicated in black). Only limbic, thalamic, hypothalamic and brainstem structures, but no neocortical structures, are part of the extended system. The frontal stippled area is involved in a rather limited control of the voice in simple vocal conditioning paradigms. The stippled area in the peri-aqueductal grey (PAG) of the midbrain delineates the matrix of the vocal system. Its destruction results in mutism in land-living vertebrates, including humans (Jürgens & Ploog, 1970, 1976).

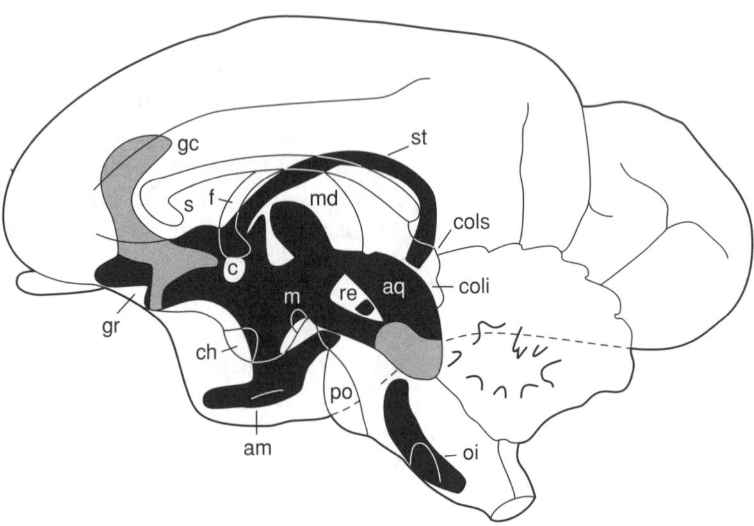

Figure 1. Vocalisation-producing brain areas in the squirrel monkey (*Saimiri sciureus*) (Jürgens & Ploog, 1976). Key: am, amygdala; aq, substantia grisea centralis; c, commissura anterior; ch, chiasma opticum; coli, colliculus inferior; cols, colliculus superior; f, fornix; gc, gyrus cinguli; gr, gyrus rectus; m, corpus mammillare; md, nucleus medialis dorsalis thalami; oi, nucleus olivaris inferior; po, griseum pontis; re, formatio reticularis tegmenti; s, septum; st, stria terminalis.

The cingulate vocalisation pathway

Combined lesion and tracer studies show that all the limbic structures from which vocalisations are elicitable converge in the PAG (Figure 2). There is one major pathway, the 'cingular vocalisation pathway' (Jürgens & Pratt, 1979), that runs from the anterior cingulate gyrus monosynaptically into the PAG. Lesions along this pathway abolish calls elicited from the anterior cingulate. The course of this tract joins the pyramidal tract in the internal capsule and follows it down to the caudal diencephalon. On leaving the pyramidal tract at the cerebral peduncle, the fibres ascend dorsally to the PAG and follow its course to its end, where they sweep laterally through the parabrachial area and descend through the lateral pons and lateral medulla oblongata to the nucleus ambiguus, which is the nucleus for the internal laryngeal motor neurones (Müller-Preuss & Jürgens, 1976). These anatomical details are important for conclusions to be drawn later.

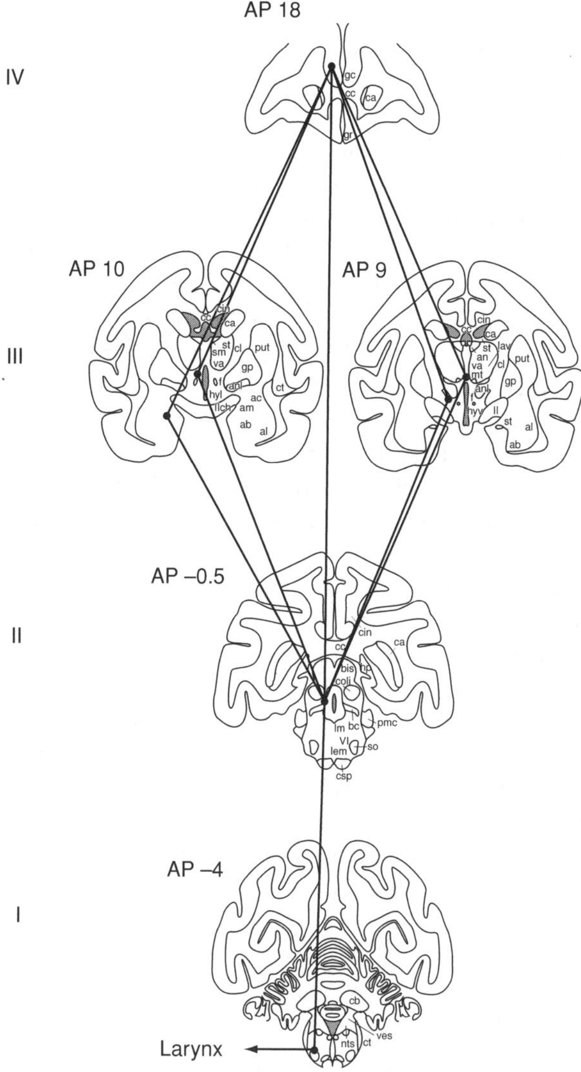

Figure 2. Scheme of hierarchical control of vocalisation. All brain areas indicated by a dot yield vocalisation when electrically stimulated. All lines interconnecting the dots represent anatomically verified direct projections (leading in rostrocaudal direction). The dots indicate in (IV) the anterior cingulate gyrus, in (III) the basal amygdaloid nucleus, dorsomedial and lateral hypothalamus and midline thalamus, in (II) the peri-aqueductal grey and laterally bordering tegmentum, and in (I) the nucleus ambiguus and surrounding reticular formation (the nucleus ambiguus itself only yields isolated movements of the vocal folds; phonation can be obtained, however, from its immediate vicinity). For further information, see the text (Jürgens & Ploog, 1981).

The cortical larynx area and its projection

Electrical stimulation of the most rostroinferior part of the neocortical face representation in the monkey (corresponding to area 6ba in the Brodmann–Vogt nomenclature) yields movements of the vocal folds. Bilateral removal of this area leaves the spontaneous vocalisations of both rhesus and squirrel monkeys unimpaired (Sutton *et al.*, 1974; Kirzinger & Jürgens, 1982). In humans, however, damage to the cortical face area in the dominant hemisphere results in a complete bilateral paralysis of the vocal cords. Such aphonia may last for months (Jürgens *et al.*, 1982). This difference between monkeys and humans can be explained by the fact that in humans, and apparently only in humans, there is a direct pathway from the laryngeal representation in the primary motor cortex to the laryngeal motor neurones of the nucleus ambiguus (Kuypers, 1958).

To clarify further the function of this part of the corticobulbar tract, Jürgens & Zwirner (1996) performed a crucial experiment, in which the PAG was pharmacologically blocked. In this condition it was not possible to elicit vocalisations from the anterior cingulate cortex. Different vocalisation sites in the forebrain could be blocked by a single PAG injection of kynurenic acid or procaine. However, that blockade had no effect on vocal fold movements elicited electrically from the cortical face area.

This finding strongly suggests the existence of two separate pathways controlling the vocal cords, namely the phylogenetically old limbic cingular vocalisation pathway and the phylogenetically young corticobulbar pathway, which is part of the pyramidal tract, a structure found only in mammals. Its entire spinal component is best developed in primates and reaches its greatest development in humans. The most important evolutionary step in its development is a group of corticospinal fibres that originate in the precentral cortex (area 4) and project monosynaptically to motor neural pools in the most ventral part of the cervical and lumbar enlargements of the spinal cord.

The neocortical vocal pathway

This corticomotoneuronal pathway emerges in mammals and develops in substance from monkeys to apes to humans, executing highly fractionated movements of the hands and relatively independent finger movements. This pathway is indispensable for the voluntary control of movements (Hepp-Reymond, 1988; Phillips, 1979). It does not seem too far-fetched to assume that this growing tendency of increasing dexterity and control in evolution is also effective in regard to the fine tuning of the vocal folds by the laryngeal motoneurones, for example in human songs, and the hypoglossal neurones for lingual articulation, which is crucial for speech but seems to be almost completely absent in the monkey's vocal communication. Again Jürgens and his group did some pioneering work in this area (Chen & Jürgens, 1995). They compared the projections of the tongue area of the primary motor cortex in the tree shrew, the tamarin and the rhesus monkey. They found that the tree shrew lacks a direct connection between motor cortex and hypoglossal nucleus. In the tamarin very few fibres could be detected there, and in the rhesus monkey there were marked terminal fibres in the hypoglossal nucleus. The authors concluded that 'there is a phylogenetic trend from lower to higher primates strengthening the cortico-hypoglossal connections'.

THE HUMAN CASE

In humans, the neocorticobulbar (pyramidal) system and the limbic vocal system co-operate inseparably, but may be separately involved in certain clinical cases.

The cingulate gyrus

A 41-year-old male patient was seen after a cerebral infarction, which was found to have affected the anterior cingulate cortex bilaterally, the left supplementary motor area, and the medial orbital cortex bilaterally (Figure 3). During the first 6 weeks post-infarction the patient remained in a state of akinetic mutism. Occasionally he responded to a painful stimulus with a moan. After 10 weeks, he could repeat long sentences in the whispering mode, articulating clearly without mistakes. Spontaneous utterances, however, remained reduced to a few monosyllabic words. After a year his speech was still restricted and monotonous. In an intonation test performed after 5 years, he was unable to attach the appropriate emotion to certain short exclamations, such as 'shut up' or 'terrific!'. The sonograms revealed that his ability to speak with emotion was greatly reduced, and he was unable to correct this deficiency voluntarily. Responsible for this lack of prosody and the inability to control intonation was the bilateral lesion of the anterior cingulate cortex (Jürgens & Cramon, 1982; Cramon & Jürgens, 1983).

The cortical face area

A second clinical example demonstrates that in humans that part of the cortex in which larynx, pharynx, tongue and mouth are represented is necessary for controlling the voice, whereas in the monkey these structures are not needed for phonation. A 52-year-old right-handed male patient had had an embolic cerebral infarction of the left middle cerebral artery (Figure 4a). He had

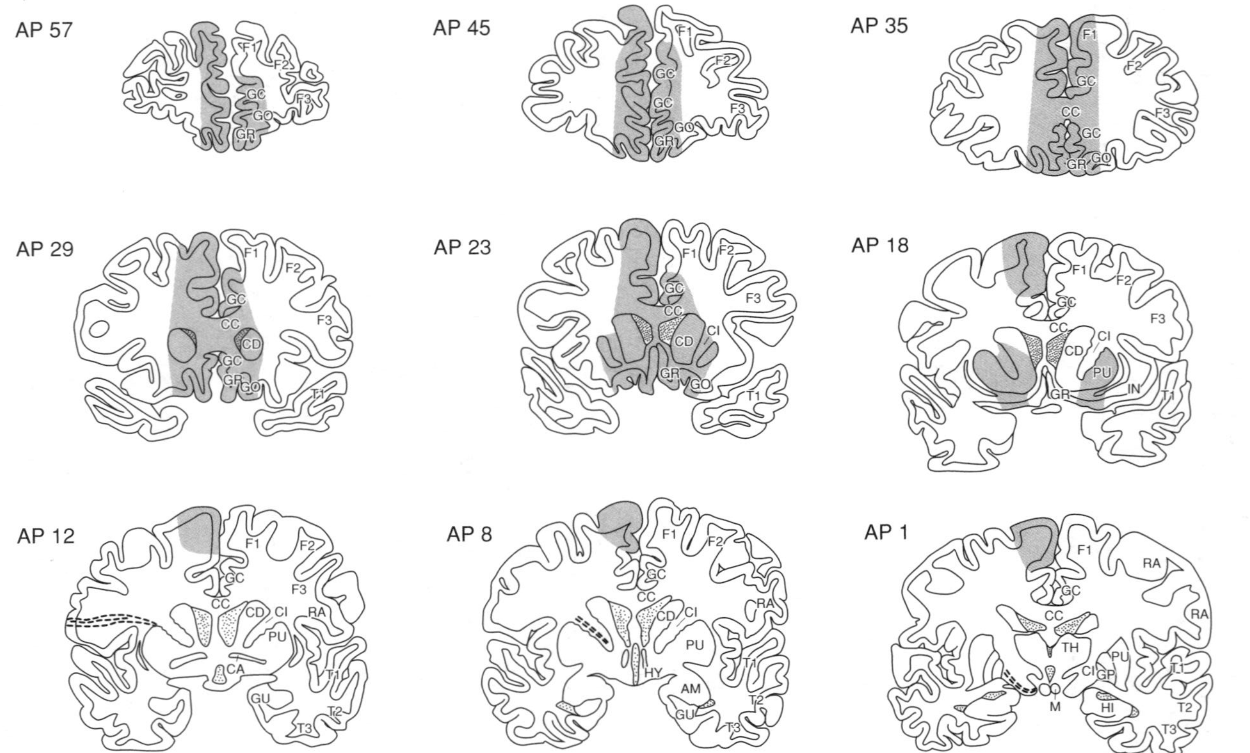

Figure 3. Schematic view of the human brain at nine levels from front (AP 57) to rear (AP 1). Dark zones: cerebral infarction sites, AP 12, AP 8, AP 1. Dashed lines: motor pathway from cortex to larynx and articulatory organs (Jürgens & Cramon, 1982).

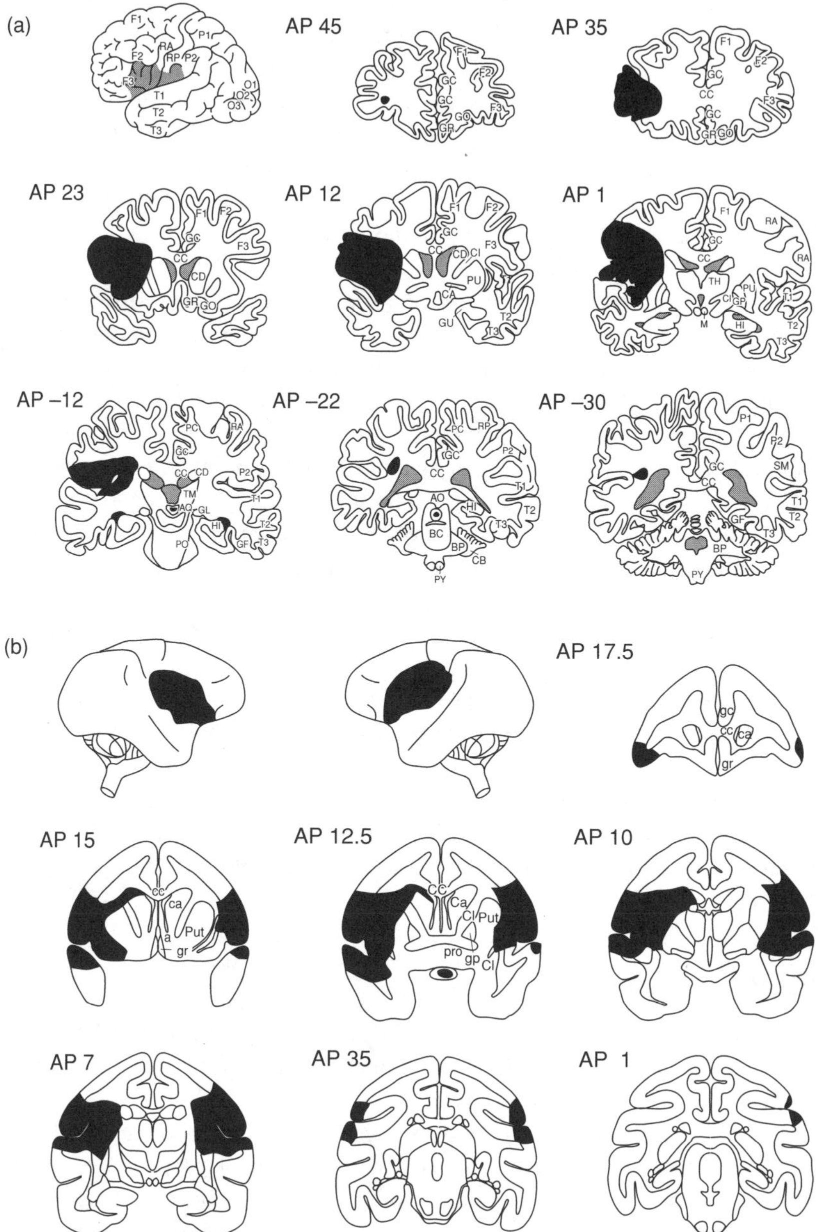

Figure 4. Frontal sections through (a) human brain and (b) monkey brain. Dark zones: (a) cerebral infarction area, (b) cerebral lesion sites (Jürgens *et al.*, 1982).

hemiplegia on the right side, a right-sided lower facial paresis, and tongue deviation to the right. After regaining consciousness, he could understand what was said to him but could not utter a single sound except that he coughed when the base of his tongue was touched. A laryngoscopic examination revealed that his vocal folds remained motionless during both respiration and attempts at phonation. This state of complete mutism and inability to phonate lasted for 11 weeks. Then, in the subsequent 2 weeks, phonation was completely restored. In contrast to the patient in the first example, this patient had considerable difficulty articulating and making oral movements.

The patient's brain lesion was experimentally modelled in the monkey (Figure 4b). In non-human primates vocalisations are elicitable from subcortical areas either in the left or in the right hemisphere. Therefore, the lesion in the monkey was carried out bilaterally. Homologous to the patient's lesion, it invaded the equivalent of Broca's area, the inferior pre- and post-central cortex, the rolandic operculum, and other structures. Although the monkey's tongue, lips and masticatory muscles were completely paralysed after the operation, its phonation remained intact. The spectrographically recorded vocalisations included all call types of the species. Consequently, the vocal folds were functioning, i.e. the central nervous patterning of calls was not impaired (Jürgens *et al.*, 1982).

The results are summarised in Figure 5. On the left side we see the limbic pathway, running from the cingulate gyrus monosynaptically to the PAG (AP –0.5), and from there to the laryngeal motor neurone (AP –4); on the right side, the corticopyramidal pathway, running from the cortical face area to the laryngeal motor neurones (AP –4). Triangles indicate sites the electrical stimulation of which produces vocalisation (AP 19) and isolated vocal fold movements, respectively. The dot at AP –0.5 indicates an injection site capable of blocking limbically, but not neocortically, induced vocal fold activity (Jürgens & Zwirner, 1996).

The two systems in tandem

It remains to be explained how these two systems, the neocortical executive system and the limbic vocal system, function inseparably in everyday life. From the neuroanatomical point of view, in the squirrel monkey there are numerous connections between the two systems, of which only a few shall be mentioned here. There are two areas that receive direct projections from the cortical larynx area, namely the anterior cingulate area and the parabrachial nuclei at the PAG, two nodal areas in the limbic vocal system (Jürgens, 1976). Conversely, projections from the limbic cingulate area into neocortical areas, for example the dorsal medial frontal cortex and Broca's area 44, are manifold (Müller-Preuss & Jürgens, 1976). The point is that the limbic part of the vocal

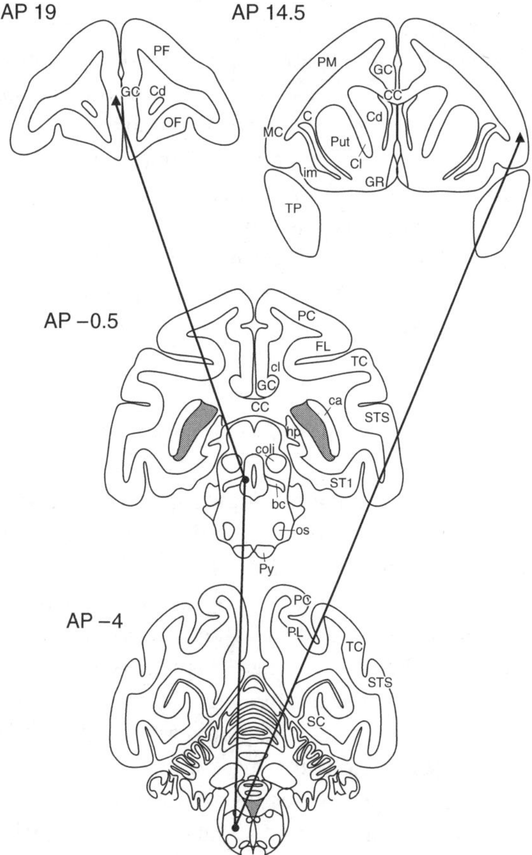

Figure 5. Scheme of a monkey brain as in Figure 2. Triangles indicate sites the electrical stimulation of which produces vocalisation (AP 19) or isolated vocal fold movements (AP 14.5), respectively. Square (AP −0.5) indicates injection site capable of blocking limbically but not neocortically induced vocal fold activity. Circle (AP −4) indicates site of laryngeal motoneurones. The connecting lines represent monosynaptic connections in the case of the projection from the anterior cingulate cortex to the peri-aqueductal grey, and polysynaptic connections to the other two cases (Jürgens & Zwirner, 1996).

system is closely tied to neocortical functions, as demonstrated in the two clinical cases.

The drastic effect of damage to the anterior cingulate area on human emotional speech is explained further by the fact that the anterior nucleus of the thalamus, which provides a large limbic source for the anterior cingulate area, has in humans substantially increased in volume over what would be expected in an ape thalamus of human brain size (Armstrong, 1986).

These brief remarks underline the indispensable participation of limbic functions in human speech. Limbic emotional vocal expressions and neocortically controlled speech work together in tandem. Only humans are able to act out their emotions with a species-specific motor system: the speech system. This explains why, for instance, even very aggressive acts are often expressed as verbal attacks and why verbal expressions of emotions often lead to relief from the respective emotional state. Talking to each other, the highest form of social communication, may cause pleasure, not talking to each other may cause discomfort. Although human social communication is dominated by the cortically guided language, the limbic structures are always involved, as clearly seen in all vocal, facial, postural and other expressions of emotion.

A HYPOTHESIS

Considering the strengthening and outgrowing of the corticopyramidal tract in the phylogeny of mammals, especially in primates, the hypothesis is advanced that the last step in the evolution of the phonatory system in the brain was the outgrowing and augmenting of the fine fibre portion of the pyramidal system that serves the direct and fast innervation of the larynx and tongue muscles via the nucleus retroambiguus and nucleus hypoglossus. For the more effective movement of the vocal cords and the tongue, more neurones of the pyramidal cell type in the increasingly larger neocortical face area may have been recruited. The direct and voluntary control of vocal behaviour allows the fractionation of species-specific vocal patterns such as cooing and babbling, transforming them into imitated and successively learnt articulated vocal gestures, i.e. words (Levelt, 1989; Ploog, 1990, 1995). In a way, this process is reflected in human ontogeny during language acquisition, where the child not only gradually gains control over its babbling, mastering increasingly more phonemes of its mother tongue, but also separating the speech motor system from the rest of the body motor system. Indicative of this stage in development are the conspicuous so-called 'associative movements' of the extremities, especially the hands, which accompany the articulatory movements (Noterdame *et al.*, 1988).

DISCUSSION

Comment: Patients who are mute and have limited cortical lesions make an effort to express themselves in other ways, for example by gesture.

Ploog: Yes, that is correct, but these people have lost their desire to communicate.

Comment: So then there is no lesion that leaves a person trapped in their brain trying to communicate but can't?

Ploog: They don't try to communicate. The drive to communicate is gone.

Comment: What did early hominids do? They wouldn't have used their hands for communication because they needed them for tool manufacture. You could turn that around and ask why tool manufacture was so static for so long and then it suddenly took off so quickly. You could reasonably argue that the reason it was so static for so long was that they needed their hands for communication.

Ploog: There must have been selective pressure for vocalisation. Why should it be since vocal communication among primates was so predominant and so forceful, why should we introduce gestures? Of course, when a chimp does a certain facial expression, it is also a kind of gesture, but it is an expression of emotion. This may mean that language and emotions are somehow related.

Comment: You mention how these primates communicate so well with lots of signals and do this on an interpersonal basis. This is also cortical and it is also fixed.

Ploog: It is not cortical; it is subcortical.

Comment: Signals that are elicited in fixed situations are not really like language.

Ploog: It is like language in that it serves as communication; but it is of course not language in that it is not grammar or syntax.

Comment: We laugh and grunt to communicate and these are not language and they are subcortical. Is this what you mean?

Ploog: Yes. Language is something different. Vocalisation evolved into speech, but it is not language. The argument cannot be turned around, however, to say that because language has syntax and grammar, it has nothing to do with vocalisation.

Comment: Yes, speech is obviously a form of language that has a lot to do with vocalisation.

References

Armstrong, E. (1986) Enlarged limbic structures in the human brain: the anterior thalamus and medial mamillary body. *Brain Research*, **362**, 394–7.

Chen, Y. & Jürgens, U. (1995) Phylogenetic trends in the projections of the cortical tongue area within primates. *European Journal of Neuroscience*, Supplement **8**, Abstract 53.04, 149.

Cramon, D. von & Jürgens, U. (1983) The anterior cingulate cortex and the phonatory control in monkey and man. *Neuroscience and Biobehavioral Reviews*, **7**, 423–6.

Hepp-Reymond, M.-C. (1988) Functional organization of motor cortex and its participation in voluntary movements. In: *Comparative Primate Biology*, Vol. 4: *Neurosciences* (eds H. D. Steklis & J. Erwin), pp. 501–624. New York: Alan R. Liss Inc.

Herzog, M. & Hopf, S. (1984) Behavioral responses to species-specific warning calls in infant squirrel monkeys reared in social isolation. *American Journal of Primatology*, **7**, 99–106.

Jürgens, U. (1976) Projections from the cortical larynx area in the squirrel monkey. *Experimental Brain Research*, **25**, 401–11.

Jürgens, U. & Cramon, D. von (1982) On the role of the anterior cingulate cortex in phonation: a case report. *Brain and Language*, **15**, 234–48.

Jürgens, U. & Lu, C.-L. (1993) The effects of periaqueductally injected transmitter antagonists on forebrain-elicited vocalization in the squirrel monkey. *European Journal of Neuroscience*, **5**, 735–41.

Jürgens, U. & Ploog, D. (1970) Cerebral representation of vocalization in the squirrel monkey. *Experimental Brain Research*, **10**, 532–54.

Jürgens, U. & Ploog, D. (1976) Ethologische Grundlagen. In: *Handbuch der Psychologie*, Vol. 8.1 (ed. L. J. Pongratz), pp. 599–633. Göttingen, Toronto, Zürich: Hogrefe.

Jürgens, U. & Ploog, D. (1981) On the neural control of mammalian vocalization. *Trends in Neurosciences*, **4**, 135–7.

Jürgens, U. & Pratt, R. (1979) The cingular vocalization pathway in the squirrel monkey. *Experimental Brain Research*, **34**, 499–510.

Jürgens, U. & Zwirner, P. (1996) The role of the periaqueductal grey in limbic and neocortical vocal fold control. *Neuroreport*, **7**, 2921–3.

Jürgens, U., Kirzinger, A. & Cramon, D. von (1982) The effects of deep-reaching lesions in the cortical face area on phonation. A combined case report and experimental monkey study. *Cortex*, **18**, 125–40.

Kirzinger, A. & Jürgens, U. (1982) Cortical lesion effects and vocalization in the squirrel monkey. *Brain Research*, **233**, 299–315.

Kuypers, H.G.J.M. (1958) Corticobulbar connexions to the pons and lower brain-stem in man. *Brain*, **81**, 364–88.

Levelt, W.J.M. (1989) *Speaking: From Intention to Articulation*. Cambridge, MA and London: The MIT Press.

Müller-Preuss, P. & Jürgens, U. (1976) Projections from the cingular vocalization area in the squirrel monkey. *Brain Research*, **103**, 29–43.

Noterdame, M., Amorosa, H., Ploog, M. & Scheimann, G. (1988) Quantitative and qualitative aspects of associated movements in children with specific developmental speech and language disorders and in normal pre-school children. *Journal of Human Movements Studies*, **15**, 151–69.

Phillips, C.G. (1979) The cortico-spinal pathway of primates. In: *Integration in the Nervous System* (eds H. Asanuma & V. J. Wilson), pp. 263–8. Tokyo: Igaku-Shoin.

Ploog, D. (1981) Neurobiology of primate audio-vocal behavior. *Brain Research Reviews*, **3**, 61–76.

Ploog, D. (1986) Biological foundations of the vocal expressions of emotions. In: *Emotion: Theory, Research and Experience*, Vol. III: *Biological Foundations of Emotion* (eds R. Plutchick & H. Kellerman), pp. 173–97. New York: Academic Press.

Ploog, D. (1988) Neurobiology and pathology of subhuman vocal communication and human speech. In: *Primate Vocal Communication* (eds D. Todt, P. Goedeking & D. Symmes), pp. 195–212. Berlin, Heidelberg: Springer.

Ploog, D. (1990) Neuroethological foundations of human speech. In: *From Neuron to Action* (eds L. Deecke, J. Eccles & V. Mountcastle), pp. 365–74. Berlin, Heidelberg, New York: Springer.

Ploog, D. (1995) Neuroethological prerequisites for the evolution of speech. *Biology International*, Special Issue, **33**, 46–9.

Ploog, D., Hupfer, K., Jürgens, U. & Newman, J.D. (1975) Neuroethologic studies of vocalization in squirrel monkeys with special reference to genetic differences of calling in two subspecies. In: *Growth and Development of the Brain* (ed. M.A.B. Brazier), pp. 231–54. New York: Raven Press.

Sutton, D., Larson, C. & Lindeman, R.C. (1974) Neocortical and limbic lesion effects on primate phonation. *Brain Research*, **71**, 61–75.

Winter, P., Ploog, D. & Latta, J. (1966) Vocal repertoire of the squirrel monkey (*Saimiri sciureus*), its analysis and significance. *Experimental Brain Research*, **1**, 359–84.

Winter, P., Handley, P., Ploog, D. & Schott, D. (1973) Ontogeny of squirrel monkey calls under normal conditions and under acoustic isolation. *Behaviour*, **47**, 230–9.

Laterality and Human Speciation

MICHAEL C. CORBALLIS

Summary. Although cerebral asymmetries abound in non-human animals, there are still reasons to suppose that there may have been a single-gene mutation producing a 'dextral' (D) allele, which created a strong bias toward right-handedness and left-cerebral dominance for language at some point in hominid evolution. The alternative 'chance' (C) allele is presumed directionally neutral, although there may be other influences producing weak population manual and cerebral asymmetries in the absence of the D allele. It is unlikely that this laterality gene is located in homologous regions of the X and Y chromosomes, as suggested by Crow (1998), but there is a case for supposing that it is located solely on the X chromosome. I argue that language evolved from manual gestures, and the D allele may have served to guarantee manual and vocal control in the same (left) hemisphere in the majority of humans. The 'speciation event' that distinguished *Homo sapiens* from other large-brained hominids may have been a switch from a predominantly gestural to a predominantly vocal form of language.

IT IS OFTEN ARGUED that one of the characteristics distinguishing humans from other apes, and perhaps other species generally, is laterality, that we are, in other words, the lopsided apes (Corballis, 1991). Of course, asymmetries abound in nature, but what may be unique to our species is the strong bias toward right-handedness and left-cerebral dominance for language. This view has been strongly endorsed by Crow (1993, 1998), who has argued for a 'speciation event' that gave rise to laterality, language itself, theory of mind, and a proneness to thought disorders such as schizophrenia. It was this event that gave rise to the emergence of our own species, *H. sapiens*, perhaps 150,000 years ago. Crow also argues that the gene responsible for these changes is located in homologous regions of the X and Y chromosomes. In this chapter, I shall be mainly concerned with the role that laterality has to play in this scenario.

Proceedings of the British Academy, **106**, 137–152, © The British Academy 2002.

ARE WE REALLY THE LOPSIDED APES?

Laterality in non-human species

Over the past decade or so there has been growing resistance to the idea that handedness and cerebral asymmetry are uniquely human attributes. In a recent review, Vallortigara *et al.* (1999) note that functional and structural asymmetries of the brain are widespread in vertebrates, including fish, reptiles, amphibians and mammals. All species so far tested that show gregarious behaviour display evidence of cerebral lateralisation at the population level, with the evidence generally consistent with a left-hemisphere specialisation for stimulus categorisation and a right-hemisphere specialisation for attack and agonistic behaviour. It is not difficult to see how these asymmetries might underlie the left-hemispheric specialisation for language and the right-hemispheric specialisation for emotional and spatial behaviour in humans. Some 40% of the species tested that do not show gregarious behaviour also show some degree of complementary lateralisation. The population bias is typically not as extreme as that of left-hemispheric speech dominance in humans, but one extreme case was recently claimed by Gannon *et al.* (1998), who reported that the left temporal planum was larger than the right in 17 out of 18 chimpanzees. This proportion is actually larger, even significantly so,[1] than that reported in humans (Geschwind & Levitsky, 1968).

Population-level preference for one or other limb, be it hand, foot or paw, also appears to be quite widespread. It has long been known that parrots show a preference for the left foot in picking up small objects (Friedman & Davis, 1938), and the population bias of about 90% is about the same as that in humans. Evidence from other species usually shows less extreme biases, and they are typically contingent on the actual activity performed with the hand or paw. For example, although Collins (1970) found no overall bias among mice for one or other paw in reaching into a glass tube for food, a more recent study has shown a right-paw preference on one test and a left-paw preference on another (Waters & Denenberg, 1994). Among primates the evidence is mixed, and somewhat controversial. MacNeilage *et al.* (1987) suggested that primates tend to be left-handed in reaching, but those that became less arboreal evolved a complementary right-hand preference for fine motor acts. Hopkins (1996) has documented evidence that about two-thirds of captive chimpanzees are consistently right-handed for a number of activities, such as extracting peanut butter from a tube, and, as shown in more recent work, gestural communication (Hopkins & Leavens, 1998). Byrne (1996) has also reported a population-

[1]The proportion in humans is about 67% (Geschwind & Levitsky, 1968), leading to an expected value of 12 out of 18. Comparing the observed value of 17 out of 18 in chimpanzees to this expected value yields a chi-square of 6.25, which is significant at $P < 0.05$.

level hand preference among gorillas preparing vegetable matter for consumption, with about two-thirds showing a right-hand preference for the more intricate components.

McGrew & Marchant (1997: 201) have expressed some scepticism as to the generality of this work; in a review of handedness in primates, they conclude that 'only chimpanzees show signs of a population bias … to the right, but only in captivity and only incompletely'. It is possible that the bias in captive chimpanzees is culturally determined. For example, there is no evidence, despite extensive observation, that chimpanzees ever use pointing gestures in the wild, but pointing has been widely adopted among the 115 captive chimpanzees studied by Hopkins & Leavens (1998), as well in other captive apes (reviewed by Leavens *et al.*, 1996). This suggests that chimpanzees readily acquire manual actions from human contact, and may also be influenced by predominance of right-handedness among their human captors.

What is unique about human laterality?

Although some of the evidence is rather conflicting, there is little doubt that there are population-level cerebral and manual asymmetries in many non-human species, including primates. We must therefore ask whether it is reasonable to conclude that there is still some aspect of laterality that is unique to humans, perhaps to the point of defining a speciation event. It was probably never realistic to suppose that human laterality appeared *de novo*, as a result of a single genetic mutation, say, and even the characteristic asymmetries of the internal organs depend on a cascade of influences rather than on the actions of a single gene (Garcia-Castro *et al.*, 2000). It is not out of the question that there *are* aspects of human laterality unique to our species, perhaps dependent on a single genetic mutation, but any such mutation would surely have operated against a background of existing asymmetries (Corballis, 1997). Let us consider, then, what aspects of human laterality might be considered unique, and what their genetic basis might be.

In most people, articulate speech is controlled by the left cerebral hemisphere, and language itself is generally considered uniquely human (Chomsky, 1980; Pinker, 1994; Calvin & Bickerton, 2000). It follows, even if only secondarily, that the left-hemispheric dominance for speech and other aspects of language is uniquely human. There are also manual activities typically carried out with the right hand, or with different contributions from the two hands, that are at least arguably uniquely human, again bestowing a uniqueness on the asymmetries themselves. One is throwing. Marzke (1996) has argued that as far back as *Australopithecus afarensis* changes in the structure of the hand, the steady bipedal stance, and control of the trunk were adapted for accurate and potentially lethal throwing. Chimpanzees, by contrast, are capable only of a crude form of underarm propulsion that is better described as a fling. Indeed, the ability to throw with precision, as a means of both attack and defence, may well have been necessary to ensure survival on the savanna, especially for a species previously restricted to a largely arboreal existence. Most people throw with the right hand, and few if any can throw equally well with either hand. Most people, too, use tools preferentially in the right hand, or with the right hand performing the critical operation while the left hand serves a holding function; for example, the left hand holds the nail while the right wields the hammer, or the left hand holds the bow while the right hand pulls back the arrow for accurate aim. Although other species have been shown to use improvised tools, the systematic manufacture of tools is generally attributed to the genus *Homo*.

Distinctively human activities like speaking, throwing, and manufacturing and using tools, may be better programmed within a cerebral hemisphere, so that interhemispheric conflict is eliminated (Corballis, 1991). Wilkins & Wakefield (1995) also point out that lateralisation might facilitate internally generated manual activities, like manipulation and throwing, by shortening the feedback from sensorimotor to motor areas. These considerations may well have favoured the selection of one or more further mutations to enhance, and perhaps guarantee, cerebral asymmetry for the programming of complex action sequences in our species.

However, it may not have been simply a matter of lateralising a number of independent activities. The importance of a consistent lateralising influence may derive from the fact that it guaranteed the lateralisation of speech and manual control in the *same* cerebral hemisphere. Calvin & Bickerton (2000) have proposed that the origins of syntax may lie, at least partly, in the neural mechanisms involved in accurate throwing; if this is so, then it would follow that the two might be located in the same hemisphere, as proposed by Calvin (1983). But there is perhaps more compelling evidence for a link between language and *gesture* (of which throwing might be considered just an example). There are cells in area F5 in the pre-motor cortex of the macaque that fire when the monkey makes specific grasping movements, and a subpopulation of these cells, known as 'mirror neurones', also fire when the animal observes a person making the same grasping movement (Rizzolatti & Gentilucci, 1988). Because this system maps a programmed movement onto the perception of the same movement, it has been suggested as a precursor to language, and as implying that language evolved from manual gestures (Rizzolatti & Arbib, 1998). There are many other reasons to suppose that language may have evolved from a system of manual gestures rather than vocal calls (Hewes, 1973; Corballis, 1991, 1999; Armstrong *et al.*, 1995; Givon, 1995).

A further reason to suppose that mirror neurones may be part of a system that is a precursor to language is that area F5 in macaques is roughly homologous to Broca's area in humans. There is an important difference, though:

mirror neurones are located bilaterally in the macaque whereas in the great majority of humans Broca's area, at least in so far as it is involved in speech, is confined to the left hemisphere. Moreover, there is also evidence that there is a system for recognising gestures in humans that is similar to that in macaques, and that it is left-hemispheric and may well overlap with Broca's area (Rizzolatti *et al.*, 1996). This suggests that, at some point in the evolution of our species, what was initially a bilateral system became predominantly unilateral, perhaps when the programming reached a certain level of complexity, and when it co-opted oral as well as manual sequencing. The enlargement of the left temporal planum in chimpanzees (Gannon *et al.*, 1998) may suggest that this had already occurred in the common ancestor of humans and chimpanzees, although there is as yet no evidence that this anatomical asymmetry is accompanied by a functional asymmetry. Moreover, an asymmetry of the temporal planum need not imply a corresponding asymmetry in the language-mediating areas of the prefrontal cortex. There is evidence that Broca's area is first discerned in hominid skulls in *Homo habilis*, suggesting that the lateralised circuits for language, whether gestural or vocal, might have evolved with the emergence of the genus *Homo* (Tobias, 1987).

According to this scenario, then, what is unique to humans may be a lateralised system for the programming of language that couples manual and vocal control. The direction of lateralisation may have been determined by pre-existing cerebral asymmetries, or perhaps by underlying developmental gradients that dictate different rates of development on the two sides of the brain (Corballis, 1991). Whatever the case, it does not seem unreasonable to suppose that it was governed by a mutation, perhaps involving a gene governing rate of growth, at some point in hominid evolution.

GENETIC THEORIES OF HANDEDNESS

The most compelling genetic theories of handedness are based on the insightful suggestion of Annett (1972), that human handedness may depend on two genetic influences, one creating a bias toward right-handedness and the other creating no disposition towards either right- or left-handedness. That is, the genetic influence is not over whether a person will be right- or left-handed, but is over whether a person will be right-handed *or not*. These may be considered as alleles of a single 'right-shift' gene, and Annett (1993) has labelled them RS+ and RS–, respectively. In her model, differences in performance between the two hands are subjected to random environmental influences, producing a normal distribution, but in those homozygotic individuals inheriting a double dose of the RS+ allele, this distribution is shifted markedly to favour the right hand. In heterozygotes inheriting one of each allele, the distribution is also shifted to the right, but to a lesser extent. In those homozygotic for the RS– allele, there is no genetic disposition towards either left- or right-handedness, although cultural pressures may induce a small shift to the right. Annett also proposes that a heterozygotic advantage in terms of fitness has resulted in balanced polymorphism, ensuring that both alleles remain in the population.

McManus (1985, 1999) has proposed a similar single-gene model, with two alleles, one labelled D for 'dextral' and the other C for 'chance'. In his scheme, handedness is defined in terms of preference rather than performance, and is considered to be a dichotomous variable. Hence *all* DD homozygotes are considered to be right-handed, while CC homozygotes are considered to be divided equally into right- and left-handers. Among CD heterozygotes, it is proposed that the proportions lie midway between those of CC and DD homozygotes, so that 75% will be right-handed and 25% left-handed.

Although rather different in terms of underlying assumptions about handedness, these two models provide good, and essentially equivalent, fits to data on the proportions of left- and right-handers born to parents of differing handedness. Both Annett and McManus also assume that the gene influences not only handedness, but also cerebral asymmetry for language. McManus (1999) points out that his model readily accounts for the observed relation between handedness and language dominance if it is assumed that the D and C alleles influence language dominance in exactly the same way but in a statistically independent fashion. Thus DD individuals are all right-handed and left-language dominant, while in CC homozygotes handedness and language dominance are assigned independently and at random, so that there are equal proportions in each combination of handedness and language dominance. DC heterozygotes are 75% right-handed and 75% left-language dominant, but the two asymmetries are independent, so that the breakdown is as follows:

right-handed and left-language dominant	56.25%
right-handed and right-language dominant	18.75%
left-handed and left-language dominant	18.75%
left-handed and right-language dominant	6.25%

It follows from these last two figures that left-handers should be more often left- than right-language dominant, although the difference is attenuated slightly by the addition of CC left-handers. The evidence has consistently shown that, among right-handers, the incidence of left-language dominance is at least 96%, while among left-handers it is lower but is still around 70% (Warrington & Pratt, 1973; Rasmussen & Milner, 1977; Pujol *et al.*, 1999), almost exactly as predicted.

Annett and McManus have both proposed that this laterality gene is uniquely human. In view of the evidence for a weak population-level right-handedness in primates (Hopkins, 1996), I have suggested that the D allele may operate against a background, not of equal proportions of left- and right-handers, but against a pre-existing bias of about 67% in favour of right-handers (Corballis, 1997). That is, CC individuals who lack the D allele may be 67% right- and 33% left-handed, and the proportions of right- and left-handers among CD heterozygotes might be recalibrated accordingly to 83% and 17%, respectively. This also accords with evidence that human asymmetries other than handedness and language dominance, such as the asymmetries of the face, the right-ear dominance in dichotic listening, or the fetal position of the final trimester in which the right hand faces towards the mother's front, are in a ratio of approximately 67 : 33 rather than 90 : 10; these and other asymmetries are summarised by Previc (1991). [Ironically, one such asymmetry is the enlargement of the left temporal planum relative to the right, which makes the more pronounced ratio (17 : 1) reported in chimpanzees (Gannon *et al.*, 1998) all the more striking, and perhaps difficult to accept as a true estimate of the population ratio.] The assumption that the background asymmetry is 67 : 33 rather than 50 : 50, when incorporated into Annett's and McManus's models, provides a slightly better fit to aspects of the data on inheritance (Corballis, 1997).

IS THE GENE ON THE SEX CHROMOSOMES?

Crow (1998) suggested that the laterality gene is not only responsible for language and theory of mind, thereby further defining *H. sapiens* as a distinct species, but that it might be located in homologous regions of the sex chromosomes. If true, this might not only explain the slight differences in cerebral asymmetry, but could also suggest that the gene was subject to sexual selection. Crow's reasons for suggesting that the gene is located on the sex chromosomes are based on certain genetic disorders. He cites evidence that people lacking an X chromosome, a condition known as Turner's syndrome, have deficits that may be described as deficits of the right hemisphere, while those with an extra X chromosome, whether XXY (Klinefelter's syndrome) or XXX, have deficits that appear to be deficits of left-hemispheric functioning. This suggests that a gene on the X chromosome influences cerebral dominance. As males, like females with Turner's syndrome, also carry only one X chromosome, yet do not show deficits associated with right-hemisphere malfunction, a gene on the Y chromosome must balance that on the X. According to Crow, then, this identifies the gene as one of the select class of X–Y homologous genes.

This theory suggests that siblings of the same handedness should be more often of the same sex than of opposite sex, and the reverse should be true of opposite-handed siblings. This follows because fathers can pass on their Y chromosomes only to sons and their X chromosomes only to daughters. The expected concordance is weak, in part because mothers pass on their X chromosomes to either sons or daughters, and in part because the C allele (or RS– allele in Annett's terminology) does not determine the direction of handedness, so that two sons, for example, might receive the C allele from the father's Y chromosome yet be of opposite handedness. Nevertheless, the concordance was demonstrated in a large-scale study of handedness in sibling pairs (Corballis *et al.*, 1996), although there was an anomaly in that the relatively small numbers of left-handed pairs were more often of opposite than of same sex.

There are, however, two serious difficulties. The first is that of explaining how the gene came to be present on both the X and the Y chromosome. As there is no recombination, there would have to be a second event, such as a transposition, for the gene to be copied over to the other chromosome. The problem may not be insuperable, as there is evidence that such transpositions do occur, and there is at least one sequence that occurs only on the X chromosome in gorillas and chimpanzees, but is duplicated with more than 99% homology on the Y chromosome in humans (Bickmore & Cook, 1987). But even if the gene exists on both chromosomes, there is still the question of how the mutation that produced the D allele could have resulted in that allele being present on both chromosomes.

The second difficulty is that polymorphisms on the Y chromosome are unstable under any selection regime, including a regime in which there is a selective advantage to heterozygotes, as postulated by Annett (1993). This is shown algebraically by Clark (1987), and is consistent with empirical evidence that Y chromosome polymorphisms are rare (Spurdle & Jenkins, 1992). A simulation also shows that if a mutation does result in the D allele occurring on both X and Y chromosomes and there is an ensuing heterozygotic advantage favouring CD genotypes, then the probability of the D allele will at first increase on both chromosomes under the selection regime. A balanced polymorphism will eventually stabilise on the X chromosome, but the alleles on the Y chromosome will eventually regress to a state in which only one of them remains (Corballis, 1997). The D allele will prevail if DD homozygotes are fitter than CC homozygotes, and this is reversed if CC homozygotes are the fitter. Polymorphisms may persist on the Y chromosome if there is no selective regime, but it is difficult to imagine why a laterality allele would appear unless it were associated with some increase in fitness.

Jones & Martin (2000) have argued that the Y chromosome may indeed carry only the C allele, leaving the X chromosome to carry both C and D alleles. The difficulty with this proposal is that it predicts too large a sex difference to fit the facts. By manipulating parameters, Jones & Martin (2000) show that, in an extreme case, the resulting estimates of the incidence of left-handedness are 13.02% for males and 8.07% for females, which they consider reasonably close to empirical estimates of 11.64% and 9.79%, respectively, as reported by McManus & Bryden (1992). But in reality it is not a good fit, because the measured sex difference is less than half that predicted, and the parameters that Jones & Martin (2000) used are unrealistic. For example, to achieve this fit they assume that the proportion of left-handedness in CC individuals is 0.21, which is far removed from the value of 0.5 assumed by McManus (1985), or even the value of 0.33 assumed in my own revised version of McManus's model (Corballis, 1997). They also assume that the D allele is dominant, so that DC individuals are all right-handed. This means that the model can no longer fit the data on relations between handedness and language dominance as described by McManus (1999) and outlined above. Moreover, if there is no phenotypic difference between DD and CC individuals, this raises questions about how the heterozygotic advantage, which is necessary to ensure polymorphism, is achieved.

Is the gene on the X chromosome only?

An alternative possibility is that the laterality gene is located solely on the X chromosome, with no counterpart on the Y chromosome. McKeever (2000) has recently reported data on the handedness of parents and their offspring in a large sample that are largely consistent with this possibility. In particular, couples in which the mother was left-handed and the father right-handed produced more left-handed offspring, especially in the case of sons, than did couples in which both were right-handed. Couples in which the mother was right-handed and the father left-handed produced more left-handed daughters than did right-handed couples, but no more left-handed sons. This last result is to be expected according to the X chromosome hypothesis, because fathers can pass on the X chromosome only to their daughters.

Rather surprisingly, perhaps, this model need not predict a sex difference in handedness. In males, there are only two genotypes, which we can label 0C and 0D, where '0' stands for the absence of a corresponding gene on the Y chromosome. We can suppose that the 0D genotype always results in right-handedness, and never in left-handedness, while the 0C genotype results in left-handedness with a probability of 0.5. If we suppose that the incidence of the C allele is c, then the overall incidence of left-handedness in males is simply $c/2$. In the case of females, there are three genotypes: CC, CD and DD. Following McManus (1985), we assume that the probability of left-handedness is 0.5 in CC individuals, 0.25 in DC individuals, and 0 in DD individuals. The overall incidence of left-handed CC females is therefore $c^2/2$, while that of left-handed CD females is $2c(1 - c)/4$, which reduces to $c/2 - c^2/2$. As none of the DD females is left-handed, we can add over CC and CD individuals to get the overall incidence of left-handedness in females, which gives $c/2$. This is exactly the same as that predicted for males. The slight sex difference might then be attributed to a higher susceptibility to birth stress in males (Bishop, 1990).

The idea that only a quarter of CD females will be left-handed also follows naturally from X inactivation. For most of the genes on the X chromosome, one of each pair of X genes in females is inactivated very early in embryonic development (Willard, 1995). This inactivation is random, so that in some cells it is the X chromosome received from the mother that is inactivated, and in some the inactivated chromosome is the one received from the father. This ensures equal dosage of the gene product in both sexes, so that in the CD genotype half of the cases will resemble the 0C male genotype and half will resemble the 0D genotype. We therefore expect the incidence of left-handedness in females inheriting the CD genotype to lie midway between those of the 0C and 0D male genotypes, in other words, a quarter.

On the face of it, this model seems more plausible than a model postulating a gene in homologous regions of the X and Y chromosomes, as the great majority of genes on the X chromosome are not paired with homologous genes on the Y. Moreover, it removes the problem of explaining how the D allele came to be located on both sex chromosomes in the absence of recombination. However, it does not square with Crow's (1998) argument that the gene must be on both chromosomes, because males do not show the 'right-hemisphere' deficits shown by 0X females with Turner's syndrome. It is possible, perhaps, that in Turner's syndrome individuals there is partial deactivation of the remaining X chromosome. Another difficulty with the X chromosome model is that it is in general not supported by large-scale studies of handedness in families, other than that reported by McKeever (2000). McKeever has noted this, and suggested ways in which other studies may have provided distorted information, but this is a matter in need of further resolution.

CONCLUSIONS

There is still reasonable support for the notion that human laterality, in its distinctive aspects, might depend in part on a single gene. It is unlikely that this gene is located in homologous regions of the X and Y chromosomes, but it is possible that it is located on the X chromosome only.

Could this gene be responsible for the 'speciation event' proposed by Crow (1998)? If it is the D allele that was instrumental in this event, then the answer is probably no, because there are individuals lacking this allele who are nevertheless undeniably human, and possessed of normal language abilities, even though the risk of language disorders might be slightly elevated (for a critical review see Bishop, 1990). It is possible that a *population* with this allele had some adaptive advantage over a population, such as the Neanderthals, that did not possess it, perhaps by virtue of the heterozygotic advantage, or perhaps simply because of a greater diversity of genotype, but one may question whether this could have represented a speciation event on the scale envisaged by Crow.

An alternative scenario, sketched by McManus (1999), is that an earlier form of the D allele, say D*, became dominant perhaps 2 million years ago, so that all members of the genus *Homo* from that point were right-handed. But then a second mutation resulted in the appearance of the present-day C allele, reducing the incidence of right-handedness to 90%. McManus suggests that this mutation may have occurred some time between 100,000 and 10,000 years ago, but it must surely have occurred prior to the migrations of *H. sapiens* from Africa, which may go back at least 125,000 years (Walter *et al.*, 2000). Perhaps it was this second mutation that defined the speciation event, which might be good news for left-handers. In invoking two mutations, however, this model is somewhat unparsimonious and is perhaps too speculative to be considered a serious possibility in the absence of further evidence.

Whatever the nature of the mutations that produced laterality, the bulk of evidence is beginning to weigh against the notion that human speciation itself involved a dramatic change. Arguments for a sudden discontinuity have been based largely on the grounds that language is quite unlike animal communication, and well beyond the capabilities of even our closest relatives, chimpanzees and bonobos (Chomsky, 1980; Bickerton, 1995). Even Bickerton, once a staunch advocate of what has been dubbed the 'big bang' theory of the emergence of syntax (Bickerton, 1995), has more recently argued that the elements of syntax might be found in reciprocal altruism in the great apes (Calvin & Bickerton, 2000). There are other scenarios more consistent with the gradual evolution of language through natural selection than with the notion that language emerged as the lucky outcome of a single mutation (Pinker & Bloom, 1990; MacNeilage, 1998).

From hand to mouth

My own view is that language probably evolved from manual gestures, and that its roots can be traced back to a system of intentional manual activity in our primate ancestors tens of millions of years ago (Hewes, 1973; Corballis, 1991, 1992, 1999; Armstrong *et al.*, 1995; Rizzolatti & Arbib, 1998). No doubt the vocal element would have assumed greater prominence as our hominid forebears gradually achieved greater cortical control over vocalisation, so that by 500,000 years ago, say, language was an approximately equal mixture of manual gesture and vocalisation. The critical event in the evolution of our own species may have been the switch to a system in which the vocal element was dominant, and carried the entire burden of syntax (Goldin-Meadow &

McNeill, 1999), although vocal language is still characteristically accompanied by gesture (McNeill, 1985). The switch from gestural to vocal language may well have been facilitated by the allele that guaranteed that manual and vocal control were located in the left cerebral hemisphere.

In terms of cognitive capacity, and even linguistic capacity, this may have been a small step, but it may have had large consequences. It would have freed the hands, allowing people to communicate freely while carrying things or tending infants, and to carry out manufacturing and other manual activities while at the same time explaining them to novices. Indeed it may have been this freeing of the hands that led to the cumulative development of technological sophistication that characterises our species. It has been suggested that this did not begin until the so-called 'evolutionary explosion' of some 35,000 years ago (Pfeiffer, 1985), as evidenced by cave drawings, the crafting of ornaments and objects displaying visual metaphor (White, 1989), and more sophisticated manufacture. But this is a Eurocentric view; the advancement of manufacture must have begun earlier than that.

For example, watercraft must have been developed to carry people from the Asian mainland to New Guinea and Australia (once joined) well over 60,000 years ago, as there is evidence that *H. sapiens* had reached south-eastern Australia by about that time (Thorne *et al.*, 1999). Evidence of a sophisticated bone industry, including the manufacture of harpoons, has been discovered in Zaire and dates from 90,000 years ago (Yellen *et al.*, 1995). It therefore seems likely that the evolutionary explosion documented in Europe actually began much earlier in Africa and expanded into Asia, and later into Europe. But it was not the result of a major speciation event, but rather a small change that gave voice to our activities. It is possible, but by no means proven, that a laterality gene had a small part to play in producing this change.

DISCUSSION

Bickerton: If you think that *Homo* had syntactic language, and if you agree with me that syntax is what enables you to have thought, why did we see so little cultural additive to this for thousands of years?

Corballis: Because manual language got in the way. Once speech emerged the hands were free for other activities.

Bickerton: Well couldn't they just stop and make something?

Corballis: This would still be inefficient. Once language was freed from manual activity, technological advance would increase cumulatively.

Questioner: Different populations have different allele frequencies. On the expectation of a single gene on the X and Y chromosomes, would you then expect to see differences in handedness across populations?

Corballis: We do see differences in the same populations with time. If the frequencies are adjusted across populations for the heterozygous advantage, one gets population differences.

Questioner: Do you require selective pressure to generate frequency differences across populations?

Corballis: It can happen by population drift.

Questioner: Yes, by population drift and since some differences in frequencies are quite high you would expect huge differences in handedness frequencies.

Crow: I'm not clear what your heterozygous advantage is for?

Corballis: It is selecting for language; the advantage is for language. A double dose of this gene gives too much pruning on the right hemisphere and too much is bad for you. You showed a couple of years ago that at the point of equality there is actually a dip in intellectual achievement (Crow *et al.*, 1998). There is a disadvantage for having a double dose of the chance allele. What is being selected for is the right dose, i.e. the heterozygous advantage. So what is being selected for is language, but if you get a double dose of it, you suffer. Marian Annett tried to document this with reading, showing that the heterozygotes could read better than extreme right-handers or extreme left-handers.

Questioner: Can you just clarify the role that brain enlargement plays in your story?

Corballis: What I am assuming is if brain enlargement started about 2 million years ago that it may have been driven by the added complexities that were involved and probably driven by the advantages of more sophisticated language.

Questioner: What about the increase in population of a half million years ago?

Corballis: I'm not sure whether the switch to vocalisation might have had an effect on civilisation … I don't know whether it goes along with my story or not.

Questioner: If there is selection for heterozygosity, I don't see how you can reach equilibrium.

Corballis: If it is 50 : 50 then there is.

Questioner: Could vocal language be at a disadvantage to gestural language?

Corballis: How do you get it off the ground? You are standing there with a creature who can't vocalise: you gesture to him. You have a lot of organisation to do to get the hand and vocalisation together.

Questioner: The difference between gesturing and signing and the difference between signing with right or left hand?

Comment: There is a difference between left- and right-hand signers: a mirror image. Do they reverse: is there mirror reversal?

Corballis: I don't think so.

References

Annett, M. (1972) The distribution of manual asymmetry. *British Journal of Psychology*, **63**, 343–58.

Annett, M. (1993) The right shift theory of a genetic balanced polymorphism for cerebral dominance and cognitive processing. *Cahiers de Psychologie Cognitive*, **14**, 427–80.

Armstrong, D.F., Stokoe, W.C. & Wilcox, S.E. (1995) *Gesture and the Nature of Language*. Cambridge: Cambridge University Press.

Bickerton, D. (1995) *Language and Human Behavior*. Seattle: University of Washington Press.

Bickmore, W.A. & Cook, H.J. (1987) Evolution of homologous sequences on the human X and Y chromosomes, outside of the meiotic pairing segment. *Nucleic Acids Research*, **15**, 6261–71.

Bishop, D.V.M. (1990) *Handedness and Developmental Disorder*. Hove: Erlbaum.

Byrne, R.W. (1996) The misunderstood ape: cognitive skills of the gorilla. In: *Reaching Into Thought* (eds A. E. Russon, K. A. Bard & S. T. Parker), pp. 111–30. Cambridge: Cambridge University Press.

Calvin, W.H. (1983) *The Throwing Madonna*. New York: McGraw-Hill.

Calvin, W.H. & Bickerton, D. (2000) *Lingua Ex Machina*. Massachusetts: Bradford Books/MIT Press.

Chomsky, N. (1980) *Rules and Representations*. New York: Columbia University Press.

Clark, A.G. (1987) Natural selection and Y-linked polymorphism. *Genetics*, **115**, 569–77.

Collins, R.L. (1970) The sound of one paw clapping: an inquiry into the origins of left handedness. In: *Contributions to Behavior–Genetic Analysis—the Mouse as a Prototype* (eds G. Lindzey & D. D. Thiessen), pp. 115–36. New York: Meredith Corporation.

Corballis, M.C. (1991) *The Lopsided Ape*. New York: Oxford University Press.

Corballis, M.C. (1992) On the evolution of language and generativity. *Cognition*, **44**, 197–226.

Corballis, M.C. (1997) The genetics and evolution of handedness. *Psychological Review*, **104**, 714–27.

Corballis, M.C. (1999) The gestural origins of language. *American Scientist*, **87**, 138–45.

Corballis, M.C., Lee, K., McManus, I.C. & Crow, T.J. (1996) Location of the handedness gene on the *X* and *Y* chromosomes. *American Journal of Human Genetics (Neuropsychiatric Genetics)*, **67**, 50–2.

Crow, T.J. (1993) Sexual selection, Machiavellian intelligence, and the origins of psychosis. *Lancet*, **342**, 594–8.

Crow, T.J. (1998) Sexual selection, timing, and the descent of man: a theory of the genetic origins of language. *Current Psychology of Cognition*, **17**, 1237–77.

Crow, T.J., Crow, L.R., Done, D.J. & Leask, S.J. (1998) Relative hand skill predicts academic ability: global deficits at the point of hemispheric indecision. *Neuropsychologia*, **36**, 1275–82.

Friedman, H. & Davis, M. (1938) 'Left handedness' in parrots. *Auk*, **55**, 478–80.

Gannon, P.J., Holloway, R.L., Broadfield, D.C. & Braun, A.R. (1998) Asymmetry of chimpanzee planum temporale: human-like brain pattern of Wernicke's area homolog. *Science*, **279**, 220–1.

Garcia-Castro, M.I., Vielmetter, E. & Bronner-Fraser, M. (2000) N-Cadherin, a cell adhesion molecule involved in establishment of embryonic left–right asymmetry. *Science*, **288**, 1047–51.

Geschwind, N. & Levitsky, W. (1968) Human brain: left–right asymmetries in temporal speech region. *Science*, **161**, 186–7.

Givon, T. (1995) *Functionalism and Grammar*. Philadelphia: Benjamins.

Goldin-Meadow, S. & McNeill, D. (1999) The role of gesture and mimetic representation in making language the province of speech. In: *The Descent of Mind* (eds M. C. Corballis & S. E. G. Lea), pp. 155–72. Oxford: Oxford University Press.

Hewes, G.W. (1973) Primate communication and the gestural origins of language. *Current Anthropology*, **14**, 5–24.

Hopkins, W.D. (1996) Chimpanzee handedness revisited: 55 years since Finch (1941). *Psychonomic Bulletin and Review*, **3**, 449–57.

Hopkins, W.D. & Leavens, D.A. (1998) Hand use and gestural communication in chimpanzees (*Pan troglodytes*). *Journal of Comparative Psychology*, **112**, 95–9.

Jones, G.V. & Martin, M. (2000) A note on Corballis (1997) and the genetics and evolution of handedness: developing a unified distributional model from the sex-chromosomes gene hypothesis. *Psychological Review*, **107**, 213–8.

Leavens, D.A., Hopkins, W.D. & Bard, K.A. (1996) Indexical and referential pointing in chimpanzees (*Pan troglodytes*). *Journal of Comparative Psychology*, **110**, 346–53.

McGrew, W.C. & Marchant, L.F. (1997) On the other hand: current issues in a meta-analysis of the behavioural laterality of hand function in nonhuman primates. *Yearbook of Physical Anthropology*, **40**, 201–32.

McKeever, W.F. (2000) A new family handedness sample with findings consistent with X-linked transmission. *British Journal of Psychology*, **91**, 21–39.

McManus, I.C. (1985) Handedness, language dominance, and aphasia: a genetic model. *Psychological Medicine Supplement*, **8**, 1–40.

McManus, I.C. (1999) Handedness, cerebral lateralization, and the evolution of language. In: *The Descent of Mind* (eds M. C. Corballis & S. E. G. Lea), pp. 194–217. Oxford: Oxford University Press.

McManus, I.C. & Bryden, M.P. (1992) The genetics of handedness, cerebral dominance and lateralization. In: *Handbook of Neuropsychology*, Vol. 6 (eds I. Rapin & S. J. Segalowitz), pp. 115–44. Amsterdam: Elsevier.

MacNeilage, P.F. (1998) The frame/content theory of the evolution of speech production. *Behavioral and Brain Sciences*, **21**, 499–546.

MacNeilage, P.F., Studdert-Kennedy, M.G. & Lindblom, B. (1987) Primate handedness reconsidered. *Behavioral and Brain Sciences*, **10**, 247–303.

McNeill, D. (1985) So you think gestures are nonverbal? *Psychological Review*, **92**, 350–71.

Marzke, M. (1996) Evolution of the hand and bipedality. In: *Handbook of Human Symbolic Evolution* (eds A. Lock & C. R. Peters), pp. 126–54. Oxford: Oxford University Press.

Pfeiffer, J.E. (1985) *The Emergence of Humankind*. New York: Harper & Row.

Pinker, S. (1994) *The Language Instinct*. New York: William Morrow.

Pinker, S. & Bloom, P. (1990) Natural language and natural selection. *Behavioral and Brain Sciences*, **13**, 707–84.

Previc, F. (1991) A general theory concerning the prenatal origins of cerebral lateralization in humans. *Psychological Review*, **98**, 299–334.

Pujol, J., Deus, J., Losilla, J.M. & Capdevila, A. (1999) Cerebral lateralization of language in normal left-handed people studied by functional MRI. *Neurology*, **52**, 1038–43.

Rasmussen, T. & Milner, B. (1977) The role of early left-brain injury in determining lateralization of cerebral speech functions. *Annals of the New York Academy of Sciences*, **299**, 355–69.

Rizzolatti, G. & Arbib, M.A. (1998) Language within our grasp. *Trends in Neuroscience*, **21**, 188–94.

Rizzolatti, G. & Gentilucci, M. (1988) Motor and visual-motor functions of the premotor cortex. In: *Neurobiology of Motor Cortex* (eds P. Rakic & W. Singer), pp. 269–84. New York: John Wiley & Sons.

Rizzolatti, G., Fadiga, L., Gallese, V. & Fogassi, L. (1996) Premotor cortex and the recognition of motor actions. *Cognitive Brain Research*, **3**, 131–41.

Spurdle, A. & Jenkins, T. (1992) The search for Y chromosome polymorphism is extended to negroids. *Human Molecular Genetics*, **1**, 169–70.

Thorne, A., Grün, R., Mortimer, G., Spooner, N.A., Simpson, J.J., McCulloch, M., Taylor, L. & Curnoe, D. (1999) Australia's oldest human remains: age of the Lake Mungo 3 skeleton. *Journal of Human Evolution*, **36**, 591–612.

Tobias, P.V. (1987) The brain of *Homo habilis*: a new level of organization in cerebral evolution. *Journal of Human Evolution*, **16**, 741–61.

Vallortigara, G., Rogers, L.J. & Bisazza, A. (1999) Possible evolutionary origins of cognitive brain function. *Brain Research Reviews*, **30**, 164–75.

Walter, R.C., Buffler, R.T., Bruggemann, J.H., Guillaume, M.M.M., Berhe, S.M., Negassi, B., Libsekal, Y., Cheng, H., Edwards, R.L., von Cosel, R., Neraudeau, D. & Gagnon, M. (2000) Early human occupation of the Red Sea coast of Eritrea during the last interglacial. *Nature*, **405**, 65–9.

Warrington, E.K. & Pratt, R.T.C. (1973) Language laterality in left handers assessed by unilateral ECT. *Neuropsychologia*, **11**, 423–8.

Waters, N.S. & Denenberg, V.H. (1994) Analysis of two measures of paw preference in a large population of inbred mice. *Behavioural Brain Research*, **63**, 195–204.

White, R. (1989) Visual thinking in the ice age. *Scientific American*, **267**, 74–81.

Wilkins, W.K. & Wakefield, J. (1995) Brain evolution and neurolinguistic preconditions. *Behavioral and Brain Sciences*, **18**, 161–226.

Willard, H.F. (1995) The sex chromosomes and X chromosome inactivation. In: *The Metabolic and Molecular Bases of Inherited Disease* (eds C. R. Scriver, A. L. Beaudet, W. S. Sly & D. Valle), 7th edn, pp. 719–35. New York: McGraw-Hill.

Yellen, J.E., Brooks, A.S., Cornelissen, E., Mehlman, M.J. & Stewart, K. (1995) A Middle Stone Age worked bone industry from Katanda, Upper Semliki Valley, Zaire. *Science*, **268**, 553–6.

When Did Directional Asymmetry Enter the Record?

JAMES STEELE

Summary. This chapter reviews evidence for the evolution of primate asymmetries in brain morphology and in behaviour, including the fossil and archaeological record of human evolution. This evidence suggests that, while morphological asymmetries are conserved features of the human brain, human functional asymmetries are derived, at least in their degree and consistency. A partial explanation is offered that takes account of allometric scaling in the evolution of brain size, neocortex size, and intra- and interhemispheric connectivity.

INTRODUCTION

LATERALITY OF FUNCTION is a well-described feature of human cognition, in language processing as well as in other tasks. Because human language has no obvious parallel in the cognitive and communication systems of other animals, it is reasonable to assume that there must be some derived feature of the organisation of the human cerebral hemispheres that reflects past selection for the language capacity. However, the issue is complicated by the existence of human polymorphisms both in aspects of behavioural laterality (such as handedness and language lateralisation), and in aspects of morphological asymmetry of the cerebral hemispheres (such as the larger side by volume, by surface area, or by some other dimension of shape in an area relevant to language processes). The rather low levels of heritability of such right–left asymmetries (best documented in the case of left-handedness) further complicate matters. There is therefore huge appeal in genetic models such as those of Annett (1995), Tim Crow (see the chapter in these *Proceedings*) and McManus (1999), which account for these polymorphisms and for observed levels of heritability of handedness while retaining selection for language (and for an optimal degree of lateralisation) as the evolutionary driving force.

Proceedings of the British Academy, **106**, 153–168, © The British Academy 2002.

At their current stage of development, however, these models all assume that human patterns of behavioural laterality and morphological brain asymmetry are unique and derived. There is an increasing body of evidence that this is not the case. If such evidence proves to be reliable, then we may need to modify these models significantly if they are to account for the appearance of human language as an adaptation. In this chapter I will review some of the most relevant comparative and evolutionary anatomical observations, and I will tentatively suggest an alternative account of the some of the more relevant aspects of human brain evolution.

FOSSIL AND COMPARATIVE DATA ON THE EVOLUTION OF HUMAN BRAIN ASYMMETRY

One of the keystones of the hypotheses of Crow, Annett and McManus is the premise that population-level right-handedness is a derived human trait reflecting an underlying brain adaptation for cognitive laterality (and, specifically, a population-level left-hemisphere dominance for speech processes). Certain key findings are cited recurrently; examples of such evidence from the period of human evolution prior to *Homo sapiens* include Holloway & de la Coste-Lareymondie's (1982) work on cerebral asymmetries in hominid skulls and endocasts, and Toth's (1985) experimental work on the preferential direction of rotation of the stone core by right-handed tool makers and his argument from archaeological evidence that this preference was characteristic of early Palaeolithic stone tool makers as well.

Although these genetic models identify lateralised brain processes as the focus of selection, their principal control data describe modern prevalence rates of right- and left-handedness. Similarly, when we conflate evidence of Palaeolithic right-handedness with that for morphological brain asymmetry, we are assuming that the one is somehow caused by the other. Studies have indeed shown a relationship between gross morphological asymmetries of the cerebral hemispheres and handedness, but the correlations are quite weak and are complicated by sex- and population-specific variation (Steele, 2000a). Steele (2000a: table 1) summarises observations of brain and skull asymmetry in human populations. It is apparent that there is a tendency for greater dimensions of the left occipital and right frontal poles, where these differ significantly from symmetry.

The clearest evidence from the fossil record relates to 'petalia' patterns. Following Cheverud *et al.* (1990: 368), petalias are defined as follows:

> the greater anterior or posterior protrusion of one cerebral lobe relative to the other. The protruding lobe may also be wider. In humans, both frontal and occip-

ital petalias are quite common, with the typical pattern consisting of a right frontal petalia and a left occipital petalia.

Steele (2000a: table 7) has summarised data from endocasts of hominid skulls. These indicate a preponderance of left occipital and right frontal width as well as length asymmetries in hominids as far back as early *Homo*, and perhaps including the later Australopithecines as well. If we accept that this pattern of brain morphology is associated with predominant right-handedness, then this too would be expected to have characterised these earlier species as long ago as 2–2.5 million years before the present.

If these brain asymmetries reflect the underlying features that cause human patterns of behavioural laterality (including handedness), then we should expect their evolutionary appearance to coincide with that of population-level right-handedness. However, with respect to the origin of this brain morphological pattern, data from studies of great apes suggest that it has an even more ancient ancestry in the primate lineage. Table 1 summarises observations of the same variables in the great apes. These data suggest that left occipital and right frontal length petalias are conservative features of the human brain, shared with the apes [however, Zilles *et al.* (1996) found no significant directional asymmetry in the length petalia patterns of chimpanzees ($n = 9$)]. The low incidence of occipital width asymmetry (and the relatively low incidence of frontal width asymmetry) in the apes in the earlier studies recorded in Table 1 is notable, as occipital width asymmetry is the dimension which seems to be most significantly associated with hand preference in humans (Steele, 2000a). However, we should note also that this contrast in the data reflects the high percentage of cases of great ape endocasts where Holloway & de la Coste-Lareymondie (1982) could recognise no width asymmetry by visual inspection, whereas LeMay *et al.* (1982) were able to recognise its presence more often using quantitative analyses of outlines of great ape endocasts (which were computer-detected from telephotographs by thresholding). The absolute size of great ape brains is about one-third by volume of that of humans, which will make such asymmetries harder to detect by eye in the great ape material. In one recent study, Hopkins & Marino (2000) found significant directional asymmetry of the human pattern in occipital and frontal width in a larger pooled sample of 19 great apes (nine chimpanzees, four orang-utans, two gorillas and four bonobos) (see also Pilcher *et al.*, 2001). Clearly there is scope for additional work on further samples of great ape brains and crania in order to resolve this issue.

Although Hopkins & Marino (2000) found no significant directional asymmetries of width in the cerebral hemispheres of Old and New World monkeys, Heilbroner & Holloway (1988) did find more localised temporal lobe asymmetries, greater length of the Sylvian fissure in the left hemisphere, in four out of

Table 1. Incidences of great ape brain asymmetry (frontal and occipital poles)

Species and study	Frontal length			Frontal width			Occipital length			Occipital width		
	R↑	=	L↑	R>	=	L>	R↓	=	L↓	R>	=	L>
(a) BRAIN ASYMMETRIES												
Pongo pygmaeus												
LeMay (1976) (brains)	–	–	–	–	–	–	3	9	0	2	8	2
LeMay *et al.* (1982) (endocasts)	5	1	0	3	3	0	0	2	4	0	4	2
Holloway & de la Coste-Lareymondie (1982) (endocasts)	–	–	–	6	13	1	4	11	5	0	19	1
Gorilla gorilla												
LeMay (1976) (brains)	–	–	–	–	–	–	5	0	2	3	1	3
LeMay *et al.* (1982) (endocasts)	4	2	0	4	2	0	0	2	4	0	2	4
Holloway & de la Coste-Lareymondie (1982) (endocasts)	–	–	–	17	21	2	3	9	28	2	31	7
Pan troglodytes												
LeMay (1976) (brains)	–	–	–	–	–	–	3	6	0	1	2	6
LeMay *et al.* (1982) (endocasts)	3	2	1	1	4	1	0	3	3	0	3	3
Holloway & de la Coste-Lareymondie (1982) (endocasts)	–	–	–	8	21	5	9	13	12	3	31	0
Pan paniscus												
Holloway & de la Coste-Lareymondie (1982) (endocasts)	–	–	–	22	15	4	10	11	20	4	36	1
Total incidence	**12**	**5**	**1**	**61**	**79**	**13**	**37**	**66**	**78**	**15**	**137**	**27**
(b) CRANIAL VAULT ASYMMETRIES												
Pongo pygmaeus												
LeMay *et al.* (1982)	3	3	2	–	–	–	4	0	4	–	–	–
Gorilla gorilla												
LeMay *et al.* (1982)	8	2	3	–	–	–	2	2	7	–	–	–
Pan troglodytes												
LeMay *et al.* (1982)	6	4	2	–	–	–	2	3	7	–	–	–
Total incidence	**17**	**11**	**7**	–	–	–	**8**	**5**	**18**	–	–	–

five Old and New World monkey species studied, where the study group consisted of large samples (n = 20–30) of formalin-fixed brain specimens. Gilissen (1992) made similar findings in capuchin monkeys, but not in spider monkeys. LeMay (1985) had earlier found a human-like pattern of asymmetry in Sylvian fissure length in great apes, but not in monkeys. These findings are relevant because the length of the Sylvian fissure is taken as an indicator of the surface area of the planum temporale, an area of the temporal lobe involved in speech processing (cf. Galaburda *et al.*, 1987). More recently, Gannon *et al.* (1998) have reported direct surface area measurements showing a larger planum temporale in the left hemisphere of 17 out of 18 chimpanzee brains studied, and Hopkins *et al.* (1998) [using magnetic resonance imaging (MRI) scan data] have reported a significant size bias towards the left hemisphere in a sample of 21 great apes (of whom 16 had greater planum temporale length in the left hemisphere). A similar result has also been reported by Gilissen *et al.* (1998) for common chimpanzees, while Gannon & Kheck (1999) also found a larger left planum temporale in the majority of a small sample of both gorilla and orangutan brains (n = 6 and n = 7, respectively). Finally, Gannon *et al.* (2000) have also reported a significant left over right asymmetry for volume of the planum temporale (defined cytoarchitectonically) in an Old World monkey (*Macaca fascicularis*).

These observations of morphological brain asymmetries in humans, in our hominid ancestors, and in other living primates suggest continuity rather than discontinuity. It is certainly hard to sustain the view that morphological brain asymmetry in these dimensions is a unique derived human trait representing a speciation event in the appearance of *H. sapiens*. Other more subtle morphological features of the human brain may indeed represent novel, derived, traits. At the macroscopic level, Holloway (1996) has suggested that the third inferior frontal convolution (which includes Broca's area) has distinctive characteristics in the human brain that can also be recognised in the endocast of KNM-ER 1470 [*Homo rudolfensis*, or *Australopithecus rudolfensis* according to Wood & Collard's (1999) taxonomic revision]. Holloway also notes that the endocast evidence for later hominids is insufficient to enable us to identify any new features of the convolutional pattern of the brain with the appearance of anatomically modern humans. However, Cantalupo & Hopkins (2001) have found human-like Broca's area asymmetry in great apes. At the microscopic level, Buxhoeveden and collaborators (Buxhoeveden *et al.*, 1996, 2001) have found differences in cortical cell-packing in human and non-human primate brains in Tpt (auditory association cortex), including differences in the degree of left–right asymmetry in cell density, which may indicate derived human patterns of connectivity. Others have found that the ratio of grey matter to neuropil is lower in the left hemispheres of human brains, suggesting that 'the dominant hand is controlled by a cortical region with a greater amount of connectivity than found in the homologous area of the other hemisphere' (Amunts *et al.*, 1997: 400). However, we should note that both cerebral convolutedness and cell packing density vary systematically with absolute size of the primate cerebral cortex, making it hard to distinguish changes in the human brain that are definitely not explicable by allometric scaling associated with size increase.

FOSSIL, ARTEFACTUAL AND COMPARATIVE DATA ON THE EVOLUTION OF HUMAN HANDEDNESS

Despite the apparent continuities in human and great ape brain morphological asymmetry, behavioural evidence of handedness in non-human primates is perplexingly equivocal. Perhaps there is a directional bias to the use of the right hand in precise manipulative tasks at the population level in primates, but even in chimpanzees this has proved extremely hard to verify (Hopkins & Morris, 1993; McGrew & Marchant, 1996, 1997; Hopkins & Fernández-Carriba, 2000). The most comprehensive attempt to demonstrate continuities between human and non-human primate behavioural laterality has come from MacNeilage and collaborators (MacNeilage *et al.*, 1987; MacNeilage, 1993). But even MacNeilage (1993: 325) acknowledges that:

> it cannot be denied that … the [functional] asymmetries are for the most part, though not entirely … inherently weaker and less consistent in their direction in other primates than in humans.

Direct evidence of the hand preferences of Palaeolithic hominids comes in two forms: skeletal modifications that reflect the unequal loading history of the two upper limbs during life, and biases in the production and use of artefacts that reflect the ergonomics of technological actions involving a preferred hand. A large number of studies in recent years have demonstrated the range of adaptive responses of the skeleton to patterns of mechanical loading *in vivo* (Steele, 2000a, b). These responses can include increases in bone strength due to increased bone density and/or cross-sectional area, increases in mechanical efficiency by shape change, and resistance to avulsion by increasing the surface area of the sites of attachment of muscles and ligaments on a bone's surface. Evidence suggests that, in any particular case, the effect of muscle strength and mechanical loading on bone mineral formation is localised to the specific site of muscle–bone interaction. Because a consistent hand preference leads to lateral asymmetry in the mechanical loading experienced by the two hands, arms and shoulders during life, we can detect handedness by studying right–left differences in the degree of skeletal response to loading strains.

Fossil hominid remains have been studied from this perspective. It is evident that predominant right-handedness extends back in time to at least the early members of our own genus *Homo*. The skeleton of the Nariokotome boy, WT-15000 (early African *Homo erectus*, also called *Homo ergaster*), has greater development of the clavicular area of attachment of the right deltoid muscle and greater length of the right ulna, consistent with right-handedness (Walker & Leakey, 1993). Humeral shaft asymmetry consistent with right-arm dominance is also prevalent in Neanderthal skeletons: of six skeletons in which the relevant measurements could be taken bilaterally, all were more robust in the right arm (Trinkaus *et al.*, 1994).

Schultz (1937) recorded asymmetries of the lengths of arm bones (humerus and radius) in a large sample of ape skeletons (including 130 gorillas, 82 chimpanzees, eight orang-utans and 21 gibbons). In marked contrast with the 722 human skeletons in his sample, he found no tendency for the right arm to be dominant in apes as assessed by this measure. He also found that the mean degree of asymmetry (unsigned) in apes was about half that found in the arm bones of humans. These findings suggest that apes do not exhibit either the population-level right-handedness seen in humans, or the degree of loading of the individually dominant side that is seen in human bones.

A variety of artefactual data is available that confirms a predominance of right-handedness throughout the Palaeolithic period. A number of aspects of object manipulation (such as direction of rotation and grip orientation) are influenced by the anatomical structure of the hand, such that use of left or right hand for these actions will follow a characteristic and asymmetric pattern. This has enabled archaeologists to infer hand preference from tool-making debris, and from wear patterns on hand-held artefacts they can establish whether the hand used in each task was the preferred or the non-preferred one. The latter type of inference is usually based on ethnographic observation and on experimental reproduction of similar artefacts or wear traces by individuals with different hand preferences.

Toth (1985) found that there was a preferential direction of rotation of the stone core by knappers, who hold it in their non-dominant hand while using the preferred hand to strike the knapping blows. This preferential direction of core rotation varies according to which hand holds the core, with the result that the rotation bias can be detected in archaeological waste material from stone tool making. The pattern which he found in right-handed knappers was also found in the stone flakes recovered from the sites of Koobi Fora, Kenya (dating to about 1.9–1.4 million years ago), and from Ambrona, in Spain (dating to about 0.3–0.4 million years ago). While this indicates a preponderance of the right-handed phenotype at these sites, it is difficult to say exactly what proportion of earlier hominid tool makers (probably genus *Homo* in both cases) was right-handed. This is not just because Toth's (1985) estimation technique is 'noisy'. It is also because more recent experimental work suggests that the proportions found of stone flakes showing traces of clockwise and counter-clockwise core rotation varies not just with hand use, but also with experience and skill level. Ludwig & Harris (1994) report that experimental knapping by experienced stone tool makers showed a much higher proportion of flakes produced by core rotation in the 'preferred' direction (*c.* 90%) than was reported either by Toth (1985) or by Ludwig & Harris (1994) in their own experiments with novices (about 60%). This is because the greater skill of the experienced knappers gives them more control over their procedures, and the ability to reduce a core much more systematically. Ludwig & Harris (1994) argue that this kind of skill level appeared in hominid stone tool-making traditions as early as 1.7 million years ago. Thus the proportions of flakes at Koobi Fora and Ambrona that were produced by core rotation in the counter-clockwise directions could indicate either a lower level of skill among right-handed tool makers, or a higher than expected prevalence of the left-handed phenotype among skilled tool makers (Pobiner, 1999). Future work may differentiate the effects of these factors; at present, while we can say that there was a preponderance of right-handed tool making in the Lower Palaeolithic, we cannot quantify the degree of such preponderance with any precision using this method.

Bermudez de Castro *et al.* (1988) discuss striations on the anterior teeth of several individuals from the Middle and early Upper Pleistocene in Europe, which were probably caused inadvertently by stone tools (used to cut meat or other material gripped between the individuals' own teeth). Of these individuals, the great majority showed a right-handed use pattern, one showed a left-handed use pattern, and one could not be assigned to either category. Cornford (1986) found evidence of handedness in stone tools found at various levels of the Cotte de St Brelade site in Jersey (dating from periods 130,000–240,000 years ago), and estimated a frequency of right-handed tool production of 71–84% (weighted average = 81.7%; Callow & Cornford, 1986: table D.1). Cornford (1986) argues that the task would strongly favour use of the preferred hand, thus making these data a good guide to the original frequency of right-handedness among successive occupants of the site. Semenov (1961; 1964: 173) found a pattern of right-handed use of bone retouchers for working blade edges on stone tools at two Neanderthal sites, Kiik-Koba (Crimea) and Teshik-Tash (Uzbekistan). He also suggested that about 80% of end-scraper tools from the Upper Palaeolithic of Europe and North Africa show wear patterns consistent with right-handed use (Semenov, 1964: 87; cf. Takeoka, 1991).

In summary, the fossil evidence, both of differences in right and left arm bones and of handedness in tool manufacture and use, supports the generalisation that humans have been predominantly right-handed since the dawn of flaked stone technology. It thus becomes puzzling that our brain asymmetry patterns should also appear in other living primates, while our behaviour shows a population-level bias towards right-handedness that is not nearly so apparent in these other living species.

HUMAN BRAIN EVOLUTION: HOW CAN WE EXPLAIN INCREASED EVIDENCE FOR LATERALITY WITHOUT SALTATIONS?

Evidence of continuity in morphological asymmetries does not tell the whole story of human brain evolution. One other major trend in this story is the absolute size increase of the hominid brain (the arguments in this section have been presented in an earlier form in Steele, 1998). We can infer a great deal from this latter trend, because comparative studies have demonstrated very significant developmental constraints on the evolution of relative sizes of the major brain structures, constraints that account for the scaling of human brain structure volumes as well as of those of other primates (Finlay & Darlington, 1995). For example, comparisons across primate genera demonstrate that in taxa with absolutely larger brains, a higher proportion of total brain volume is made up by neocortex. The following regression models illustrate this scaling relationship for each of the two major subdivisions of the primate order, strepsirhines and haplorhines. The two regression formulae (reduced major axis technique) describing these relationships are as follows (equations from Steele, 1996; data from Stephan *et al.*, 1981).

For strepsirhines (n = 12 genera)
$\log_{10}$ neocortex (mm^3) = 1.125 $\log_{10}$ (rest of brain, mm^3) – 0.460, $r^2 = 0.985$
For haplorhines (n = 25 genera, excluding *Homo*)
$\log_{10}$ neocortex (mm^3) = 1.20 $\log_{10}$ (rest of brain, mm^3) – 0.523, $r^2 = 0.992$

In each case, the slope in the regression model indicates that larger brained taxa have relatively more neocortex. *Homo sapiens* has neither more nor less neocortex than would be expected for a typical haplorhine primate of our brain size (Steele, 1996). We are the most 'neocorticalised' primate species, but only by virtue of being the largest-brained.

Let us also note another scaling constant, this time as observed more generally among mammals. This is the tendency for the relative amount of white matter to increase in larger brained taxa, and is described by a standard major axis model of the relationship between volumes of cortical grey matter and of cortical white matter in mammals (from Hofman, 1989):

$\log V_{grey}$ (cm^3) = 0.773 $\log V_{white}$ (cm^3) + 0.732
[95% confidence interval (CI) for slope = 0.741 – 0.806, r^2=0.992]

Rilling & Insel (1999) have inverted this relation by regressing volume of cerebral white matter on volume of cerebral grey matter in a separate series of primate brains, finding a slope of 1.12 (95% CI 1.05–1.19), which tells a similar story. Now if we take these grey and white matter volumes as crude surrogates for neurone and axon numbers, respectively (a naive assumption given the contribution of axons to grey matter volume, but one which may serve), then we can see that the mammalian scaling trend, for a relative increase in white matter in larger brains, is none the less insufficient to maintain the same 'percent global connectivity' among all cortical neurones (which would necessitate a parabolic increase in axon number with any increase in neurone number; cf. Deacon, 1990: 226–31; Ringo, 1991). It follows from this that as absolute brain size increased in the course of hominid evolution, there was a parallel increase in the relative volume of the neocortex, and a decrease in the overall degree of direct global connectivity among its neurones (cf. Ringo, 1991).

What are the functional implications of this? Ringo (1991) points out most specifically that the scaling of callosal area to cortical surface area in a range of smaller-brained species, and also in humans, indicates that percentage connectedness of neurones across the callosum falls with increasing brain size, and he suggests that this might relate to efficiency savings on processing power (given the increased conduction time of signals transmitted along long fibre connections). More generally, he suggests that the processing inefficiencies of retaining the same degree of interconnectedness in larger brains mean that lateral and regional specialisation of function will be strongly correlated with absolute brain size. Ringo *et al.* (1994) have developed this hypothesis, arguing that in large brains the conduction delay involved in interhemispheric transfer has important implications for the efficiency of interhemispheric integration of processing for 'time-critical' tasks. They show that in tasks like phoneme processing in speech, such a delay would critically constrain the efficiency of processing when it involved the activation of transcortical cell assemblies. The implication is that, as the absolute size of the hominid brain expanded, many cognitive and motor processes became more constrained to intrahemispheric networks due both to the allometric reduction in overall cortical connectivity, and to time delays in the integration of cell assemblies characterised by long fibre interconnections. More recently, Rilling & Insel (1999) have also reported a decrease in corpus callosum area relative to brain volume and to neocortical surface area with larger brains, in a primate series extending from squirrel monkeys to humans (see also Hopkins & Rilling, 2000). They observe that 'results suggest that interhemispheric connectivity via the corpus callosum is reduced in larger primate brains, whereas intrahemispheric connectivity is augmented' (Rilling & Insel, 1999: 1457), and that:

> in view of evidence that interhemispheric connectivity is inversely related to functional laterality, our results suggest that lateralisation of function may be an emergent property accompanying brain enlargement in primate evolution. Moreover, the human brain should be the most functionally lateralised anthropoid

> brain because it has the least amount of interhemispheric connectivity between neocortical neurons. (Rilling & Insel, 1999: 1459)

These scaling laws of human brain evolution do not account for all the pieces in the puzzle of human laterality. But they may account for the paradox whereby our brain morphological asymmetry is conserved from a distant primate common ancestor, while our behavioural asymmetries are (in their degree and consistency) evolutionarily novel. Perhaps we became more lateralised as an inevitable by-product of the need to make the most efficient use of our enlarged neocortex, with the typical pattern of right–left differences in functional specialisation simply amplifying pre-existing morphological asymmetries and wiring biases. It is extremely hard to sustain the alternative argument, that human laterality arose as a result of the evolution of novel brain morphological asymmetries restricted to the hominid line. But this resolution would raise new questions as well as provide new answers. Is there a quantitative threshold in hominid brain enlargement beyond which functional laterality was fully human, or did the degree and consistency of such functional laterality gradually increase with brain expansion throughout the Palaeolithic? What are the implications for language lateralisation? Answering these new questions will require, among other things, some very careful mathematical modelling of the properties of neural networks with differing degrees of connectivity.

DISCUSSION

McManus: Many of these asymmetries that people have been studying for years are really quite small and they are really on top of a very asymmetric body. Of course the heart of most individuals is on the left side. In fact if you take a fetal brain and pump blood into it from a highly asymmetric vascular system, it is not too surprising to also find minor asymmetries. So to demonstrate that any of the brain asymmetries are related to language, you must first demonstrate that they are independent of situs inversus (heart on right side) and secondly you need to actually demonstrate that they are related to language lateralisation. If you go back to the planum temporale data of Geschwind and Galaburda, really all they showed was that there was a large asymmetry and because there was another large asymmetry in language they assumed that the two must be related. They never demonstrated a relationship between the two and no one ever has since. There is no evidence that someone who has a larger right planum temporale has reversed language asymmetry. Thus it is not surprising that the asymmetries are now found in the great apes and other non-humans.

Wolpert: Isn't someone addressing this using scanning techniques?

Crow: Yes they are. In fact there does not seem to be a volume asymmetry of the whole hemispheres in humans. There really is not a volume asymmetry of the brain in humans.

McManus: There was just a large study done in the functional imaging lab at Queens Square where they looked at about 480 structural MRI scans done prior to functional analysis. They know that about 80 of them are left-handed and they couldn't find any differences in the structure of the brains in the right- and left-handers no matter what they measured. It is not because the techniques are not sensitive because they are finding sex differences interestingly. My strong suspicion is that there is no difference in left- and right-handers in their morphological asymmetries.

Questioner: I thought Marjorie LeMay has found differences between left- and right-handers?

McManus: Yes she did, but a number of other studies were not able to replicate this. Handedness is a relatively weak correlate of language lateralisation. What we need are studies of structural scans with corresponding functional MRI scans to see what is actually correlated.

Comment: We must not forget that brain asymmetries are widely spread throughout the animal kingdom. They are not just present in humans.

Crow: But they are not there in the chimpanzee, which is what we are really concerned with. Marchant and McGrew, who are in conflict with Hopkins, have very carefully reviewed the literature (McGrew & Marchant, 1996, 1997) and did a comprehensive study in the Gombe National Park (Marchant & McGrew, 1996). The conclusion is that there is no overall asymmetry in chimpanzees (for anatomical data on the planum temperale, see Buxhoeveden *et al.*, 2001).

Comment: There is a difference between handedness in the morphological sense and handedness in the functional sense. This is actually important because in primates vocalisations are localised on one side.

Crow: There is a problem with that literature. What you are talking about is a lesion literature which is not compatible with the literature on anatomical and functional asymmetries.

McManus: People want to find these asymmetries. Many studies are not blind. Asymmetries are publishable, symmetries are not. There was a recent publication that sheep have an asymmetric brain for recognising conspecifics and that they have right hemisphere specialisation.

Comment: Left-handedness has been associated with all kinds of things, for example autoimmunity and weakly associated with atrophy.

McManus: This was another one of Geschwind's observations, but meta-analyses don't replicate this. There are so many artefacts that we have to be very careful in reviewing these data.

Comment: What can we reasonably not doubt?

McManus: Probably, Broca's original observation that the vast majority of patients who have unilateral damage to their brain, if it's on the left, lose speech. That's the only thing I think is clear from the whole literature.

References

Amunts, K., Schmidt-Passons, F., Schleicher, A. & Zilles, K. (1997) Postnatal development of interhemispheric asymmetry in the cytoarchitecture of human area 4. *Anatomy and Embryology*, **196**, 393–402.

Annett, M. (1995) The right shift theory of a genetic balanced polymorphism for cerebral dominance and cognitive processing. *Current Psychology of Cognition*, **14**, 427–80.

Bermudez de Castro, J.M., Bromage, T.D. & Jalvo, Y.F. (1988) Buccal striations on fossil human anterior teeth: evidence of handedness in the middle and early Upper Pleistocene. *Journal of Human Evolution*, **17**, 403–12.

Buxhoeveden, D., Lefkowitz, W., Loats, P. & Armstrong, E. (1996) The linear organization of cell columns in human and nonhuman anthropoid Tpt cortex. *Anatomy and Embryology*, **194**, 23–36.

Buxhoeveden, D.R., Switala, A.E., Litaker, M., Roy, E. & Casanova, M.F. (2001) Lateralization of minicolumns in human planum temporale is absent in nonhuman primate cortex. *Brain, Behavior and Evolution*, **57**, 349–58.

Callow, P. & Cornford, J.M. (1986) Estimation of right hand preference among the La Cotte flint-knappers. In: *La Cotte de St Brelade 1961–78* (eds P. Callow & J. M. Cornford), pp. 413–4. Norwich: Geo Books.

Cantalupo, C. & Hopkins, W.D. (2001) Asymmetric Broca's area in great apes. *Nature*, **414**, 505.

Cheverud, J.M., Falk, D., Hildebolt, C., Moore, A.J., Helmkamp, R.C. & Vannier, M. (1990) Heritability and association of cortical petalias in rhesus macaques (*Macaca mulatta*). *Brain, Behavior and Evolution*, **35**, 368–72.

Cornford, J.M. (1986) Specialised resharpening techniques and evidence of handedness. In: *La Cotte de St Brelade 1961–78* (eds P. Callow & J. M. Cornford), pp. 337–51. Norwich: Geo Books.

Deacon, T.W. (1990) Fallacies of progression in theories of brain-size increase. *International Journal of Primatology*, **11**, 194–236.

Finlay, B.L. & Darlington, R.B. (1995) Linked regularities in the development and evolution of mammalian brains. *Science*, **268**, 1578–84.

Galaburda, A.M., Corsiglia, J., Rosen, G.D. & Sherman, G.F. (1987) Planum temporale asymmetry, reappraisal since Geschwind and Levitsky. *Neuropsychologia*, **25**, 853–68.

Gannon, P.J., Hof, P.R. & Kheck, N.M. (2000) Brain language area evolution: human-like pattern of cytoarchitectonic, but not gross anatomic, L > R hemispheric asymmetry of planum temporale homolog in *Macaca fascicularis*. *American Journal of Physical Anthropology*, **S30**, 155.

Gannon, P.J. & Kheck, N.M. (1999) Primate brain 'language' area evolution: anatomy of Heschl's gyrus and planum temporale in hominids, hylobatids and macaques and of planum parietale in *Pan troglodytes*. *American Journal of Physical Anthropology*, **S28**,132–3.

Gannon, P.J., Holloway, R.L., Broadfield, D.C. & Braun, A.R. (1998) Asymmetry of chimpanzee planum temporale: humanlike pattern of Wernicke's brain language area homolog. *Science*, **279**, 220–2.

Gilissen, E. (1992) Les scissures néocorticales du singe capucin (*Cebus*): mise en évidence d'une asymétrie de la scissure sylvienne et comparaison avec d'autres Primates. *Comptes Rendus de l'Academie des Sciences de Paris*, **314** Serie III, 165–70.

Gilissen, E., Amunts, K., Schlaug, G. & Zilles, K. (1998) Left–right asymmetries in the temporoparietal intrasylvian cortex of common chimpanzees. *American Journal of Physical Anthropology*, **S26**, 86.

Heilbroner, P.L. & Holloway, R.L. (1988) Anatomical brain asymmetries in New World and Old World monkeys: stages of temporal lobe development in primate evolution. *American Journal of Physical Anthropology*, **76**, 39–48.

Hofman, M.A. (1989) On the evolution and geometry of the brain in mammals. *Progress in Neurobiology*, **32**, 137–58.

Holloway, R.L. (1996) Evolution of the human brain. In: *Handbook of Human Symbolic Evolution* (eds A. Lock & C. R. Peters), pp. 74–125. Oxford: Clarendon.

Holloway, R.L. & de la Coste-Lareymondie, M.C. (1982) Brain endocast asymmetry in pongids and hominids: some preliminary findings on the paleontology of cerebral dominance. *American Journal of Physical Anthropology*, **58**, 101–10.

Hopkins, W.D. & Fernández-Carriba, S. (2000) The effect of situational factors on hand preferences for feeding in 177 captive chimpanzees (*Pan troglodytes*). *Neuropsychologia*, **38**, 403–9.

Hopkins, W.D. & Marino, L. (2000) Asymmetries in cerebral width in nonhuman primate brains as revealed by magnetic resonance imaging (MRI). *Neuropsychologia*, **38**, 493–9.

Hopkins, W.D. & Morris, R.D. (1993) Handedness in great apes: a review of findings. *International Journal of Primatology*, **14**, 1–25.

Hopkins, W.D. & Rilling, J.K. (2000) A comparative MRI study of the relationship between neuroanatomical asymmetry and interhemispheric connectivity in primates: implications for the evolution of functional asymmetries. *Behavioral Neuroscience*, **114**, 739–48.

Hopkins, W.D., Marino, L., Rilling, J.K. & MacGregor, L.A. (1998) Planum temporale asymmetries in great apes as revealed by magnetic resonance imaging (MRI). *Neuroreport*, **9**, 2913–8.

LeMay, M. (1976) Morphological cerebral asymmetries of modern man, fossil man, and nonhuman primate. *Annals of the New York Academy of Sciences*, **280**, 349–66.

LeMay, M. (1985) Asymmetries of the brains and skulls of nonhuman primates. In: *Cerebral Lateralization in Non-Human Species* (ed. S. D. Glick), pp. 233–45. New York: Academic Press.

LeMay, M., Billig, M.S. & Geschwind, N. (1982) Asymmetries of the brains and skulls of nonhuman primates. In: *Primate Brain Evolution* (eds E. Armstrong & D. Falk), pp. 263–77. New York: Plenum.

Ludwig, B.V. & Harris, J.W.K. (1994) *Handedness and Knapping Skill: Their Effects on Plio-Pleistocene Lithic Assemblage Variability. Pre-circulated papers from the World Archaeological Congress-3, New Delhi.*

McGrew, W.C. & Marchant, L.F. (1996) On which side of the apes? Ethological study of laterality of hand use. In: *Great Ape Societies* (eds W. C. McGrew, L. F. Marchant & T. Nishida), pp. 255–72. Cambridge: Cambridge University Press.

McGrew, W.C. & Marchant, L.F. (1997) On the other hand: current issues in and meta-analysis of the behavioral laterality of hand function in nonhuman primates. *Yearbook of Physical Anthropology*, **40**, 201–32.

McManus, I.C. (1999) Handedness, cerebral lateralisation and the evolution of language. In: *The Descent of Mind* (eds M.C. Corballis & S.E.G. Lea), pp. 194–217. NY: OUP.

MacNeilage, P.F. (1993) Implications of primate functional asymmetries for the evolution of cerebral hemispheric specializations. In: *Primate Laterality* (eds J. P. Ward & W. D. Hopkins), pp. 319–41. New York: Springer.

MacNeilage, P.F., Studdert-Kennedy, M.G. & Lindblom, B. (1987) Primate handedness reconsidered. *Behavioral and Brain Sciences*, **10**, 247–63.

Marchant, L.F. & McGrew, W.C. (1996) Laterality of limb function in wild chimpanzees of Gombe National Park: comprehensive study of spontaneous activities. *Journal of Human Evolution*, **30**, 427–43.

Pilcher, D.L., Hammock, E.A.D. & Hopkins, W.D. (2001) Cerebral volume asymmetries in non-human primates: a magnetic resonance imaging study. *Laterality*, **6**, 165–79.

Pobiner, B.L. (1999) The use of stone tools to determine handedness in hominids. *Current Anthropology*, **40**, 90–2.

Rilling, J.K. & Insel, T.R. (1999) Differential expansion of neural projection systems in primate brain evolution. *Neuroreport*, **10**, 1453–9.

Ringo, J.L. (1991) Neuronal interconnection as a function of brain size. *Brain, Behavior and Evolution*, **38**, 1–6.

Ringo, J.L., Doty, R.W., Demeter, S. & Simard, P.Y. (1994) Time is of the essence: a conjecture that hemispheric specialization arises from interhemispheric conduction delay. *Cerebral Cortex*, **4**, 331–43.

Schultz, A.H. (1937) Proportions, variability, and asymmetries of the long bones of the limbs and the clavicles in man and apes. *Human Biology*, **9**, 281–328.

Semenov, S.A. (1961) Traces of work on tools and evidence that Neanderthal men worked with their right hand. [In Russian] *Korotkie soobscheniya Instituta arkheologii [Short Bulletins of the Institute of Archaeology]*, **84**.

Semenov, S.A. (1964) *Prehistoric Technology* [translated by M. W. Thompson, original published in Russian in 1957]. London: Cory, Adams and Mackay.

Steele, J. (1996) On predicting hominid group sizes. In: *The Archaeology of Human Ancestry* (eds J. Steele & S. Shennan), pp. 230–52. London: Routledge.

Steele, J. (1998) Cerebral asymmetry, cognitive laterality, and human evolution. *Cahiers de Psychologie Cognitive*, **17**, 1202–14.

Steele, J. (2000a) Handedness in past human populations: skeletal markers. *Laterality*, **5**, 193–220.

Steele, J. (2000b) Skeletal indicators of handedness. In: *Human Osteology* (eds M. Cox & S. Mays), pp. 307–23. London: Greenwich Medical Media.

Stephan, H., Frahm, H. & Baron, G. (1981) New and revised data on volumes of brain structures in insectivores and primates. *Folia Primatologica*, **35**, 1–29.

Takeoka, T. (1991) Développement de la latéralité examinée à partir de l'analyse de la pierre taillée du paléolithique. *Journal of the Anthropological Society of Nippon*, **99**, 497–516.

Toth, N. (1985) Archaeological evidence for preferential right-handedness in the Lower and Middle Pleistocene, and its possible implications. *Journal of Human Evolution*, **14**, 607–14.

Trinkaus, E., Churchill, S.E. & Ruff, C.B. (1994) Postcranial robusticity in Homo. II. Humeral bilateral asymmetry and bone plasticity. *American Journal of Physical Anthropology*, **93**, 1–34.

Walker, A. & Leakey, R. (1993) The postcranial bones. In: *The Nariokotome* Homo erectus *Skeleton* (eds A. Walker & R. Leakey), pp. 95–160. Berlin: Springer-Verlag.

Wood, B. & Collard, M. (1999) The human genus. *Science*, **284**, 65–71.

Zilles, K., Dabringhaus, A., Geyer, S., Amunts, K., Qu, M., Scheicher, A., Gilssen, E., Schlang, G. & Steinmetz, H. (1996) Structural asymmetries in the human forebrain and the forebrain of non-human primates and rats. *Neuroscience and Biobehavioral Reviews*, **20**, 593–605.

Bihemispheric Language: How the Two Hemispheres Collaborate in the Processing of Language

NORMAN D. COOK

Summary. Speech production in most people is strongly lateralised to the left hemisphere (LH), but language understanding is generally a bilateral activity. At every level of linguistic processing that has been investigated experimentally, the right hemisphere (RH) has been found to make characteristic contributions, from the processing of the affective aspects of intonation, through the appreciation of word connotations, the decoding of the meaning of metaphors and figures of speech, to the understanding of the overall coherency of verbal humour, paragraphs and short stories. If both hemispheres are indeed engaged in linguistic decoding and both processes are required to achieve a normal level of understanding, a central question concerns how the separate language functions on the left and right are integrated. Relevant studies on the hemispheric contributions to language processing are reviewed, and the role of interhemispheric communications in cognition is discussed.

INTRODUCTION

The long-term goal of psychology is an understanding of the human brain such that psychological disorders can be treated at the appropriate level of intervention. Although psychological medicine is still a young science, already there are clear distinctions between organic brain diseases requiring surgical therapy, metabolic disorders that can be treated pharmaceutically, personality abnormalities that can be dealt with psychotherapeutically, and other brain problems caused by genetic defects that might some day be amenable to genetic engineering therapy. The psychological disorder that has proven most difficult to treat is schizophrenia, in which bizarre and paranoid ideation is found, often

Proceedings of the British Academy, **106**, 169–194, © The British Academy 2002.

associated with excessive and disordered speech, the schizophrenic 'word salad'. It is a disease that has no clear analogue in the animal world, but affects about 1% of the human population, regardless of cultural or racial background. For this reason, Crow (1997) has argued that schizophrenia may be as old as the divergence of *Homo sapiens* from other primate lines, is as characteristic of humans as the potential for learning and using language, and can be considered as the 'price' that we, as a species, pay for having language capabilities. Because functional hemispheric specialisation is one of the most unusual characteristics of the human brain, relative to other primates, and various abnormalities of cerebral dominance have been implicated in schizophrenia (e.g. Crow, 1998; Gur, 1999), a proper understanding of the mechanisms underlying cerebral lateralisation is arguably a prerequisite for understanding the psychopathology of schizophrenia.

Since the 1960s, the lateralised functions of the cerebral hemispheres have been the focus of much research that has led gradually to a more comprehensive understanding of hemispheric differences, if not yet a full understanding of hemispheric interactions. The changing emphasis on various laterality themes can be briefly outlined, as in Figure 1.

Following reports on the sometimes very different cognition of the two hemispheres in patients who had undergone severance of the corpus callosum (Sperry, 1968; Sperry *et al.*, 1969), the fact that all people contain two potentially independent 'brains' in one skull became widely known, and attention was focused on precisely how these 'two brains', and possibly these 'two personalities', differ. While some of the speculation prompted by the split-brain research was perhaps excessive, the core findings on the split-brain patients have stood the test of time: such patients do show signs of internal

Emphasis on left hemispheric dominance
for speech and handedness
(→1950s)

Focus on left and right hemisphere independence
as a consequence of the split-brain studies
(1960–1970s)

Focus on right hemispheric specialisations
in normal and brain-damaged subjects
(1980s)

Consideration of
hemispheric interactions
(1990→)

Figure 1. The recent evolution of the main themes concerning human laterality.

contradictions that can be traced to the loss of corticocortical connections between the cerebral hemispheres. It can therefore be inferred that the corpus callosum in the intact brain acts to resolve contradictions between the hemispheres and integrate the cognition of the left and right to produce a more or less unified self.

Subsequent to the initial split-brain work, neuropsychological studies on brain-damaged patients provided numerous examples of hemispheric specialisation. Related, if generally much weaker, results in normal subjects using tachistoscopic and dichotic techniques were also reported, and gradually a host of methodological issues have been addressed. There has even been a resurgence in interest in the 'two personalities' of the cerebral hemispheres and the implications for psychotherapy (Schiffer, 1998). Most recently, brain-imaging techniques have made it possible to measure directly the cortical activity in normal subjects, and these new methodologies have again invigorated laterality research.

Many of the bold dichotomies of hemisphere function proposed in the 1960s and 1970s have found their way into the textbooks, but virtually none has survived the harsh glare of empirical research. 'Verbal and visuospatial' remains the single most popular summary of left and right hemispheric specialisations, but studies of unilateral brain damage, continued investigation of the split-brain patients, behavioural studies of normal subjects, and the brain-imaging work of the 1990s are unanimous in showing that both hemispheres process both verbal and visuospatial information in their own ways. Since the earliest PET studies, a consistent finding has been approximately bilateral activation of the cerebral hemispheres during language processing. Given the rapid changes in regions of cortical activation (as found in EEG studies) and the extreme localisation of functions (as seen in fMRI studies), it can be said that if any consensus has been provided by recent brain imaging it lies in the idea that hemispheric functional asymmetries need to be considered as dynamic, rather than as static, processes. Evidence indicating that both hemispheres are actively engaged in language processing is reviewed below.

COMPLEMENTARY LANGUAGE PROCESSING

The analysis of language processes in normal subjects and brain-damaged patients has been undertaken from the level of the smallest segments (morphemes, graphemes and phonemes) through small units (words and phrases) to complete utterances and coherent messages (sentences, jokes, short-stories, etc.). As summarised below, at each level a fairly consistent pattern of hemispheric functional asymmetry has been found and indicates the active involvement of both hemispheres at multiple levels of language processing (Figure 2).

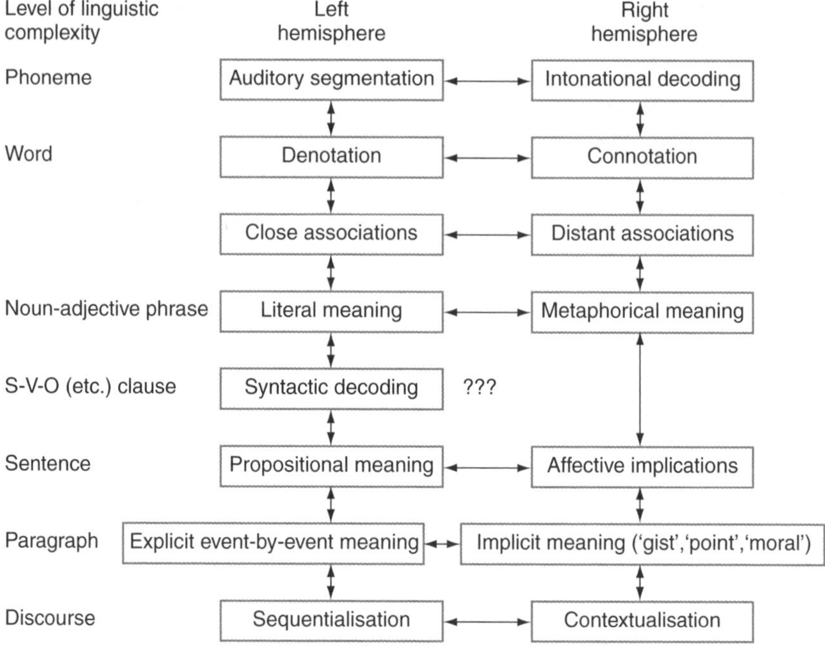

Figure 2. A schema of the multiple levels of bilateral language processing. The horizontal arrows indicate levels at which the cerebral hemispheres may interact via the corpus callosum. The vertical arrows indicate the bottom-up sequential processing from smaller to larger linguistic units, as well as the top-down effects from larger to smaller units. Whether or not there are RH functions comparable to and interacting with the syntactic decoding of the LH is uncertain.

The loss of the affective prosody of speech following RH damage is well-known clinically (Weniger, 1984). As in music perception/production, the two principal dimensions of the prosody of speech concern the temporal dimension, i.e. the rhythm and timing of speech output, particularly consonant stops, and the pitch dimension, i.e. the fluctuation in auditory frequency of particularly vowel sounds. Abnormalities of timing and fine temporal discrimination are found following LH damage, and may be a key factor responsible for dyslexia (Tallal *et al.*, 1996). Following RH damage, the pitch dimension appears to be most disturbed (Behrens, 1989; Ross *et al.*, 1997); callosal damage has similar effects (Klouda *et al.*, 1988). The range of pitch fluctuations is reduced, the frequency of changes in the direction of pitch intervals (melodiousness) decreases (Schirmer *et al.*, 2001), and what prosody there is often seems inappropriate to the linguistic content. Although sometimes dismissed as 'para' linguistic, the production and understanding of prosody is clearly important for normal verbal communication and is a function for which the RH is dominant.

In dichotic listening experiments on normal subjects, Bulman-Fleming & Bryden (1994) and Grimshaw (1998) have studied intonation using a design that allows the measurement of affective and linguistic understanding simultaneously. They demonstrated the superiority of the RH in detecting emotional prosody (happy, sad, etc.) and the LH in detecting linguistic meaning. In studies on patients with unilateral brain lesions, Van Lancker & Sidtis (1992) showed a double dissociation between pitch and rhythm perception, and Alcock *et al.* (2000) reported similar effects (unrelated to speech) that suggest LH specialisation for rhythm and RH specialisation for pitch. Ross *et al.* (1981, 1997) have defended the idea that there are types of aprosodia due to focal lesions of the RH that are as specific as the varieties of aphasia that occur with focal lesions of the LH. Most importantly, they present clinical evidence for a distinction between two forms of aprosodia related to the understanding and to the production of prosody, a distinction that mirrors sensory and motor aphasia. Finally, Zatorre *et al.* (1994, 2002) have repeatedly reported evidence of RH involvement in pitch perception and have emphasised the role of the RH in both speech prosody and music.

In line with the idea that the LH and RH functions can be summarised as verbal and visuospatial, respectively, the possibility that concrete, easily visualised, words might be more competently processed in the RH, and abstract, less easily visualised, words by the LH, has often been studied. Results have been mixed, probably reflecting differences in stimulus materials. In a recent fMRI study designed to examine specifically the abstract/concrete aspects of hemispheric processing, Kiehl *et al.* (1999) found (1) bilateral activation in temporal, parietal and frontal regions during all verbal processing, with (2) more activation of right temporal cortex for abstract words, and (3) more activation of left temporal cortex for concrete words. This pattern is the exact opposite of what many would predict, and is an indication that the concrete/abstract dimension is problematical for distinguishing between LH and RH processing.

With the notable exception of the so-called function words, most nouns, adjectives and verbs have connotative meanings, in addition to their dictionary denotations. The connotation is related to the affective state and the larger cognitive context within which the words are typically used, and normally has implications beyond the literal meaning of the phrase or sentence. As both cerebral hemispheres in most individuals will be exposed to the same words in the same contexts for an entire lifetime, it is of extreme interest that unilateral brain damage can lead to deficits of linguistic processing at either the denotative or the connotative level. In a classic study by Brownell *et al.* (1984), the understanding of words by the LH and RH was compared by having patients with unilateral brain damage group words according to their similarity. The dimension along which 'similarity' was to be determined was not specified, but the test design forced a choice between denotative and connotative grouping.

Patients with an intact RH but a damaged LH preferred metaphoric or connotative pairings, whereas patients with an intact LH but a damaged RH preferred antonymic or denotative pairings. Gainotti *et al.* (1983) also found specific lexical semantic deficits in patients with RH damage. More recently, Taylor *et al.* (1999) showed qualitative hemispheric differences in semantic category matching in normal subjects. The dimensions of RH semantic processing appear to be complex and may reflect individual differences rather than neuropsychological universals, but it cannot be said that semantics is exclusively a LH function.

The idea of a semantic network within which the words known to an individual are organised along various semantic dimensions has a long history in both psychology and artificial intelligence, and the possibility that the two hemispheres contain similar lexicons, but are organised differently, has frequently been studied. Particularly in light of the demonstration of the independence of the hemispheres in simple word recognition tasks in both normal subjects (Iacoboni & Zaidel, 1996) and split-brain patients (Zaidel, 1985), it is of interest to know if the semantic organisation of the LH and RH differs. In a split-visual field study, Rodel *et al.* (1994) found the LH to favour close associations, and the RH to favour distant associations. Using an ERP technique, Kiefer *et al.* (1998) examined the hemispheric response to closely related words, distantly related words and unrelated words. Both hemispheres responded to closely related words, but only the RH responded to distantly related words. Such findings are viewed as support of the idea that the RH maintains more associations than the LH, with the latter focusing on one of several possible trains of thought.

Judgements of metaphoric meaning show a similar laterality, with the RH preferring the metaphoric over the literal and the LH preferring the reverse. In a study by Winner & Gardner (1977), patients were asked to select one of four pictures that depicts the meaning of a phrase, such as 'lending a hand'. With one picture depicting a literal loaning of a disembodied hand and another picture depicting an individual giving help to another, the LH-damaged patients more often chose the metaphoric meaning, and the RH-damaged patients more often the literal meaning. Similar results were obtained by Van Lancker & Kempler (1987) and Anaki *et al.* (1998). Bottini *et al.* (1994) compared literal and metaphoric sentences in a PET study using normal subjects. The relevant comparison revealed right-sided frontal, temporal and parietal activations during the metaphoric sentences and left-sided activations during the literal sentences. Finally, Burgess & Chiarello (1996) have provided evidence indicating that an intact RH is essential for metaphor comprehension.

The construction of a coherent paragraph consisting of several, individually coherent (grammatically and semantically), sentences necessitates the sequencing of the sentences in the semantically correct order. In a comparison of the abilities of brain-damaged patients to do so, Gardner *et al.* (1983) found the RH-damaged patients to perform more poorly. Schneiderman *et al.* (1992) found that RH damage significantly disrupts patients' ability to arrange sentences into coherent paragraphs. The understanding of short stories requires one to grasp not only the individual actions of story participants, but also the consistency, overall coherence and sequence of events. Wapner *et al.* (1984) presented short stories to groups of brain-damaged subjects and had them retell the stories immediately upon completion of presentation. The stories were constructed with various logical anomalies, including temporally or causally anomalous events, and counter-intuitive actions. Regardless of the type of anomaly, RH-damaged patients were generally capable of retelling the story including its main elements, but failed to detect the nature of the anomaly. In contrast, the LH-damaged patients with intact RH, despite more diverse language problems, were capable of detecting the unusualness of the stories. These and related results have led Gardner *et al.* (1983) to consider the RH as an 'anomaly detector'. In related work on the understanding of jokes, Winner *et al.* (1998) have reported deficits in understanding following RH damage, and have argued that it represents a loss of second-order mental states.

In one of Geschwind's (1982) most neglected papers, he noted that the single most common consequence of diffuse RH damage, as seen in the neurological clinic, is the 'confusional state'. He defined this condition as one in which the ability for speech production is normal, but the coherence of verbal output is degraded, leading to unwitting humour, paramnesias and an inability to carry a train of thought to its logical conclusion. Being based on clinical observations, Geschwind's (1982) argument that the RH normally prevents 'confusion' must be considered anecdotal, but raises the interesting question of what state we are in when we are 'not confused'. Whether in conversation or in a monologue of speaking or writing, when ideas fall into place and lead to coherent conclusions, it might be said that each word, thought or statement is 'in context', and that cognition as a whole is contextually grounded. If such coherency is a function of the RH, it may be that the highest level contribution of the RH to language functions is the construction or maintenance of cognitive contexts. [Note, however, that Leonard *et al.* (1997) have failed to demonstrate a contextual role of the RH in brain-damaged patients. This might be attributable to the syntactic nature of the task (the resolution of ambiguous pronouns) but, in any case, highlights the need for a more precise definition of 'context'.]

What is significant about the above findings is that they indicate that, within the linguistic realm, both cerebral hemispheres are engaged in information processing at approximately the same level of complexity, but with apparently different strategies. Unlike hemispheric dichotomies in which very unlike processes are contrasted (verbal/visuospatial, etc.), the multiple levels of bilateral language processing summarised in Figure 1 suggest a complementary competence of the 'two brains'. Both hemispheres process linguistic information but manage *not* to duplicate their processing, despite the fact that their life-long experience of all language input is identical. The complementarity, as distinct from dissimilarity, of the two modes of cognition has been a recurring theme in the laterality literature (Landis *et al.*, 1979; Kinsbourne, 1982). Bogen (1997), in particular, has been a persistent defender of the idea that the RH is capable of high-level cognition, and has shown convincingly that both hemispheres in most split-brain patients have linguistic competence, provided only that one does not insist on a definition of 'linguistic competence' that is syntax-based.

For further discussion of individual experiments and laterality models that focus on the bilaterality of language, two edited volumes can be recommended: *Right Hemisphere Language Comprehension* (Beeman & Chiarello, 1998) and *Language and the Brain: Representation and Processing* (Grodzinsky *et al.*, 2000). An older review by Code (1987), *Language, Aphasia and the Right Hemisphere*, and a more recent discussion of metaphoric and figural understanding by Burgess & Chiarello (1996), are also noteworthy.

EFFECTS OF CALLOSOTOMY ON LANGUAGE FUNCTIONS

The results concerning the language specialisations listed in Figure 2 have come predominantly from patients with unilateral brain damage, but a remarkable fact is that callosal damage alone can produce effects similar to those following RH damage [for example the loss of affective intonation (Klouda *et al.*, 1988; Ross *et al.*, 1997) and infrequent use of affect-related words following callosal section (TenHouten *et al.*, 1985)]. In general, the language abnormalities of split-brain patients are mild when tested in a non-lateralised fashion, but already in the earliest discussions of these patients, Sperry (1968; Sperry *et al.*, 1969) noted that their spontaneous speech was affectively flat or inappropriate, and unusually concrete with a tendency toward literalism. These comments are particularly noteworthy as they were made before most of the neuropsychological studies on the affective, contextual and higher-order contributions of the RH to language understanding.

Another remarkable acute effect of callosal section is mutism. Cutting the corpus callosum results in the complete loss of speech for days, weeks or months in most callosotomy patients (Ross *et al.*, 1984). The effect is not permanent, but remains unexplained. Why would the speech-competent LH require input from the RH to initiate speech? RH damage itself does not normally produce mutism, indicating that, when the integrity of the RH is compromised, the LH is *not* prevented from acting on its own. Paradoxically, following severance of the corpus callosum when the RH is intact and capable of normal information processing, the presence of two functioning cerebral hemispheres that have been suddenly disconnected means that the LH can no longer undertake its most usual and perhaps least effortful behaviour, speech. The implication is that the LH, prior to callosotomy, normally *awaits* cognitive input from the RH before initiating verbal behaviour.

The phenomenon of mutism is perhaps not so surprising in light of the effects summarised in Figure 2. That is, if the multilevel hemispheric division of labour shown in the figure is typical of the normal brain, then mutism can be understood as a consequence of the loss of the cognitive—connotative, metaphorical, contextual—input that motivates most *normal* speech behaviour. It is relevant to note that, in response to perceived speech, even when the literal meaning is entirely clear, normal people do not necessarily respond if the 'point' of the speech is not perceived. When contextual information, implications and underlying meaning are missing, many normal people are *not* talkative and are reticent about engaging in 'meaningless' verbal discourse except for reasons of social politeness. Such an argument concerning post-callosotomy mutism remains speculative, but a loss of spontaneous speech would not be paradoxical if the various 'meanings' that normally drive verbal behaviour reside in the RH and are disconnected from their normal outlet through the LH.

In so far as callosal damage produces linguistic deficits similar to those of RH damage, the obvious inference is that, even with both hemispheres fully functional, RH cognition does not affect behaviour if information is not sent across the corpus callosum for use by the talkative 'dominant' LH. In other words, for the purposes of motor utilisation of the 'para-linguistic' information of the RH, callosal connections are essential (or, if not essential, at least the most efficient route over which information can flow between the left and right cerebral cortices). Less clear is the influence that an intact corpus callosum has on RH language understanding. That is, does the RH need the syntactical decoding of the LH to understand correctly even propositional speech? 'The boy kissed the girl', 'The girl kissed the boy' and 'The boy was kissed by the girl' (etc.) might sometimes be construed as providing the same affective 'young love' message, but these sentences might also be understood as emotionally quite different, depending on the situation. In the 'young love' context, precisely who was kissed by whom is not important, but consider a situation where the boy had been chasing the girl relentlessly for weeks, and at some point the relationship developed into a kiss! The syntactic information available from the verb form could provide the key information to distinguish between an act of unwanted harassment or one of reciprocated love. Clearly two distinct types of RH affect might be the result, depending crucially on syntactic information. In that case, does the syntactic information of the active/passive verb form, presumably processed in the LH, play a role in determining the polarity of the RH affect? If so, the processing of the RH would make use of input from the LH to deduce the actual affective state, whereas, if the RH is a 'coarse processor' (Beeman *et al.*, 1994), the summation of the connotations of the three words, 'boy', 'girl' and 'kiss', tells the entire 'young love' story; in that case, the information contained in the verb form would presumably not play a decisive role in determining the affect perceived by the RH, and the flow of information from left to right would not be an important aspect of RH language processing.

Whatever the case may be with regard to hemispheric co-operation during language understanding, what is known about language *expression* is that the RH does *not* act as a language processor capable of independent action. It relies on the LH for verbal expression, and when its access to the LH is prevented by callosal section, RH information is simply not expressed verbally. The affective state of the RH may be 'leaked' through limbic mechanisms (blushing and giggling) or somatically through gestures or facial expressions, but the RH remains verbally silent if direct transfer to the LH is not possible. The only apparent exception to this rule is verbal expression through singing. Although brain-damage studies support the idea that the RH is capable of singing but incapable of normal speech, the clearest demonstration of this effect comes from unilateral anaesthesia (the Wada test). Following left carotid artery injection of sodium amytal, the LH is temporarily incapacitated, but singing is not disrupted. Contrarily, right-sided anaesthesia has little effect on language production but disrupts singing. The capability of the RH to sing provides an interesting insight into the nature of RH language capabilities. While it seems likely that the prosody and pitch contour of the song aids the RH in its verbal expression, what is most remarkable about RH singing is that, with the help of the melody, the RH is capable of correct pronunciation, correct syntax and appropriate timing of speech output. Propositional speech may not be its strength, but the RH is *not* non-verbal!

INTEGRATION

While the ability to respond literally to simple questions and to produce syntactically coherent propositional statements is a prerequisite to more complex language usage, verbal exchanges among normal people rarely remain at the literal level. If you don't laugh at my jokes, don't respond appropriately to my metaphors, don't pick up on the 'gist' of my argument or if you giggle in response to my unhappy news, we do not 'understand' one another in the sense that we normally use the word 'understand'. It may be the case that literal language use and non-metaphoric information exchanges constitute the foundation on which metaphoric language is built, but the syntactic and literal semantic issues that have been the primary topic of traditional linguistics, and are the linguistic strengths of the LH, are closer to the starting point than the completion of an understanding of characteristically human communications. In an extensive review of the cognitive psychology of non-literal language use, Gibbs (1994) has argued that:

> Metaphor, metonymy [part-whole metaphors], irony and other tropes [figures of speech] are not linguistic distortions of literal mental thought but constitute basic schemes by which people conceptualize. (Gibbs, 1994: 1)

> Metaphor is a fundamental mental capacity by which people understand themselves and the world through the conceptual mapping of knowledge from one domain onto another. (Gibbs, 1994: 207)

Clearly, in so far as we are engaged in verbal communication more complex than asking directions to the nearest bus stop, the understanding of language requires the contributions of both literal and metaphoric/connotative/affective processes. As linguistic and paralinguistic information must be brought together to obtain the benefits of literal and non-literal modes, the question of 'integration' is an important issue still facing cognitive psychology. This general point has been understood for many years, and felt acutely by researchers in artificial intelligence who have been able to implement a variety of literal language-understanding processes and logical inference mechanisms, but have utterly failed to build intelligent machines. Given the nature of psychological research and the underlying assumptions of a scientific methodology, it is inevitable that definition of the identifiable components of cognition should precede discussion of the integration of those components, but the gap between robotic language processing and the level of normal human language use is as great as ever. From a neuropsychological perspective, the bridging of the gap between the realm of literal language and that of non-literal language means addressing questions of the relationship between the language functions of the LH and RH. Unfortunately, most neuropsychological arguments about LH and RH 'capabilities', 'specialisations' and 'information-processing modules' still conclude with statements concerning the nature of the differences or unilateral superiorities, and fail to address the next issue, that of interaction.

A recent example is the remarkable book by Ivry & Robertson (1998), *The Two Sides of Perception*. They defended the idea that the LH and RH are specialised for, respectively, high and low frequency information-processing in both the auditory and visual domains. The book constitutes a coherent, reductionist, argument about laterality and, if not the final word, it is certainly a worthy attempt to delineate a core mechanism that might account for a host of hemispheric functional differences, including language capabilities. On the penultimate page of the monograph, however, the authors note explicitly that the possibility of hemispheric interaction has not been dealt with:

> In our current development of the … theory we have emphasized how processing within each of the hemispheres considered in isolation can account for laterality effects in a variety of task domains. Our current lack of consideration of inter-hemispheric communication is an obvious weakness of the theory … (Ivry & Robertson, 1998: 276)

When dealing with 'low-level' psychophysical phenomena, the omission of callosal effects may be justified (although some would challenge the very notion of low-level), but when the discussion turns to 'high-level' language processing and the functions of association cortex where the density of callosal fibres is greatest, it is far from obvious that a 'hemispheres in isolation' perspective will have any validity. Callosally connected cortical regions are *not* isolated, and it remains an open question whether or not specialised functions are influenced by contralateral input.

Despite the continuing preference for treating the cerebral hemispheres as if they were disconnected brains, there are clear indications that hemispheric interaction does occur at various levels in the normal brain, and indeed that a failure of integration of functioning components is one possible clinical syndrome following brain damage. The clearest example of such failure of integration is the disconnection syndrome seen in the split-brain patients. Each hemisphere actively processes information, but, with the corpus callosum absent, each is ignorant of what the other hemisphere knows. Unfortunately, the split-brain studies have had the unintended effect of emphasising the independence of the LH and RH. Without a corpus callosum, *independent* and somewhat different information-processing can occur in the split-brain patients, but, however important that insight might be concerning the possibility for radically different modes of hemispheric processing in such patients, the intact brain necessarily has, in addition, the opportunity for collaboration, interaction and integration across the corpus callosum. While perhaps no one any longer advocates the idea that the corpus callosum does nothing of interest psychologically, the bulk of current theorising about the cerebral hemispheres still emphasises dominance and potential independence, rather than what it may mean to have two somewhat different hemispheric processes communicating with one another in the normal condition. Final answers may not yet be possible, but the general question of hemispheric interaction clearly requires some attention.

HERA

Starting in 1993, several laboratories reported a consistent, but unexpected, prefrontal asymmetry of activation in various short-term memory tasks; Tulving *et al.* (1994) labelled this phenomenon 'the hemispheric encoding/retrieval asymmetry' (HERA). The basic effect is that LH prefrontal regions are relatively active during the encoding phase of stimulus memorisation, whereas RH prefrontal regions are activated in the recall or retrieval phase. This was found using various brain-imaging techniques, including EEG, ERP, PET and fMRI, and reported by diverse groups. Debate continues regarding the influence of the nature of the stimuli and whether or not recall success and/or effortfulness are important factors (Nyberg, 1998), but the reality of the effect using meaningful verbal stimuli is not in doubt (Fletcher *et al.*, 1998a, b; Heun *et al.*, 1999). Many issues remain unsettled, but the asymmetry of cortical activation during verbal information-processing in short-term memory tasks is an unambiguous indication of some form of hemispheric collaboration.

In a typical HERA experiment, word-pairs, such as category–exemplar combinations (for example, furniture–bookcase, tool–hammer, fruit–papaya), are presented during the encoding phase, and retrieval of the exemplars in response to the category label is demanded in the retrieval phase. The familiarity of the words, their concreteness and ease of visualisation are factors that might influence the strength of activation, but the most robust effects have been found in tasks requiring a verbal response to a verbal stimulus (notably, the use of semantically 'empty' stimuli, i.e. pronounceable non-words do not elicit the HERA effect; Lee *et al.*, 2000). Although activation of the LH during the encoding of verbal material is unremarkable and expected solely on the basis of LH dominance for language, the activation of the RH during recall to produce a verbal response is a nonsensical, inexplicable effect if the possibility of interhemispheric communication is not considered. That is to say, in an extreme 'independent-hemispheres' model, HERA simply cannot be explained: if the information is initially encoded in the LH, recall should not involve RH activation at all, much less activation more robust than that of the LH regions involved in encoding. Presumably, the prefrontal RH activation during recall is indicative of retrieval of information in response to the verbal (category) stimulus, and transfer of that information to the LH to initiate the appropriate speech response. That the corpus callosum (CC) may be involved in this interhemispheric communication is the obvious first consideration, so that a working hypothesis might be summarised as follows.

The encoding phase:

> **sensory organs** (→ posterior LH) → **prefrontal LH** (→ CC → prefrontal RH)

Followed by the retrieval phase:

> **sensory organs** (→ posterior LH → prefrontal LH →CC) → **prefrontal RH** (→ CC → prefrontal LH) → **speech organs**

The structures in bold type are known to be involved, while the involvement of the structures in parentheses is empirically uncertain, but theoretically necessary, whenever the basic HERA effect is obtained. That is, if retrieval occurs from structures to which encoding must necessarily have first taken place, then we must postulate a relatively 'silent' (or delayed) involvement of the RH subsequent to LH encoding. Similarly, the involvement of the LH during retrieval must be assumed, even if not evident in brain images, initially to register the category to which an exemplar must be matched, and subsequently to provide the exemplar speech response. Interhemispheric communication is thus a *necessary* component of any viable explanation of the HERA phenomenon. This prefrontal asymmetry is found in the normal adult brain with intact corpus callosum in response to bilateral (auditory or visual) stimulus presentation, thus making an 'independent-hemisphere' model extremely unlikely. As the LH is clearly dominant for speech output in most individuals and notably superior at syntactic functions, its importance for both decoding input and encoding output seems clear. Despite the fact that the RH is remarkably *unable* to control the organs of speech for normal propositional speech output, it is known to be involved in various language processes, but that fact alone does not explain the HERA effect. The most obvious possibility is that the RH contributes to language processing by communicating with the LH across the corpus callosum, rather than using its information for direct control over the relevant somatic musculature. While the more difficult questions of neuronal mechanisms remain to be explored, this first-order understanding of the flow of language information in the human brain can be summarised in what I refer to as a 'central dogma' (Cook, 1986, 1989, 2002) for human neuropsychology:

$$\text{RH} \leftrightarrow \text{LH} \rightarrow \text{striate musculature}$$

The 'dominance' of the LH is due to its role in, particularly, speech output. Of course, both cerebral hemispheres are heavily connected to sensory and motor organs, but, at least with regard to language processing, the RH does *not* actively control the midline organs of speech. Moreover, in the absence of the LH, the RH is incapable of understanding even mildly complex propositional speech, despite its rich lexicon. In other words, although the RH contributes in characteristic ways to language understanding and production, it is not 'independent'. Instead, a large part of the flow of linguistic information from the RH seems to go through the LH. This fact about the neuropsychology of language is as frequently misunderstood as it is widely acknowledged. On the one hand, the cerebral hemispheres are each virtually complete information-processing neural structures: if either hemisphere is removed at a young age, the other hemisphere alone is capable of a nearly normal range of sensory, cognitive, affective and motor processing, including language. On the other hand, in the normal intact brain, the hemispheres become specialised during the developmental process to such an extent that, subsequent to acute brain damage in adulthood, the undamaged RH often cannot take over lost LH functions, and vice versa. It is this paradox of equipotentiality, but functional specialisation (and consequent asymmetry of information flow), that lies at the heart of the 'central dogma.'

Analogous to the central dogma of molecular biology, this psychological dogma does *not* address the important issues of mechanisms, i.e. the neurophysiology of intracerebral information flow (i.e. a 'brain code' with a scientific clarity comparable to the nucleotide base-pairing of the genetic code), and is essentially nothing more than a highly simplified flow chart. For focusing on the most important processes in language input and output, however, it is a valid summary of the division of labour that the human brain appears to employ. That there may be other flow charts for non-linguistic processes or for the cognitive processes of other species is likely, but for the all-important usage of language (and possibly tools) the human brain has evolved a functional asymmetry that entails not only functional differences, but a specific pattern of asymmetrical information flow (Figure 3).

Regardless of the label given to this pattern of hemispheric interaction, its general validity with regard to linguistic processing has been partially known since the mid-1800s, when the dominance of the LH for speech became established. Precisely what the RH does during language processing has remained more of a puzzle, but most of the confusion about hemispheric specialisation has come from attempts to summarise all types of hemispheric functions with a single dichotomy of a psychological nature: verbal/non-verbal, logical/emotional, or whatever. Unfortunately, even when there is strong empirical support for a given dichotomy in a given context, the dichotomy is defined by the nature

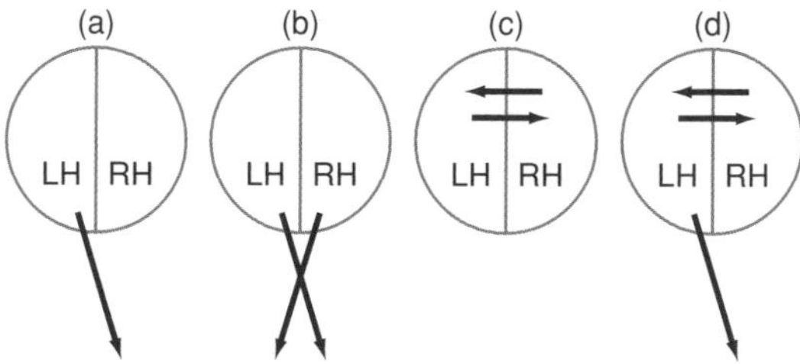

Figure 3. The evolution of ideas on human laterality. An emphasis on LH 'dominance' for speech and handedness was the main theme of early (pre-1950) laterality studies (a), while hemispheric independence and the distinct modes of LH and RH cognition were the focus of most split-brain and split-brain-motivated studies (1960s–1980s) (b). Hemispheric interaction has more recently (1990s) been researched and debated (c), but hemispheric interaction models still need to be reconciled with traditional ideas of dominance, particularly with regard to language (d).

of stimulus materials and does not necessarily generalise to other modalities. What is needed, therefore, is not the perfect set of philosophical or psychological adjectives, but a neuronal mechanism, or a small set of neuronal mechanisms, that can be understood to operate on various kinds of cortical information.

The establishment of mechanisms to replace (and explain) a host of psychological dichotomies may seem unachievable at the present time, but the relative simplicity of callosal anatomy (largely homotopic connections between cortical areas that process information of the same modality or modalities) and the two-dimensional organisation of the cortical surface greatly constrain the types of callosal mechanisms that must be considered (Reggia *et al.*, 1998; Cook, 1999; Shkuro *et al.*, 2000). Already prior to 1986, all four of the most obvious possibilities (i.e. assuming predominantly excitatory or inhibitory synaptic effects, and predominantly diffuse or focal callosal fibre termination) had been defended in the laterality literature as 'the crucial mechanism' underlying human laterality (reviewed in Cook, 1986). Since then several variations on the basic excitatory and inhibitory models have been advocated (for a review see Burgess & Lund, 1998; for discussion of the relative merits of the models see Querné *et al.*, 2000; for some clarity on the various versions of interhemispheric inhibition see Chiarello & Maxfield, 1996). When taken as the-one-and-only mechanism of hemispheric interaction, however, no single hypothesis alone can account for the diverse nature of interhemispheric information flow, but the validity of each model taken on a cortical module-by-module basis may well be demonstrable both behaviourally and in brain-imaging studies. Final answers are not at hand, but the HERA pattern of cortical activation provides an indication that something as simple as the 'central dogma' flow chart for interhemispheric communications may be a useful framework into which the neuronal mechanisms (specific cortical circuitry, neurotransmitters, etc.) will eventually need to be plugged.

NEURONAL MECHANISMS

From the point of view of cognitive psychology, neurones are rather simple things: as far as is known, they have significance for cognition only when they generate an action potential that leads to the release of neurotransmitters at synapses. The strength of the synaptic effect and the frequency of firing can vary continuously from small to large values, but the synaptic polarity is fixed as either excitatory or inhibitory (Dale's law). The complexity of neuronal 'information-processing' therefore lies in (1) the spatial configuration of neurones (numbers of neurones, numbers of synapses and pattern of connectivity) and (2) the temporal structure of neuronal firing. Arguably the single most important addition to the basic concepts of neurophysiology in general since the 1950s has been the realisation that the temporal dimension, i.e. synchronisation of neuronal firing, may be important for information processing (Singer, 1993; Crick, 1994; Singer & Gray, 1995). That is to say, not only does the frequency of neuronal firing have influence on cognition, but the temporal relationships are also relevant as a means for 'binding' cognitive operations that occur in diverse locations in the nervous system. Synchronisation thus adds an important *temporal* dimension to the basic concepts of neurophysiology that were established in the first half of the twentieth century, and promises to play a central role in the elucidation of so-called higher cognitive activity.

The dimensions along which information is organised in early sensory and late motor cortex are known and give rise to the retinotopic, somatomotor, etc., topographical mapping of large portions of the neocortex. Despite some indication of the semantic organisation of association cortex obtained by direct stimulation (Penfield, 1959), the simplicity of sensory and motor cortex mapping is not found, and individual differences in the organisation of association cortex may be large. Assuming only that the association cortex also has some sort of meaningful two-dimensional organisation, it can be concluded that the interaction between homologous regions in the LH and RH will be influenced primarily by the anatomical connectivity between them, and the nature of the synaptic effects. As mentioned above, the main classes of interaction are easily summarised and simulated (Reggia *et al.*, 1998; Cook, 1999; Shkuro *et al.*, 2000). If a relatively fine-grained topographic connectivity between cortical regions is assumed, then the transfer of information (either excitatory or inhibitory) from one hemisphere to the other is possible. In contrast, if callosal effects are relatively diffuse, the possibilities for the transmission of detailed information decrease, and the corpus callosum will act to alter the hemispheric balance of arousal and attention.

Attentional models of hemispheric functions have been advanced by Kinsbourne (1970, 1982), Heilman & Van Dan Abell (1979) and Guiard (1980), and more recently by Banich (1998) and Liederman (1998). These attempts to explain laterality effects on the basis of asymmetrical 'arousal' or 'attention' are psychologically plausible, but suffer from terminological problems. Commonly used phrases such as 'the dynamic allocation of attentional resources' have no obvious physiological meaning and thus lack the specificity to bridge the gap between psychological phenomena and neuronal mechanisms. Interestingly, recent advances in explaining arousal, attention and awareness on the basis of the synchronisation of neuronal firing mean that the psychological concepts of the attention theories, such as 'resources,' 'spotlights' and 'bottlenecks', might be translatable into the language of neuronal activity (Singer, 1993; Crick, 1994).

The importance of the synchronisation hypothesis for the issue of human laterality is that it has the potential to replace a host of plausible, but inherently fuzzy, descriptions of hemispheric relations with an explicit neuronal mechanism that allows distant (including bihemispheric) cortical modules to collaborate without requiring a cortical region at which all cognitive 'results' are integrated. While explanation of bilateral cognition on the basis of synchronisation has not yet been achieved, progress has been made in defining the relationship between synchronisation and arousal, attention and awareness, the relevant frequencies of oscillation have been studied in various animal species, and possible neuronal mechanisms have been explored in artificial neural nets (Engel & Singer, 2001). Hypotheses concerning the synchronisation of LH and RH neuronal activity during language processing might eventually replace or augment ideas of topographical information transfer.

As mentioned in the introduction, the paradox of the duality of the nervous system and yet the unity of subjective consciousness remains unsolved. It remains a paradox primarily because the problem of subjective feeling itself is one of the so-called 'hard problems' in consciousness studies (Searle, 1997; Shear, 1998). That is, why do we have subjective feeling at all? Why, in addition to the 'information-processing' of neurones, do we feel that there is something like direct 'experience' that is fundamentally different from cognition? Here, as well, the synchronisation hypothesis provides an essential connection between the realm of cognition and the various issues of subjective consciousness.

Briefly, the argument concerning the relationship between synchronised neuronal firing and subjective consciousness (Cook, 2000, 2002) is as follows. The neuronal membrane is normally closed to the diffusion of ions, but at the time of the action potential there occurs (1) a localised, transient permeability of the membrane, (2) several hundred thousand ion channels are opened, and (3) about 10^8 ions per channel per second flow between the intra- and extracellular fluids as a consequence of the transmembrane potential gradient (Koch, 1999). The action potential is in effect a means by which the neurone directly 'experiences' the electrotonic state of its environment and re-equilibrates its own ionic concentrations to concentrations more similar to those of the extracellular fluid. Synchronisation of the firing of neurones is essentially the temporal correlation of many action potentials, such that there occurs a co-ordinated pattern of the inflow of ions to many neurones at diverse locations throughout the nervous system.

The behaviour of neurones is well understood and not controversial. What is unusual in this account is only the emphasis placed on the membrane dynamics of the action potential. Instead of seeing the action potential simply as a means of impulse discharge (the mechanism by which the neuronal cell body sends a message to its axonal terminals), I maintain that it is the very fact of opening ion channels and allowing the physiochemical diffusion of ions across the normally closed cell membrane that is the key phenomenon (Figure 4).

The philosophical argument is therefore that the momentary opening of the cell membrane at the time of the action potential is the single-cell protophenomenon (MacLennan, 1998) underlying 'subjectivity', literally, the opening up of the cell to the surrounding biochemical solution and a brief, controlled, breakdown of the barrier between cellular 'self' and the external world. The synchronisation of the action potentials of many neurones produces a pattern of 'openness' of the nervous system as a whole, arguably a simple by-product of the temporal co-ordination of neuronal firing that is *needed* for feature 'binding' in cognition. In this view, the normal ebb and flow in the strength of subjective feeling is real, and a direct consequence of the variable number of neurones participating in synchronous firing. When synchronisation occurs interhemispherically, not only is there a co-ordinated activation of the information in both hemispheres, and simultaneous participation in the cognition of the organism as a whole, but there occurs an associated simultaneous 'feeling' of awareness in both hemispheres (Cook, 1999, 2000).

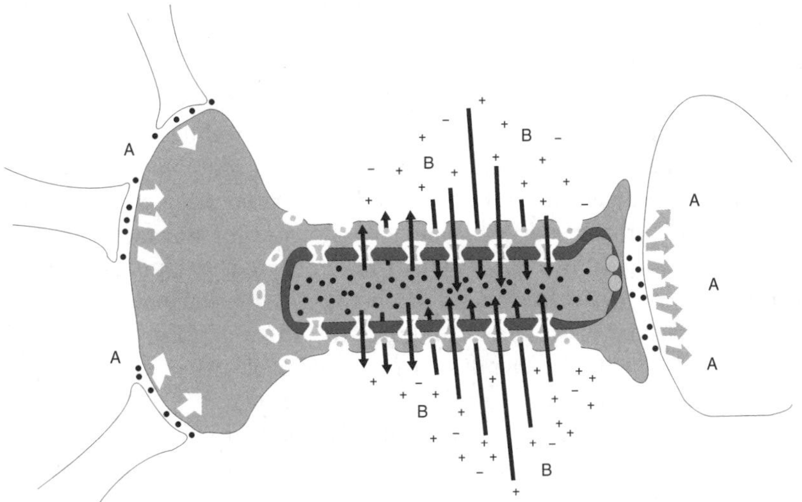

Figure 4. A cartoon of the two modes of contact between a neurone and its environment: (A) synaptic transmission and (B) ion exchange. Synaptic transmission is the functional unit of cognition, whereas ion-exchange during the action potential is the means by which the neurone makes direct contact with the extracellular fluid and, as such, is the functional unit of subjective consciousness (Cook, 2002). Both functions attain significance for the organism as a whole only when many neurones are active within synchronised neuronal networks. Synchronisation itself must occur through appropriate neuronal circuitry (i.e. synaptic effects), but means that neurones distributed throughout the nervous system can contribute to the same cognitive and conscious phenomena.

If the synchronised opening of the neuronal membrane is the basis for a feeling of openness to the external world, then the only thing 'illusory' about subjective awareness is that we tend to think mistakenly that we feel the 'outer world' directly, a world that, in fact, exists millimetres or miles outside our own skin. In contrast, it is *not* an illusion that individual nerve cells are momentarily in direct contact with their extracellular environments, with each cell briefly sampling the electrostatic state of the surrounding ion solution. This is to say that the dilemma of the hard problem is real (Searle, 1997) and not merely a linguistic trick of armchair philosophers! Already at the cellular level, there is indeed an explanatory gap between computational cognition and incomputable consciousness, between quantifiable synaptic communications that lead to behaviour and diffuse membrane dynamics that result in an unquantifiable 'feeling' with no direct behavioural implications. The gap is real, but not inexplicable, and the problem is perhaps not so hard, if it can be understood as a direct consequence of the two modes of neuronal interaction with the 'external' biochemical world (Cook, 2002) (Figure 4).

Synaptic transmission is widely believed to be the functional unit of cognition: in the act of neurotransmitter release and the exertion of inhibitory or excitatory effects on other neurones, a neurone participates in small-network logical functions, the sum of which is cognition, but a single neurone does not 'think' anything beyond its own function of inhibition or facilitation. Similarly, at the moment of the action potential, the neurone exchanges ions with its environment and participates in the organism's overall feeling of the external world, but a single neurone is not 'aware' of anything other than its own relative activation or quiescence. Of course, the action potential is the causal trigger leading to neurotransmitter release at the synapse, but transmembrane ion flow and neurotransmitter release are distinct phenomena and contribute to two different types of organism-level psychology: in addition to whatever small role the neurone may play in the cognition of the whole organism, it also makes a small contribution to the overall subjective feeling of openness of the organism through contact with its extracellular environment.

By focusing on the temporal synchronisation of neurones, solutions to two fundamental problems in consciousness studies come into view: the binding problem of cognition (including the binding of simultaneous LH and RH cognitive activity) and the subjective feeling problem of consciousness. Both of these core topics can be understood in relation to protophenomena *at the neuronal level.* Synaptic communications lead ultimately to the cognition of multicellular nervous systems, and the transmembrane flow of ions leads ultimately to the feeling of awareness of an 'external' world, i.e. subjective consciousness. Questions about subtypes of consciousness and cognition that depend on the connectivity of the multineurone networks are not addressed by this synchronisation argument, but by identifying the relevant cellular mechanisms the construction of strictly *neuronal* theories of both consciousness and cognition may be possible, and obviate the need for introducing otherworldly philosophical or quantum mechanical postulates.

CONCLUSION

The pattern of results concerning the contributions of the cerebral hemispheres in language processing clearly points away from the old ideas of 'unilateral dominance' and towards the conclusion that both hemispheres are involved in multiple subtasks during most language functions. Because the LH is essential for the motor functions underlying speech, it may be appropriate to refer to it as 'language dominant', but most findings over the past 20 years indicate that the RH is far from irrelevant, and is actively involved in processing linguistic information in ways that are distinct from those of the LH.

In addition to proposals concerning the flow of information between the cerebral hemispheres, recent work on neuronal synchronisation provides a 'non-information flow' means of interhemispheric communication. In effect, the synchronisation argument provides a mechanism for interhemispheric 'binding' and thus gives an explicit neuronal mechanism for one variety of attentional model. So doing, the need for the verbal gymnastics of most previous attentional models can be avoided and the relevant psychological phenomena can be translated into the terminology of neurophysiology. Details of the neuronal mechanisms and the important oscillatory frequencies (by species and by task) still need to be clarified, but the concept of synchronisation may achieve, in one bold step, a neurone-level elucidation of cognition, consciousness, attention and hemispheric interactions.

Modelling of hemispheric interactions across the corpus callosum remains of fundamental interest for human psychology in general, not only for clarification of the nature of hemispheric specialisation, but more importantly for answering the question why functional asymmetry is associated with the three quintessentially human forms of behaviour, language, tool usage and music.

ACKNOWLEDGEMENT

This work was supported by the 'Research for the Future Program,' administered by the Japan Society for the Promotion of Science (Project No. JSPS-RFTF99P01401).

DISCUSSION

Crow: Your model does not seem to depend on the anterior/posterior torque? Also, Is it unidirectional?

Cook: No, the direction is not one way, it is mutual. I am only talking about a small portion of the torque, the frontal association cortex. In particular, transfer of sensory information, auditory, visual, etc., would probably be excitatory. The inhibitory function is the most interesting cognitively and that is what I was trying to focus on there.

Questioner: Do you have any information about corpus callosal agenesis?

Cook: Most of these individuals are actually subnormal. They are often detected because of headaches. The brain is quite plastic and so they manage to compensate somehow. Even some of the split-brain patients have developed language in the right hemisphere.

Questioner: Do acallosal subjects develop schizophrenia?

Cook: Yes, they turn up more than would be expected in the brain scan literature.

Questioner: There are aspects of language which are located in the right hemisphere, like understanding metaphors or understanding jokes. These aspects of language are important in terms of being able to trust an individual. If someone doesn't share a set of jokes with you, these are essentially warning signals. So I wonder whether there isn't a right hemisphere aspect of language that is present in primates, and then a left hemisphere set of syntactical functions that are subject to sexual selection.

Comment: We can't do metaphor unless we have syntactic rules to do it by, so the two interact.

McManus: 10% of the population have language on the right side of the brain. If you look at the pattern of brain organisation, we actually have a brain polymorphism here. To try and give information about the mean is misleading when it is actually the variance that we ought to be looking at. Means are going to mislead us. There are always simple stories if we want to say, this is left and this is right, but half of the data are based on patients with lesions.

Cook: I agree that we are talking here about population means and we can not ignore population differences. Nevertheless, lateralisation is an issue and it keeps coming up.

Comment: We have to realise that evolution is working on the entire distribution. It is the whole 'lot' that is selected, not just the mean of the population.

References

Alcock, K.J., Wade, D., Anslow, P. & Passingham, R.E. (2000) Pitch and timing abilities in adult left-hemisphere-dysphasic and right-hemisphere-damaged subjects. *Brain and Language*, **75**, 47–65.

Anaki, D., Faust, M. & Kravetz, S. (1998) Cerebral hemispheric asymmetries in processing lexical metaphors. *Neuropsychologia*, **36**, 353–62.

Banich, M.T. (1998) The missing link: the role of interhemispheric interaction in attentional processing. *Brain and Cognition*, **36**, 128–57.

Beeman, M. & Chiarello, C. (1998) *Right Hemisphere Language Comprehension*. New Jersey: Erlbaum.

Beeman, M., Friedman, R.B., Grafman, J., Perez, E., Diamond, S. & Lindsay, M.B. (1994) Summation priming and coarse semantic coding in the right hemisphere. *Journal of Cognitive Neuroscience*, **6**, 26–45.

Behrens, S.J. (1989) Characterizing sentence intonation in a right hemisphere-damaged population. *Brain and Language*, **37**, 181–200.

Bogen, J.E. (1997) Does cognition in the disconnected right hemisphere require right hemisphere possession of language? *Brain and Language*, **57**, 12–21.

Bottini, G., Corcoran, R., Sterzi, R., Paulesu, E., Schenone, P., Scarpa, P., Frackowiak, R.J. & Frith, C. (1994) The role of the right hemisphere in the interpretation of figurative aspects of language. A positron emission tomography activation study. *Brain*, **117**, 1241–53.

Brownell, H.H., Potter, H.H., Michelow, D. & Gardner, H. (1984) Sensitivity to lexical denotation and connotation in brain damaged patients: a double dissociation. *Brain and Language*, **22**, 253–65.

Bulman-Fleming, M.B. & Bryden, M.P. (1994) Simultaneous verbal and affective laterality effects. *Neuropsychologia*, **32**, 787–97.

Burgess, C. & Chiarello, C. (1996) Neurocognitive mechanisms underlying metaphor comprehension and other figurative language. *Metaphor and Symbolic Activity*, **11**, 67–84.

Burgess, C. & Lund, K. (1998) Modeling cerebral asymmetries in high-dimensional space. In: *Right Hemisphere Language Comprehension* (eds C. Burgess & C. Chiarello), pp. 215–44. New Jersey: Erlbaum.

Chiarello, C. & Maxfield, L. (1996) Varieties of interhemispheric inhibition, or how to keep a good hemisphere down. *Brain and Cognition*, **30**, 81–108.

Code, C. (1987) *Language, Aphasia and the Right Hemisphere*. New York: John Wiley.

Cook, N.D. (1986) *The Brain Code: Mechanisms of Information Transfer and the Role of the Corpus Callosum*. London: Methuen.

Cook, N.D. (1989) Toward a central dogma for psychology. *New Ideas in Psychology*, **7**, 1–18.

Cook, N.D. (1999) Simulating consciousness in a bilateral neural network: 'nuclear' and 'fringe' awareness. *Consciousness and Cognition*, **8**, 62–93.

Cook, N.D. (2000) On defining awareness and consciousness: the importance of the neuronal membrane. *Proceedings of the Tokyo-99 Conference on Consciousness* (ed. K. Yasue), pp. 19–20. Singapore: World Scientific.

Cook, N.D. (2002) *Tone of Voice and Mind: The Connections Between Intonation, Emotion, Cognition and Consciousness*. Amsterdam: Benjamins.

Crick, F. (1994) *The Astonishing Hypothesis: The Scientific Search for the Soul*. New York: Simon and Schuster.

Crow, T.J. (1997) Schizophrenia as failure of hemispheric dominance for language. *Trends in Neurosciences*, **20**, 339–43.

Crow, T.J. (1998) Why cerebral asymmetry is the key to the origin of *Homo sapiens*: how to find the gene or eliminate the theory. *Cahiers de Psychologie Cognitive*, **17**, 1237–77.

Engel, A.K. & Singer, W. (2001) Temporal binding and the neural correlates of sensory awareness. *Trends in Cognitive Science*, **5**, 16–25.

Fletcher, P.C., Shallice, T. & Dolan, R.J. (1998a) The functional roles of prefrontal cortex in episodic memory. I. Encoding. *Brain*, **121**, 1239–48.

Fletcher, P.C., Shallice, T., Frith, C.D., Frackowiak, R.S.J. & Dolan, R.J. (1998b) The functional roles of prefrontal cortex in episodic memory. II. Retrieval. *Brain*, **121**, 1249–56.

Gainotti, G., Caltagirone, C. & Miceli, G. (1983) Selective impairment of semantic–lexical discrimination in right-brain-damaged patients. In: *Cognitive Processing in the Right Hemisphere* (ed. E. Perecman), pp. 149–67. New York: Academic Press.

Gardner, H., Brownell, H., Wapner, W. & Michelow, D. (1983) Missing the point: the role of the right hemisphere in the processing of complex linguistic materials. In: *Cognitive Processing in the Right Hemisphere* (ed. E. Perecman), pp. 169–91. New York: Academic Press.

Geschwind, N. (1982) Disorders of attention: a frontier in neuropsychology. *Philosophical Transactions of the Royal Society of London (Biological)*, **298**, 173–85.

Gibbs, R.W. (1994) *The Poetics of Mind: Figurative Thought, Language and Understanding*. New York: Cambridge University Press.

Grimshaw, G.M. (1998) Integration and interference in the cerebral hemispheres: relations with hemispheric specialization. *Brain and Cognition*, **36**, 108–27.

Grodzinsky, Y., Shapiro, L. & Swinney, D. (2000) *Language and the Brain: Representation and Processing*. New York: Academic Press.

Guiard, Y. (1980) Cerebral hemispheres and selective attention. *Acta Psychologica*, **46**, 41–61.

Gur, R.E. (1999) Is schizophrenia a lateralized brain disorder? *Schizophrenia Bulletin*, **25**, 7–9.

Heilman, K.M. & Van Dan Abell, T. (1979) Right hemispheric dominance for mediating cerebral activation. *Neuropsychologia*, **17**, 315–21.

Heun, R., Klose, U., Jessen, F., Erb, M., Papassotiropoulos, A., Lotze, M. & Grodd, W. (1999) Functional MRI of cerebral activation during encoding and retrieval of words. *Human Brain Mapping*, **8**, 157–69.

Iacoboni, M. & Zaidel, E. (1996) Hemispheric independence in word recognition: evidence from unilateral and bilateral presentations. *Brain and Language*, **53**, 121–40.

Ivry, R.B. & Robertson, L.C. (1998) *The Two Sides of Perception*. Massachusetts: MIT Press.

Kiefer, M., Weisbrod, M., Kern, I., Maier, S. & Spitzer, M. (1998) Right hemisphere activation during indirect semantic priming: evidence from event-related potentials. *Brain and Language*, **64**, 377–408.

Kiehl, K.A., Liddle, P.F., Smith, A.M., Mendrek, A., Forster, B.B. & Hare, R.D. (1999) Neural pathways involved in the processing of concrete and abstract words. *Human Brain Mapping*, **7**, 225–33.

Kinsbourne, M. (1970) The cerebral basis of lateral asymmetries in attention. *Acta Psychologica*, **33**, 193–201.

Kinsbourne, M. (1982) Hemispheric specialization and the growth of human understanding. *American Psychologist*, **37**, 411–20.

Klouda, G.V., Robin, D.A., Graff-Radford, N.R. & Cooper, W.E. (1988) The role of callosal connections in speech prosody. *Brain and Language*, **35**, 154–71.

Koch, C. (1999) *Biophysics of Computation: Information Processing in Single Neurons*. New York: Oxford University Press.

Landis, T., Assal, G. & Perret, E. (1979) Opposite cerebral hemispheric superiorities for visual associative processing of emotional faces and objects. *Nature*, **278**, 739–40.

Lee, A.C.H., Robbins, T.W., Pickard, J.D. & Owen, A.M. (2000) Asymmetric activation during episodic memory: the effects of stimulus type on encoding and retrieval. *Neuropsychologia*, **38**, 677–92.

Leonard, C.L., Waters, G.S. & Caplan, D. (1997) The use of contextual information by right brain-damaged individuals in the resolution of ambiguous pronouns. *Brain and Language*, **57**, 309–42.

Liederman, J. (1998) The dynamics of interhemispheric interaction and hemispheric control. *Brain and Cognition*, **36**, 193–208.

MacLennan, B. (1998) The elements of consciousness and their neurodynamical correlates. In: *Explaining Consciousness: The Hard Problem* (ed. J. Shear), pp. 249–66. Massachusetts: MIT Press.

Nyberg, L. (1998) Mapping episodic memory. *Behavioral Brain Research*, **90**, 107–14.

Penfield, W. (1959) *Speech and Brain Mechanisms*. New Jersey: Princeton University Press.

Querné, L., Eustache, F. & Faure, S. (2000) Interhemispheric inhibition, intrahemispheric activation, and lexical capacities of the right hemisphere: a tachistoscopic, divided visual-field study in normal subjects. *Brain and Language*, **74**, 171–90.

Reggia. J., Goodall, S. & Shkuro, Y. (1998) Computational studies of lateralization of phoneme sequence generation. *Neural Computation*, **10**, 1277–97.

Rodel, M., Cook, N.D., Regard, M. & Landis, T. (1994) Hemispheric dissociation in judging semantic relations: complementarity for close and distant associates. *Brain and Language*, **43**, 448–59.

Ross, E.D., Harney, J.H., deLacoste-Utamsing, C. & Purdy, P.D. (1981) How the brain integrates affective and propositional language into a unified behavioral function. *Archives of Neurology*, **38**, 745–8.

Ross, E.D., Thompson, R.D. & Yenkosky, J. (1997) Lateralization of affective prosody in brain and the callosal integration of hemispheric language functions. *Brain and Language*, **56**, 27–54.

Ross, M.K., Reeves, A.G. & Roberts, D.W. (1984) Post-commissurotomy mutism. *Annals of Neurology*, **16**, 114–23.

Schiffer, F. (1998) *Of Two Minds: The Revolutionary Science of Dual-Brain Psychology*. New York: Free Press.

Schirmer, A., Alter, K., Kotz, S.A. & Friederici, A.D. (2001) Lateralization of prosody during language production. A lesion study. *Brain and Language*, **76**, 1–17.

Schneiderman, E.I., Murasugi, K.G. & Saddy, J.D. (1992) Story arrangement ability in right-brain damaged patients. *Brain and Language*, **43**, 107–20.

Searle, J.R. (1997) *The Mystery of Consciousness*. New York: New York Review Books.

Shear, J. (1998) *Explaining Consciousness: The Hard Problem*. Massachusetts: MIT Press.

Shkuro, Y., Glezer, M. & Reggia, J.A. (2000) Interhemispheric effects of simulated lesions in a neural model of single-word reading. *Brain and Language*, **72**, 343–74.

Singer, W. (1993) Synchronization of cortical activity and its putative role in information processing and learning. *Annual Review of Physiology*, **55**, 349–75.

Singer, W. & Gray, W.C.M. (1995) Visual feature integration and the temporal correlation hypothesis. *Annual Review of Neuroscience*, **18**, 555–86.

Sperry, R.W. (1968) Hemisphere deconnection and unity in conscious awareness. *American Psychologist*, **23**, 723–33.

Sperry, R.W., Gazzaniga, M.S. & Bogen, J.E. (1969) Interhemispheric relationships: the neocortical commissures; syndromes of hemisphere disconnection. In: *Handbook of Clinical Neurology* (eds P. J. Vinken & G. W. Bruyn), Vol. 4, pp. 273–90. Amsterdam: North-Holland.

Tallal, P., Miller, S.L., Bedi, G., Byma, G., Wang, X., Nagarajan, S.S., Schreiner, C., Jenkins, W.M. & Merzenich, M.M. (1996) Language comprehension in language-learning impaired children improved with acoustically modified speech. *Science*, **271**, 81–4.

Taylor, K.I., Brugger, P., Weniger, D. & Regard, M. (1999) Qualitative hemispheric differences in semantic category matching. *Brain and Language*, **70**, 119–31.

TenHouten, W.D., Hoppe, K.D. & Bogen, J.E. (1985) Alexithymia and the split-brain. I–III. *Psychotherapy and Psychosomatics*, **43**, 202–8; **44**, 1–5; **44**, 89–94.

Tulving, E., Kapur, S., Markowitsch, H.J., Craik, F.I.M., Habib, R. & Houle, S. (1994) Neuroanatomical correlates of retrieval in episodic memory: auditory sentence recognition. *Proceedings of the National Academy of Science of the USA*, **91**, 2012–5.

Van Lancker, D. & Kempler, D. (1987) Comprehension of familiar phrases by left but not by right hemisphere damaged patients. *Brain and Language*, **32**, 265–77.

Van Lancker, D. & Sidtis, J.J. (1992) Identification of affective-prosodic stimuli by left and right hemisphere damaged subjects: all errors are not created equal. *Journal of Speech and Hearing Research*, **35**, 963–70.

Wapner, W., Hamby, S. & Gardner, H. (1984) The role of the right hemisphere in the apprehension of complex linguistic materials. *Brain and Language*, **14**, 15–33.

Weniger, D. (1984) Dysprosody as part of the aphasic language disorder. *Advances in Neurology*, **42**, 41–50.

Winner, E., Brownell, H., Happe, F., Blum, A. & Pincus, D. (1998) Distinguishing lies from jokes: theory of mind deficits and discourse interpretation in right hemisphere brain-damaged patients. *Brain and Language*, **62**, 89–106.

Winner, E. & Gardner, H. (1977) The comprehension of metaphor in brain-damaged patients. *Brain*, **100**, 719–27.

Zaidel, E. (1985) Language in the right brain. In: *The Dual Brain: Hemispheric Specializations in Humans* (eds D. F. Benson & E. Zaidel), pp. 205–31. New York: Guilford Press.

Zatorre, R.J., Belin, P. & Penhune, V.B. (2002) Structure and function of auditory cortex: music and speech. *Trends in Cognitive Science*, **6**, 37–46.

Zatorre, R.J., Evans, A.C. & Meyer, E. (1994) Neural mechanisms underlying melodic perception and memory for pitch. *Journal of Neuroscience*, **14**, 1908–19.

III

THE SEARCH FOR A CRITICAL EVENT

Sex chromosomal rearrangements as putative speciation events in hominid evolution

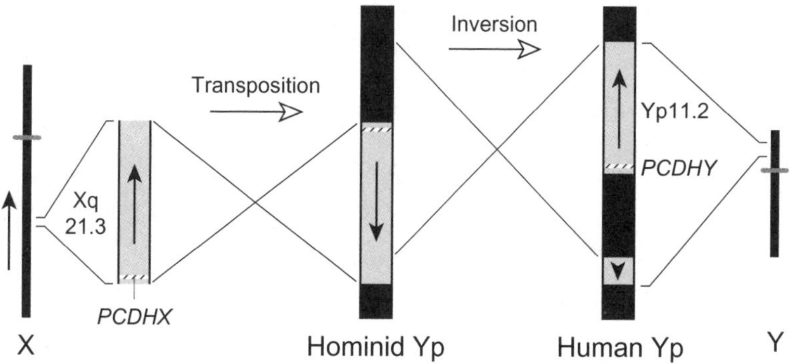

The origin of the Xq21.3/Yp regions of homology in *Homo sapiens* by a reduplicative translocation dated at between 3 and 2 million years BP and a subsequent (and presently undated) paracentric inversion that split the translocated block on a hominid Y short arm. Adapted from Schwartz A. et al (Reconstructing hominid Y evolution: X-homologous block, created by X-Y transposition, was disrupted by Yp inversion through LINE-LINE recombination. Human Molecular Genetics 1998; 7: 1-11) to show the location of the protocadherin XY gene (PCDHX and PCDHY - see chapter by Sargent et al).

Sexual Selection, Timing and an X–Y Homologous Gene: Did *Homo sapiens* Speciate on the Y Chromosome?

TIM J. CROW

Summary. A theory of the speciation of modern *Homo sapiens*, that a single gene played a critical role in the transition from a precursor species, is founded upon the following. (1) The premise that hemispheric asymmetry is the defining feature of the human brain and the only plausible correlate of language. (2) An argument for a specific candidate region (the Xq21.3/Yp11.2 region of homology) based upon the reciprocal deficits associated with the sex chromosome aneuploidies, and the course of chromosomal change in hominid evolution (supported by a weak linkage to handedness). A gene (protocadherinXY) identified within this region is expressed in the brain with the potential to account for a sex difference. (3) A particular evolutionary mechanism (sexual selection acting on an X–Y-linked gene) to account for species-specific modification of what initially was a saltational change (in this case a chromosomal rearrangement). These postulates relate to the case of modern *H. sapiens*; on the basis of the recent literature it is argued that (3) has general significance as a mechanism of speciation.

THE ORIGINS OF THE CONCEPT

THE CONCEPT (Crow, 1998a, b, 2000a) developed from two origins.

1. The theory of M. Annett, who has argued for a number of years (Annett, 1978, 1985, 1995) that her single gene theory of the genetics of cerebral asymmetry (the right-shift theory) is the key to the evolution of the human characteristic of language.
2. Arguments concerning the aetiology of psychosis, specifically that the most puzzling feature of schizophrenic psychoses (the central paradox; Crow, 2000b) is that these illnesses occur with approximately the same

incidence in all populations in spite of being associated with a substantial biological (fecundity) disadvantage. It seems that the condition is intrinsic (and therefore in some sense 'genetic') in origin. Why are the relevant genes not selected out of the population?

In 1984 I concluded that a previous environmental theory (contagion; Crow, 1983) of the aetiology of schizophrenic psychosis was untenable. What was required to explain the universal incidence and the brain changes associated with the disease was genetic diversity related to the trajectory of human brain development (Crow, 1984). On the basis that we had found that a component of the brain change was asymmetrically distributed to the hemispheres (Brown *et al.*, 1986), I formulated the hypothesis that the relevant genetic variation was that associated with the asymmetry gene that Annett had argued was responsible for the specifically human developments of the anatomy of the brain. Further evidence (Crow, 1986, 1990) for deviations in asymmetry of the structural changes accumulated, but I was puzzled by a number of sex differences [for example in age of onset, and the tendency towards same-sex concordance, i.e. affected relatives are more frequently of the same sex than would be expected, as had been noted earlier by Penrose (1942) and Rosenthal (1970)]. These influences of sex appeared to require a genetic explanation, but given the observation that illness may be transmitted from an affected father to a child of either sex, a conventional sex-linked pattern of inheritance of predisposition was excluded. To overcome this problem I proposed (Crow, 1988; see also Crow, 1987) a locus within the pseudoautosomal region. Within this region there is recombination between X and Y chromosomes in male meiosis, with the consequence that there is strict homology of genes on the X and Y. This provided an explanation of same-sex concordance (Crow *et al.*, 1989) but, as I now appreciate, on account of the strict X–Y sequence homology, it provided no explanation of a sex difference.

A more radical hypothesis was required. This came from consideration of the psychological impairments associated with the sex chromosome aneuploidies; these deficits are consistent with the location of a gene for cerebral asymmetry within a region of X–Y homology. While I had first proposed (Crow, 1989) a pseudoautosomal locus (i.e. for cerebral dominance as well as psychosis), a more satisfactory theory (Crow, 1993), because it could account for the sex differences associated with psychosis and cerebral asymmetry, was that the gene was located within one of several sex-specific (i.e. non-pseudoautosomal) regions of X–Y homology that had by that time been well described (Page *et al.*, 1984) and located with respect to primate phylogeny (Lambson *et al.*, 1992).

CEREBRAL ASYMMETRY AS THE SPECIES-DEFINING FEATURE

In writing a commentary on Annett's right-shift theory (Crow, 1995), it occurred to me that the theory could be more forcefully stated. If her claim, that the right shift was the defining characteristic of the human brain, was sustained, the conclusion had implications for speciation theory: the genetic change that generated the right shift should be considered as a speciation event. Annett (1995) had argued that population-based directional asymmetry of handedness was specific to *H. sapiens*, and had presented evidence that it was absent in the gorilla (Annett & Annett, 1991). The conclusion was substantially reinforced by the studies of Marchant & McGrew (1996) of chimpanzees in the Gombe National Park (and by the cross-species studies of Holder, 1999). In reviewing the primate literature, McGrew & Marchant (1997) concluded that:

> nonhuman primate hand function has not been shown to be lateralised at the species level—it is not the norm for any species, task or setting, and so offers no easy model for the evolution of human handedness.

The contrast is illustrated by the comparison of hand usage for the everyday range of activities in the chimpanzee and human (Figure 1).

There is a discontinuity. Directional asymmetry is present in the human but absent in the great ape population. When can this discontinuity have arisen? Clearly sometime between the separation of the hominid and chimpanzee lineages, i.e. between approximately 5 million and 100,000 years ago (the minimal estimate for the origin of modern *H. sapiens*; Stringer & McKie, 1996). The change must have had a genetic basis. The juxtaposition places Annett's theory

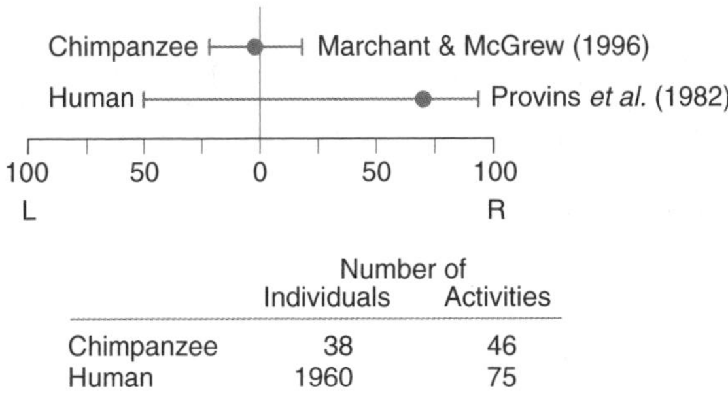

	Number of Individuals	Number of Activities
Chimpanzee	38	46
Human	1960	75

Figure 1. Hand preference in chimpanzees and humans compared. Data for chimpanzees refer to a community of wild chimpanzees (*Pan troglodytes schweinfurthii*) observed in the Gombe National Park by Marchant & McGrew (1996). Data for *Homo sapiens* were collected by questionnaire from populations of undergraduate psychology students in Scotland and Australia by Provins *et al.* (1982). Medians and boundary values (95%) have been extracted from the graphs of the original publications.

within an evolutionary context, albeit one, the nature of speciation, associated with controversy.

Thus formulated, the case of *H. sapiens* raises questions for speciation theory (see also the introduction to these *Proceedings*). Both the concept that the gene is associated with persisting and in part disadvantageous variation (that I was trying to account for in the case of psychosis, and Annett argued for in her heterozygote advantage hypothesis), and a location for the gene on the sex chromosomes, such as I was arguing for, have implications for the mechanism of transition from a precursor species. What was the origin of variation maintained at an apparently constant rate against a selective disadvantage? What were the implications of a sex difference in a novel species characteristic? Did the sex chromosomes have a special status with respect to the genetics of speciation?

WHAT DOES HAND SKILL PREDICT?

In her heterozygote advantage hypothesis, Annett argued that if the right-shift factor was the key change in *H. sapiens*, a relationship with cognitive ability would be expected. This has been a controversial prediction (Table 1).

By the balance of papers the consensus is negative, cognitive ability is unrelated to hand skill. But this consensus hides a multitude of methodological differences between studies. Some that reached positive conclusions (for example

Table 1. Does hand skill predict cognitive ability?

Positive	Equivocal	Negative
Levy (1969)		
Miller (1971)		
		Calnan & Richardson (1976)
		Hardyck *et al.* (1976)
	Harshman *et al.* (1983)	McManus & Mascie-Taylor (1983)
Annett & Kilshaw (1984)		
Annett & Manning (1990)		Bishop (1990)
	Whittington & Richards (1991)	
		McManus *et al.* (1993)
		Palmer & Corballis (1996)
		Resch *et al.* (1997)
Crow *et al.* (1998)		
		Cerone & McKeever (1999)

Levy, 1969; Miller, 1971) used samples of modest size. A number of studies that reached negative conclusions (Calnan & Richardson, 1976; McManus & Mascie-Taylor, 1983; Bishop, 1990; Whittington & Richards, 1991) did so on the basis of the large sample (more than 12,000 individuals tested at the age of 11 years) included in the UK National Child Development study. But these studies made use of the records on the 'hand that the child writes with' rather than a quantitative index.

We (Crow *et al.*, 1998) re-examined the issue in the same data set but constructed a quantitative index of 'relative hand skill' based upon the numbers of squares the child ticked with each hand in 1 minute. This analysis revealed deficits not so much at the extremities of hand skill, as Annett had suggested, but close to the point of equal hand skill (ambidexterity), as Orton (1937) and Zangwill (1960) predicted. Those around the point of 'hemispheric indecision' were impaired on verbal (Figure 2) and other abilities relative to those who were more strongly lateralised either to the right or to the left.

Thus lateralisation determines a component of ability close to the core of language. A dimension of variation persists in the population that is related to the characteristic that defines the species. Moreover, the dimension can be accounted for by a relatively simple genetic influence.

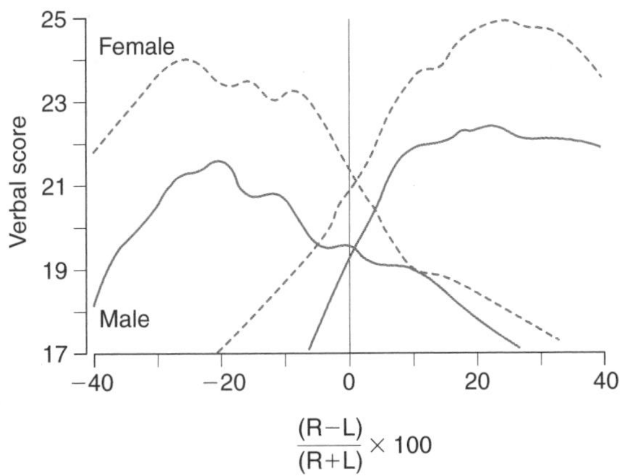

Figure 2. Relationship between verbal ability and relative hand skill (Relhand) in left- and right-hand writers (plotted separately) in 11,700 individuals in the National Child Development cohort for whom data at age 11 years were available. The curves have been smoothed with a Lowess function and truncated at Relhand values of −40 and +40, or where the numbers of individuals at a particular Relhand value discrepant with writing hand became too small to sustain the function (reproduced from Crow *et al.*, 1998).

THE CASE FOR AN X–Y HOMOLOGOUS GENE

Where is the gene? Sex differences for verbal ability are documented (Maccoby & Jacklin, 1975; Halpern, 1992) and illustrated in Figure 2: females have an advantage over males. There is also a sex difference for degrees of handedness: females are more strongly right-handed than males and are less likely to be left-handed (Annett, 1985; McManus, 1991; Crow *et al.*, 1998). It is plausible that these sex differences are related and that both in turn are related to the sex difference in brain growth: brain development is faster in females than in males (Figure 3).

The key to the genetics of asymmetry lies in the neuropsychological deficits associated with the sex chromosome aneuploidies (Crow, 1993). Individuals who lack an X chromosome (XO or Turner's syndrome) have relative deficits of non-dominant hemisphere capacity (performance IQ), while individuals with an extra X chromosome (XXY or Klinefelter's syndrome and XXX syndrome) have relative deficits of dominant hemisphere capacity (verbal IQ) (Netley & Rovet, 1982; Netley, 1986). As XXY individuals are male and XXX individuals are female, these effects cannot be attributed to gonadal hormones.

These findings indicate that a gene on the X chromosome influences the relative development of the hemispheres. The fact that deficits comparable to those in Turner's syndrome are not present in normal males, who, like Turner's syndrome individuals, have only one X chromosome, indicates that the gene must also be present on the Y chromosome. This argument generates the hypothesis that the asymmetry factor belongs to the class of X–Y homologous genes. Consistent with the theory, a tendency for handedness to be associated

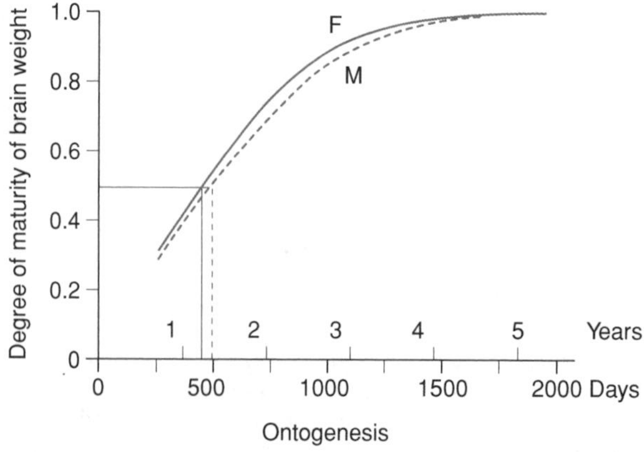

Figure 3. The sex difference in brain growth. From Kretschmann *et al.* (1979).

within sibships with sex (same-sex concordance) was observed in a collection of 15,000 families (Corballis *et al.*, 1996). Whereas the earlier version of the hypothesis (Crow, 1989), that the gene was within the pseudoautosomal region, did not explain a sex difference (such as those noted above), the later hypothesis (Crow, 1993), that it is located in a sex-specific region of homology, does so because the lack of recombination within such a region allows sequence variation on the Y to occur independently of that on the X chromosome.

McKeever (2000) has presented an analysis of the literature and a new family sample in support of a gene on the X chromosome, but discounts a locus on the Y chromosome on the grounds of Corballis' (1997) argument that a stable polymorphism on the Y would not be maintained within the population. However, the evidence summarised above from sex chromosome aneuploidies, the presence of a same-sex concordance effect and, given an X chromosomal locus, the fact of father to son transmission, all support such a locus. An alternative to Corballis' assumption that the variation on the Y chromosome is related to the DNA sequence (the basis of his rejection of the X–Y theory) is presented below.

THE SIGNIFICANCE OF THE Xq21.3 TRANSLOCATION AND THE Yp PARACENTRIC INVERSION

Genes that are present in homologous form on the X and the Y chromosomes were predicted to account for the phenomena of Turner's syndrome by Ferguson-Smith (1965). Their presence has been formally demonstrated within the pseudoautosomal region (Rouyer *et al.*, 1986; Rappold, 1993) and within the sex-specific region of the Y (Page *et al.*, 1984). In general, these regions have arisen as a result of translocations to the Y chromosome from the X. The evolutionary time–course of these chromosomal rearrangements has been charted by Lambson *et al.* (1992) (see the later chapter in these *Proceedings* by Carole Sargent *et al.*).

The regions of greatest interest with respect to evolutionary developments in humans are those that have been subject to change between the chimpanzee and *H. sapiens*. Two regions, the Xq21.3/Yp region of homology (Sargent *et al.*, 1996; Mumm *et al.*, 1997) and the 0.4 Mb pseudoautosomal region (PAR 2) at the telomeres of the long arms of the X and Y, stand out. Both representations on the Y chromosome were established after the separation of the hominid and chimpanzee lineages. Of these two, a gene within the Xq21.3/Yp region more readily explains a sex difference on the basis that sequence divergence can take place within the sex-specific regions of the X and Y but not within PAR 2.

The time–course of these changes is of great interest. The translocation from Xq21.3 to Yp has been estimated, on the basis of X–Y sequence divergence, at approximately 3 million years (Sargent *et al.*, 1996; Schwartz *et al.*, 1998). This was followed by a paracentric inversion, which has not been dated, that split and reversed the block in Yp, and by a series of changes within the homologous region around DXS214 (for details see the chapter by Carole Sargent *et al.*; Sargent *et al.*, 2001). It will be important to determine the temporal sequence of these changes in so far as this can be reconstructed. One question that can be asked about each variation is whether the change is universal in extant human populations. Where this is the case, as for the original translocation, the paracentric inversion and the changes at DXS214, the change has potential relevance to species-defining characteristics. One must assume that in each case the change has been selected. Beyond that it occurred after the separation of the chimpanzee and hominid lineages, no information has so far been obtained for the origin of pseudoautosomal region 2.

NEOTENY AND THE PLATEAU OF BRAIN GROWTH

How could these changes have influenced the evolution of language and *H. sapiens*? As Chaline *et al.* (1998) have pointed out, the morphological differences between the skulls of the great apes and those of the hominid series can be accounted for by a series of changes: (1) retreat and verticalisation of the face; (2) an increase in cranial capacity; and (3) tilting of the foramen magnum. These changes are apparent in landmark comparisons of skull outlines for the great ape and *Australopithecus* with *Homo erectus* and *H. sapiens* (Figure 4).

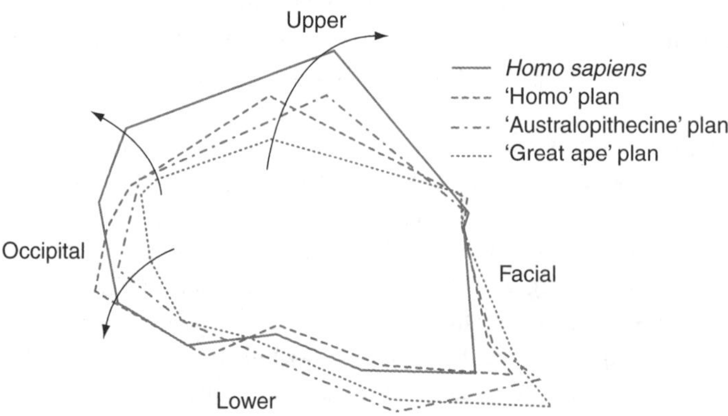

Figure 4. Comparisons of landmark-based skull outlines in *Homo sapiens*, hominids, Australopithecines and great apes. From Chaline *et al.* (1998).

The relative complexity of the differences between the adult skull outlines is much reduced in a comparison of the juvenile morphology of the great ape with *H. sapiens* (Figure 5).

The contrast is consistent with Bolk's (1926) concept that *H. sapiens* evolved by a process of neoteny, the prolongation into adult life of some features that are characteristic of infancy in a precursor species. In *H. sapiens* the topology of the skull, including the recessed conformation of the face (relative to that of the adult ape), is retained into the adult form. As Bolk suggested, it seems likely that this occurred by a process of 'heterochrony', a change in the relative timing of the components of development. In *H. sapiens* the development of the skull relative to the soma (into the topology of the adult chimpanzee) does not take place.

What selective factor could account for such a delay? It is implausible (and it does not seem to have been seriously argued) that there is selection for the facial features of infancy. What is surely more likely is that the facial features of the infant ape are retained in humans as a consequence of selection for some other characteristic, and that this characteristic relates to the brain itself rather than to its casing.

Holt *et al.* (1975) drew attention to the fact that in different primate species brain weight increases *pari passu* with body weight and then reaches a plateau. Across species the trajectory of brain growth is similar but the point of plateau differs, it is delayed in the chimpanzee relative to the macaque and in humans relative to the chimpanzee. A genetic change that accounted for the transition from one plateau to another could be at the core of the difference between two species (Figure 6).

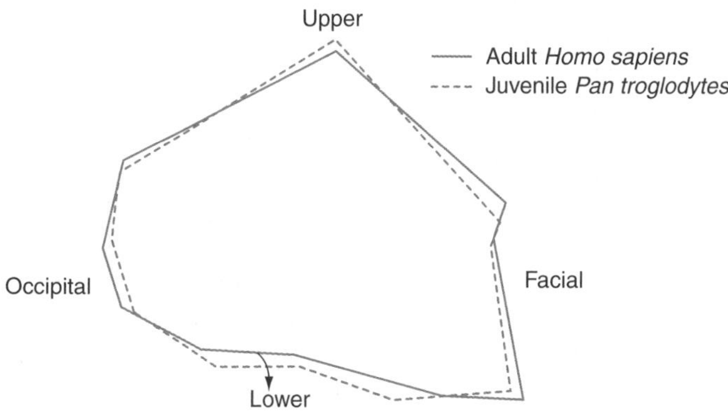

Figure 5. From Chaline *et al.* (1998).

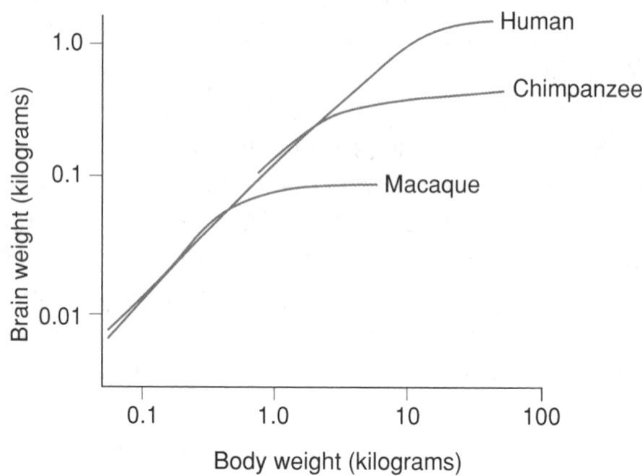

Figure 6. Brain weight in relation to body weight. From Holt *et al.* (1975).

SEXUAL SELECTION AS A MODIFIER

A critical change on the sex chromosomes, specifically one that it is located within a sex-specific region of homology between the X and the Y, introduces a new consideration, that the change will be differentially represented in the two sexes. There are two reasons why this is the case. First, in the case of a translocation from the X to the Y, such as has occurred with the Xq21.3/Yp block, the dosage in males will be double that in females (and also that in males who lack the translocation) because genes on one X chromosome are normally inactivated (see below). Secondly, lack of recombination between the X and the Y opens up the potential for sequence divergence. Either mechanism could lead to a sex difference in gene expression. Whatever its origin (gene dosage or sequence difference), such a sexual dimorphism is open to differential modification by mate choice.

Sexual selection can modify the primary change, first in males and then in females, and this modification proceeds along the dimension of variation that has been introduced by the initial change. Thus a novel sexual dimorphism becomes the focus of differential selection in the two sexes. The potential for quantitative change, constrained by the nature of the genetic innovation, enables a search for a new equilibrium. In the case of a rate-regulating process, for example the point of plateau in brain development, the attainment of the new equilibrium may be associated with a series of secondary adjustments, for example to facial morphology, as exemplified by neoteny in humans.

The hypothesis that speciation takes place as a consequence of sex chromosomal change followed by sexual selection overcomes two of the problems that saltational theories encounter. The first difficulty is explaining how an organism that has been subject to a discontinuous change can acquire a mate. According to the present theory the change occurs first in one individual in one sex, a male, and the innovation is then propagated to the progeny of that individual by a process of selection by females. Later there is a process of modification of the critical sequence or its expression in females as the males carrying the new sequence on their Y chromosomes increase in the population. Successive modification of the gene or its expression first in males and then in females appears to conform to the process of runaway sexual selection envisaged by Fisher (1930) and Lande (1981).

The second difficulty for saltational theories is to explain how a single discontinuous change can become integrated into a coherent and viable new form (the 'hopeful monster' problem; Goldschmidt, 1940). The present theory accounts for such integration on the basis that the single change (chromosomal or genetic) within a region of homology on the Y chromosome initiates a phase of sexual selection and heterochronic adjustment that equilibrates in a new plateau of maturation. This plateau will be associated with a changed suite of developmental trajectories (orientated around the core process) and its own sexual dimorphism. Thus a saltational change in a single male is selected by females and the effects (expressed through both X and Y copies of the gene) are modified by mate choice in both sexes. In this way a primary change in a rate-determining gene on the Y chromosome allows mate selection to determine a new playing field of sexual encounter.

A key question is the genetic consequence of the change on the Y chromosome for the original gene on the X chromosome. Genes on one X chromosome in females are subject to X inactivation, the process (dosage compensation) whereby the quantitative expression of genes on the X is equalised in males and females. A gene in a block on the Y that has transposed from the X is in an unusual situation, it escapes from X inactivation and is expressed in double dosage in the male. In general, one must suppose that an abrupt change in gene dose will be disadvantageous and that the great majority of such chromosomal rearrangements are rapidly selected out of the population. But in the case we are considering the relevant characteristic is positively selected and the gene on the Y is retained. In general it is observed that genes that are also present on the Y are protected from inactivation on the X (Davies, 1991), but the mechanism by which such protection is achieved is obscure. One possibility, that pairing of X and Y sequences in male meiosis plays a role (Crow, 1991; but see Burgoyne & McLaren, 1985), implies that the direction of the sequence on the Y relative to that on the X is significant, and therefore that in this case the paracentric inversion, which resulted in the realignment of X and Y sequences, is relevant. In discussing the differences between mammalian orders in the pattern of inactivation on the X of genes common to X and Y chromosomes, Jegalian & Page (1998) propose a mechanism that depends on successive changes (their figure 4) in response to selective pressures (unspecified) on first male and then female fitness. This sequence could reflect the role of sexual selection in the course of mammalian speciation. Therefore it is possible that epigenetic modification of critical sequences on the X chromosome is a necessary component of the process of sexual selection that follows the rearrangements of the X and the Y chromosome that distinguish related species.

SEXUAL SELECTION, SPECIATION AND SEX LINKAGE

The general concept that speciation and sexual selection are related has some support in the speciation literature. In attempting to explain the diversity of species of *Drosophila* on the Hawaiian archipelago, Kaneshiro & Boake (1987) concluded that the characteristics that differentiate species are those in which a change in a sexually dimorphic feature has become subject to mate choice. Similar arguments have been developed to account for the rapid speciation and diverse coloration of cichlid fish in the lakes of East Africa by Dominey (1984) and McKaye (1991), and in relation to song, morphology and plumage in birds by Price (1998). In each case it is argued that sexual selection has a role in generating pre-mating isolation of a new species from its precursor. Language in *H. sapiens* can be considered as an examplar of this sequence (Crow, 1996).

Other authors (Ringo, 1977; West-Eberhard, 1983; Turner & Burrows, 1995; Zink, 1996; Kaneshiro, 1997; Carson, 1997) have discussed aspects of the relationship between sexual selection and speciation. The thread of continuity in these discussions is that change in some identifiable characteristic associated with one sex initiates a process of sexual selection that separates the species. Here it is proposed that the genetic foundation of this sequence (as applied to mammals) is that a primary change in the Y copy of an X–Y homologous gene (1) generates a novel sexual dimorphism and (2) is subject to female choice, and that the X–Y difference becomes the target of a process of sexual selection with the runaway characteristics described by Fisher (1930).

PATERSON'S SPECIFIC MATE RECOGNITION CONCEPT

These conclusions are relevant to a definition of a species that casts new light on both the isolation or 'biological' species concept and the saltational theory: Paterson's (1985) specific mate recognition concept, the notion that what defines a sexually reproducing species, and differentiates one species from another, is the mechanism for recognising a mate (Lambert & Spencer, 1994). Paterson (1985) defined a species as 'that most inclusive population of individual, bi-parental organisms which share a common fertilisation system'. The specific mate recognition concept has been described as a 'fundamental property of a species, established at the time of speciation, and essentially unaltered thereafter' (Turner, 1994).

Sex chromosomal change has relevance to the concept. A change on the sex chromosomes can introduce novelty into the mate recognition system that is open to rapid and differential modification in the two sexes. In discussing the role of chromosome change in speciation in Hawaiian drosophilids, Carson (1970) considered the possibility that 'the founder event may be accomplished by a single propagule, probably a single fertilized female', although he later (Carson, 1997) modified this view in judging that 'genetic effects might produce a species by one or a few catastrophic steps ... to be an unrealistic expectation'. According to the concept outlined here, the primary change is on the Y chromosome, and the founder event occurs in a male.

Paterson (according to Carson, 1995) argued that reproductive success within each species is well served by a:

> specific mate recognition system comprising a number of co-adapted stages. The co-adaptation referred to is intersexual: a signal from one potential sex partner evokes a particular response from the other; these events may then culminate in a chain of alternating signals between the individuals resulting in the success or failure of copulation.

Such a description may be applied, in the case of *H. sapiens*, to language.

CONCLUSIONS

The asymmetry of the human brain is a feature that is not shared by other primates. It is variable between individuals, and this variation influences the acquisition of words, arguably the core feature of the species-defining characteristic of language.

On the basis of the neuropsychological deficits associated with the sex chromosome aneuploidies, an X–Y homologous gene for asymmetry was proposed, and is supported by the transmission of handedness within families. Such a gene can explain the sex difference in verbal and other aspects of cognitive ability and the faster development of the female brain; it will be subject to sexual selection. According to this concept the human brain evolved by a process of neoteny (delayed maturation) under the influence of mate choice for ability to communicate.

The speciation of modern *H. sapiens* is proposed as an instance of sexual selection acting on recent sex chromosomal change to establish a new 'specific mate recognition system': language. A change in a region of X–Y homology introduces novelty into the mate recognition system that is open to rapid and differential modification in the two sexes. A candidate region in *H. sapiens* is that block of sequences within Xq21.3 that transposed to the Y chromosome short arm after the separation of the chimpanzee and hominid lineages and was subject to a subsequent paracentric inversion in Yp; a candidate gene is protocadherinXY, expressed in different forms on the X and the Y chromosomes. The role for X–Y homology postulated in sexual selection and speciation draws attention to epigenetic regulation of such genes as relevant to the understanding of speciation-related variation and pathologies.

DISCUSSION

Questioner: There seems to be a selective advantage to the choice of right-hand lateralisation: one hand for eating and one for cleaning the body. Are there any cross-cultural studies of handedness?

Crow: The best cross-cultural studies are by Perelle & Ehrman (1994). There are some problems with these data, but they conclude that the population-based direction of the right-handedness is consistent across cultures.

Skuse: As you know, we've been interested in sex chromosome aneuploidies and we've studied 100 females with Turner's syndrome. According to Marian Annett's theory, the proportion of left-handers should be somewhat greater than 10% in the general population. The proportion of left-handers in the Turner's girls is exactly the same as that with females with normal X chromosomes.

Crow: I'm not supporting all aspects of Marian Annett's theory, but what you also show is the same finding that has been in the literature from the work of Money (1964, 1993), that is deficits of spatial ability in Turner's individuals. You find that irrespective of whether the X chromosome comes from the mother or the father.

Skuse: No, that's not exactly right.

Crow: That's strong evidence that there is an asymmetry factor there.

Skuse: I wouldn't interpret it that way. You are absolutely right, there are verbal advantages relative to performance deficits, it's about 80%, not 100%. In Klinefelter's syndrome most people would say there is an overall disadvantage, but the level of IQ is somewhat lower than what you've shown there.

Crow: But it's the relative IQ (see also Netley, 1998).

Skuse: Yes, you're absolutely right. There's something to be explained there. But the way you have explained it is that there's a speciation event which involved a block of X-linked genes transferred to the Y and then an inversion occurring a half million years ago. But how can something happening on the recombining part of Y chromosome explain what is happening on the X chromosome? How does what has led to asymmetry in the male lead to the equivalent change in women?

Crow: That's where, if it works at all, sexual selection has to come in. So you've got a new situation where there is a desirable characteristic selected for first by females and then by males.

Questioner: Are you saying there are two separate events: one on the Y and one also occurring on the X?

Crow: Yes. I accept that because I think the prediction is that not only should there be divergence on the Y but the gene on the X should now diverge from the gene on the primate X with which we can compare it. The divergence may be in the sequence or it may be in the epigenetic modification of its expression.

Wolpert: Within the terminology of Bickerton, there is protolanguage and language. Would you say that within these specific mate recognition systems it is necessary to have a language system rather than a protolanguage? We are talking here about the emergence of linguistic abilities and it seems unclear why a language rather than a protolanguage would be a preferential mate recognition system.

Crow: I don't see that this is a problem; both of them could be selected for.

Questioner: Are you saying that one pre-mutation in a stone age man gave us language? No evolutionary biologist would explain why echolocation is present in bats on the basis of one pre-mutation. No evolutionary biologist would explain sonar contact in whales in the same manner. There have to be selection pressures.

Crow: Let me explain the implications of a chromosomal event such as this. Clearly it has been selected. This is not on the Y in a chimpanzee. It is on the Y in every human male. It is possible that there are no genes in this region, in which case it will be completely neutral. Then it would be like heterochromatin on the Y and of no consequence at all. But if there are genes in this region then the translocation and paracentric inversion must have affected their control and expression first in males then in females as I have described. At some stage there was positive selection. It is extremely interesting that this translocation is what got selected. What we imagine is that many such changes on the sex chromosomes may occur and then get lost because they are not advantageous and are selected out. But this is exceptional; somewhere within this block the sequence is noteworthy because that is what has survived.

Wolpert: So are you saying that the first person to speak was a male and that this person was a famous person in that way?

Crow: Yes.

References

Annett, M. (1978) *A Single Gene Explanation of Right and Left Handedness and Brainedness.* Coventry: Lanchester Polytechnic.

Annett, M. (1985) *Left, Right. Hand and Brain: The Right-shift Theory*. London: Lawrence Erlbaum.

Annett, M. (1995) The right-shift theory of a genetic balanced polymorphism for cerebral dominance and cognitive processing. *Current Psychology of Cognition*, **14**, 427–80.

Annett, M. & Annett, J. (1991) Handedness for eating in gorillas. *Cortex*, **27**, 269–75

Annett, M. & Kilshaw, D. (1984) Lateral preference and skill in dyslexics: implications of the right-shift theory. *Journal of Child Psychology and Psychiatry*, **25**, 357–77.

Annett, M. & Manning, M. (1990) Reading and a balanced polymorphism for laterality and ability. *Journal of Child Psychology and Psychiatry*, **31**, 511–29.

Bishop, D.V. (1990) *Handedness and Developmental Disorder*. London: MacKeith.

Bolk, L. (1926) *Das Problem der Menschwerdung*. Jena: Gustav Fischer.

Brown, R., Colter, N., Corsellis, J.A.N. *et al.* (1986) Postmortem evidence of structural brain changes in schizophrenia. Differences in brain weight, temporal horn area, and parahippocampal gyrus compared with affective disorder. *Archives of General Psychiatry*, **43**, 36–42.

Burgoyne, P.S. & McLaren, A. (1985) Does X–Y pairing in male meiosis protect the paired region of the X chromosome from subsequent X-inactivation? *Human Genetics*, **70**, 82–3.

Calnan, M. & Richardson, K. (1976) Developmental correlates of handedness in a national sample of 11-year-olds. *Annals of Human Biology*, **3**, 329–42.

Carson, H.L. (1970) Chromosome tracers of the origin of species. *Science*, **168**, 1414–28.

Carson, H.L. (1995) Fitness and the sexual environment. In: *Speciation and the Recognition Concept* (eds D. M. Lambert & H. G. Spencer), pp. 123–37. Baltimore: John Hopkins University Press.

Carson, H.L. (1997) Sexual selection: a driver of genetic change in Hawaiian *Drosophila*. *Journal of Heredity*, **88**, 343–52.

Cerone, L.J. & McKeever, W.F. (1999) Failure to support the right-shift theory's hypothesis of a 'heterozygote advantage' for cognitive abilities. *British Journal of Psychology*, **90**, 109–23.

Chaline, J., David, B., Magniez-Jannin, F. *et al.* (1998) Quantification of the morphologic evolution of the Hominid skull and heterochronies. *CR Academy of Sciences Paris*, **326**, 291–8.

Corballis, M.C. (1997) The genetics and evolution of handedness. *Psychological Review*, **104**, 714–27.

Corballis, M.C., Lee, K., McManus, I.C. & Crow, T.J. (1996) Location of the handedness gene on the X and Y chromosomes. *American Journal of Medical Genetics (Neuropsychiatric Genetics)*, **67**, 50–2.

Crow, T.J. (1983) Is schizophrenia an infectious disease? *Lancet*, **342**, 173–5.

Crow, T.J. (1984) A re-evaluation of the viral hypothesis: is psychosis the result of retroviral integration at a site close to the cerebral dominance gene? *British Journal of Psychiatry*, **145**, 243–53.

Crow, T.J. (1986) Left brain, retrotransposons, and schizophrenia. *British Medical Journal*, **293**, 3–4.

Crow, T.J. (1987) Pseudautosomal locus for psychosis? *Lancet*, **2**, 1532.

Crow, T.J. (1988) Sex chromosomes and psychosis. The case for a pseudoautosomal locus. *British Journal of Psychiatry*, **153**, 675–83.

Crow, T.J. (1989) Pseudoautosomal locus for the cerebral dominance gene. *Lancet*, **2**, 339–40.

Crow, T.J. (1990) Temporal lobe asymmetries as the key to the etiology of schizophrenia. *Schizophrenia Bulletin*, **16**, 433–43.

Crow, T.J. (1991) Protection from X inactivation. *Nature*, **353**, 710.

Crow, T.J. (1993) Sexual selection, Machiavellian intelligence and the origins of psychosis. *Lancet*, **342**, 594–8.

Crow, T.J. (1995) The case for an X–Y homologous gene, and the possible role of sexual selection in the evolution of language. *Current Psychology of Cognition*, **14** , 775–81.

Crow, T.J. (1996) Language and psychosis: common evolutionary origins. *Endeavour*, **20**, 105–9.

Crow, T.J. (1998a) Sexual selection, timing and the descent of man: a genetic theory of the evolution of language. *Current Psychology of Cognition*, **17**, 1079–114.

Crow, T.J. (1998b) Why cerebral asymmetry is the key to the origin of *Homo sapiens*: how to find the gene or eliminate the theory. *Current Psychology of Cognition*, **17**, 1237–77.

Crow, T.J. (2000a) Did *Homo sapiens* speciate on the Y chromosome? *Psycoloquy*, **11**, 1.

Crow, T.J. (2000b) Schizophrenia as the price that *Homo sapiens* pays for language: a resolution of the central paradox in the origin of the species. *Brain Research Reviews*, **31**, 118–29.

Crow, T.J., Crow, L.R., Done, D.J. & Leask, S.J. (1998) Relative hand skill predicts academic ability: global deficits at the point of hemispheric indecision. *Neuropsychologia*, **36**, 1275–82.

Crow, T.J., DeLisi, L.E. & Johnstone, E.C. (1989) Concordance by sex in sibling pairs with schizophrenia is paternally inherited. Evidence for a pseudoautosomal locus. *British Journal of Psychiatry*, **155**, 92–7.

Davies, K. (1991) The essence of inactivity. *Nature*, **349**, 15–6.

Dominey, W.J. (1984) Effects of sexual selection and life histories on speciation: species flocks in African cichlids and Hawaiian *Drosophila*. In: *Evolution of Fish Species Flocks* (eds A. A. Echelle & I. Kornfield), pp. 231–49. Maine: Orino Press.

Ferguson-Smith, M.A. (1965) Karyotype–phenotype correlations in gonadal dysgenesis and their bearing on the pathogenesis of malformations. *Journal of Medical Genetics*, **2**, 142–55.

Fisher, R.A. (1930) *The Genetical Theory of Natural Selection*. Oxford: Oxford University Press.

Goldschmidt, R. (1940) *The Material Basis of Evolution*, reprinted 1982. New Haven: Yale University Press.

Halpern, D.F. (1992) *Sex Differences in Cognitive Abilities*. New Jersey: Lawrence Erlbaum.

Hardyck, C., Petrinovich, L.F. & Goldman, R.D. (1976) Left handedness and cognitive deficit. *Cortex*, **12**, 266–79.

Harshman, R.A., Hampson, E. & Berenbaum, S.A. (1983) Individual differences in cognitive abilities and brain organization. I. Sex and handedness differences in ability. *Canadian Journal of Psychology*, **37**, 144–92.

Holder, M.K. (1999) Influences and constraints on manual asymmetry in wild African primates: reassessing implications for the evolution of human handedness and brain lateralization (Dissertation Abstracts International Section A). *Humanities and Social Sciences*, **60**, 0470.

Holt, A.B., Cheek, D.B., Mellits, E.D. & Hill, D.E. (1975) Brain size and the relation of the primate to the non-primate. In: *Fetal and Post-Natal Cellular Growth: Hormones and Nutrition* (ed. D B. Cheek), pp. 23–44. New York: J. Wiley.

Jegalian, K. & Page, D.C. (1998) A proposed mechanism by which genes common to mammalian X and Y chromosomes evolve to become X inactivated. *Nature*, **394**, 776–80.

Kaneshiro, K.Y. (1997) RCL Perkins' legacy to evolutionary research on Hawaiian Drosophilidae (Diptera). *Pacific Sciences*, **51**, 450–61.

Kaneshiro, K.Y. & Boake, C.R.B. (1987) Sexual selection: issues raised by Hawaiian *Drosophila*. *Trends in Ecology and Evolution*, **2**, 207–11.

Kretschmann, H.F., Schleicher, A., Wingert, F., Zilles, K. & Loeblich, H.-J. (1979) Human brain growth in the 19th and 20th century. *Journal of the Neurological Sciences*, **40**, 169–88.

Lambert, D.M. & Spencer, H.G. (1994) *Speciation and the Recognition Concept: Theory and Application*. Baltimore: Johns Hopkins University.

Lambson, B., Affara, N.A., Mitchell, M. & Ferguson-Smith, M.A. (1992) Evolution of DNA sequence homologies between the sex chromosomes in primate species. *Genomics*, **14**, 1032–40.

Lande, R. (1981) Models of speciation by sexual selection on polygenic traits. *Proceedings of the National Academy of Sciences of the USA*, **78**, 3721–62.

Levy, J. (1969) Possible basis for the evolution of lateral specialisation of the human brain. *Nature*, **224**, 614–5.

Maccoby, E.E. & Jacklin, C.N. (1975) *The Psychology of Sex Differences*. Oxford: Oxford University Press.

McGrew, W.C. & Marchant, L.F. (1997) On the other hand: current issues in and meta-analysis of the behavioral laterality of hand function in nonhuman primates. *Yearbook of Physical Anthropology*, **40**, 201–32.

McKaye, K.R. (1991) Sexual selection and the evolution of the cichlid fishes of Lake Malawi, Africa. In: *Behaviour, Ecology and Evolution* (ed. M. H. A. Keenleyside), pp. 241–57. London: Chapman & Hall.

McKeever, W.F. (2000) A new family handedness sample with findings consistent with X-linked transmission. *British Journal of Psychology*, **91**, 21–39.

McManus, I.C. (1991) The inheritance of left-handedness. In: *Biological Asymmetry and Handedness (CIBA Foundation Symposium 162)* (eds Q. R. Bock & J. Marsh), pp. 251–81. Chichester: Wiley.

McManus, I.C. & Mascie-Taylor, C.G.N. (1983) Biosocial correlates of cognitive abilities. *Journal of Biosocial Science*, **15**, 289–306.

McManus, I.C., Shergill, S. & Bryden, M.P. (1993) Annett's theory that individuals heterozygous for the right-shift gene are intellectually advantaged: theoretical and empirical findings. *British Journal of Psychology*, **84**, 517–37.

Marchant, L.F. & McGrew, W.C. (1996) Laterality of limb function in wild chimpanzees of Gombe National Park: comprehensive study of spontaneous activities. *Journal of Human Ecology*, **30**, 427–43.

Miller, E. (1971) Handedness and the pattern of human ability. *British Journal of Psychology*, **62**, 111–12.

Money, J. (1964) Two cytogenetic syndromes. I. Intelligence and specific factor quotients. *Journal of Psychiatric Research*, **2**, 223–31.

Money, J. (1993) Specific neurocognitional impairments associated with Turner (45, X) and Klinefelter (47, XXY) syndromes: a review. *Social Biology*, **40**, 147–51.

Mumm, S., Molini, B., Terrell, J., Srivastava, A. & Schlessinger, D. (1997) Evolutionary features of the 4Mb Xq21.3 X–Y homology region revealed by a map at 60-kb resolution. *Genome Research*, **7**, 307–14.

Netley, C.T. (1986) Summary overview of behavioural development in individuals with neonatally identified X and Y aneuploidy. *Birth Defects Original Article Series*, **22**, 293–306.

Netley, C.T. (1998) Sex chromosome aneuploidy and cognitive developments. *Current Psychology of Cognition*, **17**, 1190–7.

Netley, C.T. & Rovet, J. (1982) Atypical hemispheric lateralization in Turner syndrome subjects. *Cortex*, **18**, 377–84.

Orton, S.T. (1937) *Reading, Writing and Speech Problems in Children*. New York: Norton.

Page, D.C., Harper, M.E., Love, J. & Botstein, D. (1984) Occurrence of a transposition from the X-chromosome long arm to the Y-chromosome short arm during human evolution. *Nature*, **331**, 119–23.

Palmer, R.E. & Corballis, M.C. (1996) Predicting reading ability from handedness measures. *British Journal of Psychology*, **87**, 609–20.

Paterson, H.E.H. (1985) The recognition concept of species. In: *Species and Speciation* (ed. E. S. Vrba), pp. 21–9. Pretoria: Transvaal Museum Monograph.

Penrose, L.S. (1942) Auxiliary genes for determining sex as contributory causes of mental illness. *Journal of Mental Science*, **88**, 308–16.

Perelle, I.B. & Ehrman, L. (1994) An international study of human handedness: the data. *Behaviour Genetics*, **24**, 217–27.

Price, T. (1998) Sexual selection and natural selection in bird speciation. *Philosophical Transactions of the Royal Society of London (Biological)*, **353**, 251–60.

Provins, K.A., Milner, A.D. & Kerr, P. (1982) Asymmetry of manual preference and performance. *Perceptual and Motor Skills*, **54**, 179–94.

Rappold, G.A. (1993) The pseudoautosomal regions of the human sex chromosomes. *Human Genetics*, **92**, 315–24.

Resch, F., Haffner, J., Parzer, P., Pfueller, U., Strehlow, U. & Zerahn-Hartung, C. (1997) Testing the hypothesis of relationships between laterality and ability according to Annett's right-left theory: findings in an epidemiological sample of young adults. *British Journal of Psychology*, **88**, 621–35.

Ringo, J.M. (1977) Why 300 species of Hawaiian *Drosophila*? The sexual selection hypothesis. *Evolution*, **31**, 694–6.

Rosenthal, D. (1970) *Genetic Theory and Abnormal Behaviour*. New York: McGraw-Hill.

Rouyer, F., Simmler, M.-C., Johnsson, C., Vergnaud, G., Cooke, H.J. & Weissenbach, J.A. (1986) A gradient of sex linkage in the pseudoautosomal region of the sex chromosomes. *Nature*, **319**, 291–5.

Sargent, C.A., Boucher, C.A., Blanco, P. *et al.* (2001) Characterization of the human Xq21.3/Yp11 homology block and conservation of organization in primates. *Genomics*, **73**, 77–85.

Sargent, C.A., Briggs, H., Chalmers, I.J., Lambson, B., Walker, E. & Affara, N.A. (1996) The sequence organization of Yp/proximal Xq homologous regions of the human sex chromosomes is highly conserved. *Genomics*, **32**, 200–9.

Schwartz, A., Chan, D.C., Brown, L.G., Alagappan, R., Pettay, D., Disteche, C., McGillivray, B., de la Chapelle, A. & Page, D.C. (1998) Reconstructing hominid Y evolution: X-homologous block, created by X–Y transposition, was disrupted by Yp inversion through LINE–LINE recombination. *Human Molecular Genetics*, **7**, 1–11.

Stringer, C. & McKie, R. (1996) *African Exodus: The Origins of Modern Humanity*. London: J. Cape.

Turner, A. (1994) The species in palaeontology. In: *Speciation and the Recognition Concept: Theory and Application* (eds D. M. Lambert & H. G. Spencer), pp. 57–70. Baltimore: Johns Hopkins University.

Turner, G.F. & Burrows, M.T. (1995) A model of sympatric speciation by sexual selection. *Proceedings of the Royal Society of London B*, **260**, 287–92.

West-Eberhard, M.J. (1983) Sexual selection, social competition and speciation. *Quarterly Review of Biology*, **58**, 155–83.

Whittington, J.E. & Richards, P.N. (1991) Mathematical ability and the right-shift theory of handedness. *Neuropsychologia*, **29**, 1075–82.

Zangwill, O.L. (1960) *Cerebral Dominance and its Relation to Psychological Function*. Edinburgh: Oliver & Boyd.

Zink, R.M. (1996) Species concepts, speciation, and sexual selection. *Journal of Avian Biology*, **27**, 1–6.

What Can the Y Chromosome Tell Us about the Origin of Modern Humans?

CHRIS TYLER-SMITH

Summary. Crow has proposed that a change to an X–Y homologous gene was an important event in human speciation. This chapter reviews how our current understanding of the human Y chromosome can contribute to an evaluation of this hypothesis. Human and ape Y lineages are generally believed to have split about 5–7 million years ago, while extant human Y lineages trace back to a common ancestor that probably lived between 40 and 200 thousand years ago. Between these dates, two substantial segments of DNA on the Y chromosome were duplicated on the Y: the Yq pseudoautosomal region and the Xq/Yp homology region. The former does not contain any good candidate speciation genes but the latter may, and these are discussed by Carole Sargent *et al.* later in these *Proceedings*. The Xq–Yp transposition probably occurred soon after the ape–human split and, at the same time or subsequently, was divided in two by an inversion. An exhaustive evaluation of the genes contained in this region provides the best way to test Crow's hypothesis.

INTRODUCTION

Why consider the Y chromosome?

DURING THE LAST DECADE, Crow has developed the hypothesis that a change in the expression of a pair of genes on the Y and Y chromosomes was a critical event in the evolution of language and the speciation of modern humans; these ideas are outlined in Tim Crow's chapter in these *Proceedings*. If the relevant genes had already been identified, there would be no need for the present chapter. The most direct test of the hypothesis is to search for such genes, and this approach is described by Carole Sargent *et al.* in the next chapter of these *Proceedings*. In the absence of clear evidence for the crucial genes, the present chapter reviews relevant aspects of our knowledge of the human Y chromosome and considers what this indirect information can contribute to the debate.

Proceedings of the British Academy, **106**, 217–229, © The British Academy 2002.

Genetic properties of the Y chromosome

Humans have 46 chromosomes, consisting of one pair of sex chromosomes (designated X and Y) and 22 other pairs (designated autosomes). The Y chromosome carries a gene, *SRY* (*S*ex-determining *R*egion of the *Y* chromosome; Sinclair *et al.*, 1990), that provides the primary sex-determining signal and directs development away from the default female pathway to the male pathway; thus in females the karyotype is 46,XX, while in males it is 46,XY. The primary sex-determining role of the Y chromosome has several important consequences for its genetics and evolution, some obvious but others less so. *SRY* must be haploid (present in only one copy per genome) in order for this sex-determining mechanism to work. It therefore has no homologue and so cannot recombine. However, recombination is required for successful meiosis, and so the Y chromosome itself must recombine. This paradox is resolved by dividing the Y into distinct sections: pseudoautosomal, which recombines with the X chromosome, and Y-specific, which carries *SRY* and does not recombine. A priori, it would seem that one pseudoautosomal region and one Y-specific region would be sufficient, and their sizes cannot be predicted. In fact, the human Y has two pseudoautosomal regions of 2.4 Mb and 0.3 Mb, one at each end, but the majority of the chromosome, the size of which is conventionally estimated at about 60 Mb, shows Y-specific inheritance (Figure 1a). In addition to *SRY*, a considerable number of other genes are located in the Y-specific region. For several reasons, it is difficult to give the exact number of these: despite the progress of the human genome sequencing project, our knowledge is incomplete and new genes are still being identified. Some genes are duplicated or are members of multigene families, and it is not always clear whether a gene is active or an inactive pseudogene. Furthermore, polymorphisms are found in the population so that a gene may be present in some individuals and absent from others, or the number of copies in a multigene family may vary between individuals. Nevertheless, about 20 different protein-coding genes with diverse functions have been described and are shown in Figure 1d. This is, however, a small number compared with other chromosomes. Chromosome 22, for example, is smaller but at least 545 genes were predicted from the sequence (Dunham *et al.*, 1999).

A consequence of the lack of recombination over most of the Y chromosome and the low gene density, is that large-scale rearrangements are tolerated more readily than on other chromosomes. These rearrangements can take forms such as duplications, deletions, inversions and translocations. Because of the low gene density, the rearrangements themselves are unlikely to disrupt the expression of crucial genes. In the absence of recombination, changes in the position of the centromere do not lead to the generation of acentric or dicentric chromosomes.

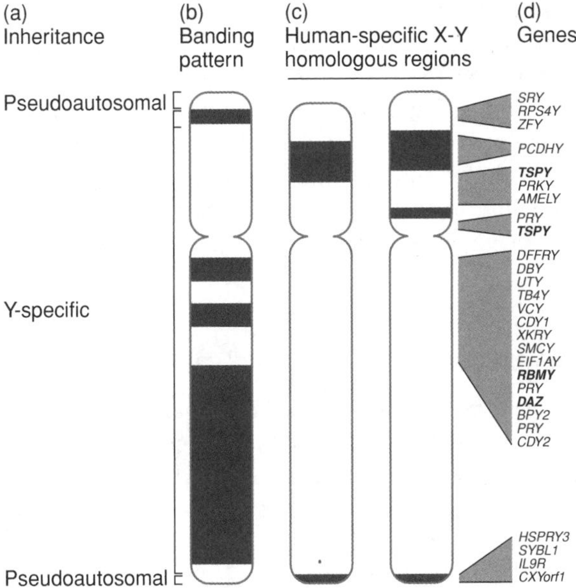

Figure 1. (a) Inheritance of different sections of the Y chromosome. (b) Conventional representation of the banding pattern. (c) X–Y homologous regions specific to the human Y lineage, indicated in black. The proximal Yp region is absent from some males (left). (d) Genes from the Y-specific and Yq pseudoautosomal regions (Lahn & Page, 1997; Jobling & Tyler-Smith, 2000). Loci shown in bold are present in multiple copies. Several genes are known from the Yp pseudoautosomal region, but are not listed here.

A consequence of the haploid state of the Y chromosome is that there are fewer copies of the Y present in the population than of the X or any autosome: a couple have, between them, four copies of each autosome and three copies of the X, but only one copy of the Y. This means that changes in gene frequency due to chance (genetic drift) will occur more rapidly on the Y than on other chromosomes, and so variant Y chromosomes will have a better chance of being fixed in the population. All genomes or sections of genomes that do not recombine must be able to trace their ancestry back to a single DNA molecule. The time taken is called the time to the most recent common ancestor (TMRCA), divergence time or coalescence time. If everything else is equal, this time will depend on the effective population size, and so the Y coalescence time will be one-third of that for a region on the X, or one-quarter of that of an autosome.

A final important property is also a consequence of the lack of recombination. If a variant Y chromosome has a selective advantage in the population, it will tend to increase in frequency and may become fixed. If this occurs, all polymorphisms that happen to be present on the selected Y will also be fixed, and all other variants will be lost from the population, even though these extra variants themselves may be neutral. This is sometimes referred to as a selective sweep or hitch-hiking. If such an event had occurred in the recent past, an unexpectedly low amount of variation would be seen on current Ys.

Thus, overall, the Y chromosome is distinguished from the rest of the genome by its ability to tolerate large-scale rearrangements and the higher probability of such changes being fixed. In effect, it is reinventing itself more rapidly.

What can Y chromosomal studies contribute?

Studies of the genomes, including the Y chromosomes, of contemporary humans and apes can contribute to our understanding of three relevant areas: (1) the timing of the divergence between apes and humans; (2) the changes that have occurred to the Y chromosome along the human lineage; and (3) the timing of the divergence of modern human Y chromosomes (Figure 2). The next three sections will consider these areas.

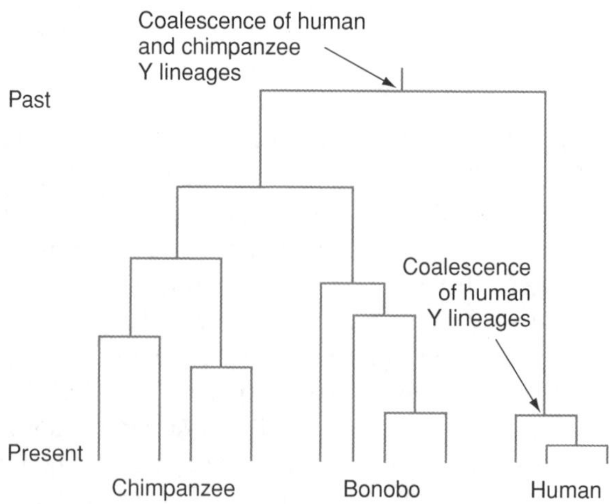

Figure 2. Schematic representation of ape and human Y lineages. Representation is schematic because no suitable data are available for ape Ys.

THE APE–HUMAN Y CHROMOSOMAL COALESCENCE

Comparison of ape and human Y chromosomes

So far, our knowledge of the overall organisation of ape and human Y chromosomes is derived mainly from cytogenetic studies. Comparisons of high-resolution (late prophase) G-banded chromosomes were described by Yunis & Prakash (1982). They observed large differences in the heterochromatic part of the Y, but judged the euchromatic part to be homologous between the different species. However, a more recent comparison, carried out using *in situ* hybridisation (Archidiacono *et al.*, 1998), which allows the positions of specific sequences on the chromosome to be visualised, showed that the situation was more complex. Sequences from the short arm of the human Y were homologous to the long arm of the chimpanzee Y. In addition, several differences in the copy number and order of the sequences were also detected, but Archidiacono *et al.* (1998) did not attempt to propose an evolutionary scenario for the chromosome.

Timing of the ape–human coalescence

The date of the split between apes and humans can be estimated by measuring the DNA sequence difference (or an indirect measure of this, such as protein structure difference) and translating this into years using an event in the fossil record that is considered well-calibrated. The initial molecular dating of the ape–human split used antisera to serum albumins (Sarich & Wilson, 1967) and an assumed divergence between hominoids and Old World monkeys of 30 million years ago. It produced a date of about 5 million years ago, which initially caused some surprise, but has subsequently been supported by a large number of other studies, and a date of 5–7 million years ago is now generally accepted. One exception is the work of Arnason *et al.* (1996), who estimated a divergence time of about 13.5 million years ago using mitochondrial DNA sequence information and the split between whales and cows at 60 million years ago as their standard.

CHANGES ON THE HUMAN Y LINEAGE

Although their structures are incompletely understood, human and chimpanzee Y chromosomes have clearly accumulated many gross differences (Archidiacono *et al.*, 1998). Half of these have probably occurred on the human lineage, and the human-specific changes can be identified by comparison with additional apes such as gorilla or orang-utan. Among them, the differences resulting in new X–Y homologous regions are of particular interest here.

The Xq–Yp translocation and Yp inversion

The focus of the Crow hypothesis is on a region of Xq that has transposed onto Yp since the divergence between apes and humans (Page *et al.*, 1984). This region is being studied intensively by Affara and colleagues, and their work is described in the next chapter. Here, I will comment only on three aspects.

First, when did the transposition occur? It is present in all human Y chromosomes that have been examined, and no chimpanzee Ys, so must have taken place after the divergence between the human and chimpanzee Y lineages and before the TMRCA of modern human Ys (Figure 2). This would place it after 5–7 million years ago and before 40–200 thousand years ago (see below). These, however, are wide limits. An alternative, and to some extent independent, date can be obtained from the extent of sequence divergence between the X and Y copies. This date is only partially independent because it may be calibrated against the same events in the fossil record as the ape–human divergence. Schwartz *et al.* (1998) measured the X–Y similarity in about 5 kb of sequence at 99.3 ± 0.2% and translated this into a date of *c.* 3–4 million years ago using an estimate of 0.2% divergence per million years (Shimmin *et al.*, 1993). A comparison of 100 kb of sequence from two bacterial artificial chromosome (BAC) clones (AC004388 from the X and AC010722 from the Y) reveals a similarity of 99.0%, which would correspond to a slightly older date of *c.* 5 million years ago using the same divergence rate. This suggests that the translocation occurred soon after the split.

Secondly, the Xq/Yp homology region forms one contiguous block on the X, but two distinct blocks on the Y, one distal and one proximal. It is thought that the Y copy has undergone an inversion with one endpoint within the homology region and the other elsewhere on Yp (Schwartz *et al.*, 1998). When did this inversion occur? It cannot have happened before the transposition, but could have been contemporaneous with it, or have occurred at any time between then and the modern human Y TMRCA. It is difficult to see how any analysis of modern Ys can narrow this interval.

Thirdly, the proximal Yp homology region is entirely absent from some males as a result of a deletion polymorphism (Figure 1c) (Santos *et al.*, 1998). The individuals carrying this deletion are quite rare (two out of 350 in the initial study, 0.6%) and their phenotypes have not been studied in detail, but they were detected in a survey of normal males and there is no evidence of any abnormalities. Thus it is unlikely that a crucial gene lies within this region.

The Xq–Yq pseudoautosomal region

A second relevant difference between ape and human Y chromosomes is that the human Y has a long arm pseudoautosomal region (Yq PAR; Figure 1) but ape Ys do not. Like the Xq/Yp homology region, the Yq PAR was formed between the times of the ape–human Y divergence and the human Y coalescence. However, unlike the Xq/Yp region, it continues to undergo exchange with the X (Freije *et al.*, 1992) and thus has not diverged in sequence, so that the timing of its origin cannot be refined using sequence information. The region has now been completely sequenced (Ciccodicola *et al.*, 2000) and is 330 kb long. Four genes have been predicted or detected: *HSPRY3*, *SYBL1*, *IL9R* and *CXYorf1*. *HSPRY* and *SYBL1* appear to be inactive on the Y, and are thus poor candidates for genes leading to speciation through their presence on the Y. This conclusion must, however, be considered provisional because gene activity has only been measured in a small number of cell types, and it remains possible that one or both of these Y genes are active in cells that have not yet been tested. *IL9R* and *CXYorf1* are probably both active on the Y, but additional factors make them poor candidates for speciation genes. *IL9R* is a growth factor for some of the cells in the blood, and thus seems excluded by its function. The function of *CXYorf1* is unknown, and its sequence provides no clues, but highly homologous genes are present on at least six other chromosomes, and it seems unlikely that such a small increase in gene number (and gene product level) could have a large phenotypic effect.

Thus the Yq PAR provides a human-specific region of X–Y homology, but no good candidate genes for speciation events.

THE HUMAN Y CHROMOSOMAL COALESCENCE

As outlined in the introduction, all copies of the non-recombining section of the Y chromosome must have descended from a single individual. Any change to the Y that contributed to speciation must have been present in this individual. It is therefore important to establish when this individual lived: the coalescence time of surviving Y chromosomes.

Variation on the human Y chromosome

Y DNA variation takes many forms, but the variants that have been most useful have been binary polymorphisms and microsatellites (also called short tandem repeats; STRs). Binary polymorphisms derive their name from the finding that they have just two alleles; the most abundant are single nucleotide polymorphisms (SNPs), where one nucleotide is replaced by another (for example

T by C), but they also include insertions or deletions of a few nucleotides and insertions of retroposon sequences. Microsatellites consist of small units (for example GATA) that are repeated in tandem. The number of copies varies between individuals: for example 11 on one Y chromosome and 12 on another. Binary polymorphisms have low mutation rates, which have been estimated from comparisons of human and chimpanzee DNA at 1.2×10^{-9} per nucleotide per year, assuming no selection and a split 5 million years ago (Thomson *et al.*, 2000). If the generation time is 25 years, this would correspond to 3×10^{-8} per nucleotide per generation. In contrast, microsatellites have much higher mutation rates, measured in modern families at 2.1×10^{-3} per locus per generation (Heyer *et al.*, 1997).

Timing of the human Y coalescence

There have been several estimates of the Y chromosomal coalescence time and these are summarised in Figure 3, where they are compared with autosomal and X chromosomal coalescence times. Six of the first seven estimates were around 150 thousand years ago, with wide confidence limits (Hammer, 1995; Whitfield *et al.*, 1995; Tavare *et al.*, 1997; Underhill *et al.*, 1997; Hammer *et al.*, 1998; Karafet *et al.*, 1999). The exception was estimate 2 (Whitfield *et al.*, 1995). However, reconsideration of estimate 2 revealed that it was an estimate of the coalescence time of five chromosomes, and that the coalescence time of the entire population of Y chromosomes would be significantly older (estimate 3) (Tavare *et al.*, 1997); thus estimates 1–7 are all consistent. They contrast with estimates 8 and 9 (Pritchard *et al.*, 1999; Thomson *et al.*, 2000), which suggest a more recent date of about half of the earlier estimates, but again with wide confidence limits. Estimate 8 is based on microsatellite variation, but estimate 9, like 1–7, is based on binary variation.

Which estimate is correct? This apparently esoteric question is important, because it raises the possibility that selection may have acted on the Y. According to simple population genetics models, in the absence of selection coalescence time is proportional to effective population size. As the ratio of Y chromosomes : X chromosomes : autosomes is 1 : 3 : 4, coalescence times should also show this ratio. Y coalescence times 1–7 are in reasonable agreement with this expectation; 8 and 9 less so. Several explanations are possible. The confidence limits are very wide, so there may be no significant departure from neutral expectation to explain. The effective population size for the Y may be smaller than assumed, or mutation rates may differ between loci in ways that are not understood. Alternatively, the Y may have undergone a selective sweep. According to this scenario, an advantageous Y variant arose about 60,000 years ago and spread through the population because of selection. Could this hypothetical variant have increased language ability?

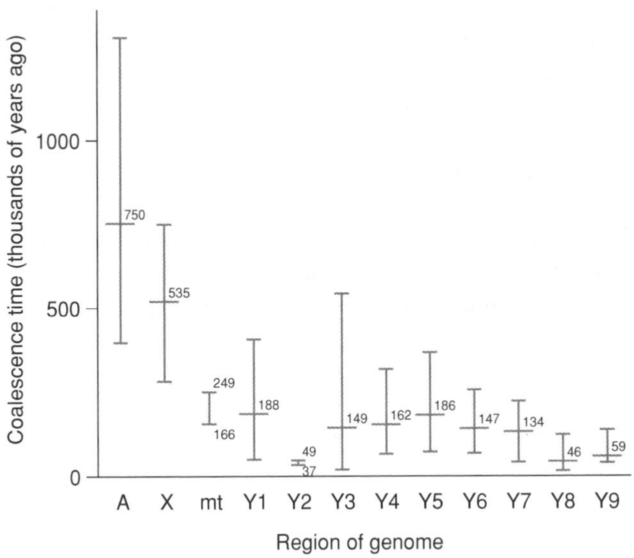

Figure 3. Published coalescence times for autosomal (A), X chromosomal (X), mitochondrial DNA (mt) and Y chromosomal (Y1–Y9) lineages. In most cases the estimate is shown, together with the 95% confidence limits given by the authors, or two times the standard error as an approximation to these. For mt and Y2, only the limits are given. Sources of data are: A, Harding *et al.* (1997); X, Kaessmann *et al.* (1999); mt, Vigilant *et al.* (1991); Y1, Hammer (1995); Y2, Whitfield *et al.* (1995); Y3, Tavare *et al.* (1997); Y4 and Y5, Underhill *et al.* (1997); Y6, Hammer *et al.* (1998); Y7 Karafet *et al.* (1999); Y8, Pritchard *et al.* (1999); Y9, Thomson *et al.* (2000).

A major reason for the different coalescent times is that estimates 1–7 assumed a constant population size, while 8 and 9 assumed population growth. As the autosomal and X chromosomal estimates also assumed constant population size, and all the different loci have been present in the same world-wide population, it may be necessary to compare estimates 8 and 9 with autosomal and X chromosomal coalescence times calculated assuming population growth. Thus the data and analysis are still too preliminary to show whether or not there has been a selective sweep.

CONCLUSIONS

There have been substantial changes to the human Y after the split from apes, including the addition of genes from the X, and some of these are thus expressed from both the X and the Y in humans but not in apes. These changes occurred after the ape–human split at 5–7 million years ago and before the common ancestor of modern human Ys at *c.* 40–200 thousand years ago. The complete sequences of the X–Y homologous regions should be available shortly and will provide an excellent starting point for the identification of all the genes they contain.

The origin of the unique characteristics of modern humans must involve changes to a small number of genes, but the nature of the changes and the number and identity of the genes are unknown. Few specific testable hypotheses have been suggested. The Crow hypothesis proposes that one of the human-specific X–Y homologous genes influences linguistic ability and contributed to speciation. We have the tools to test this hypothesis and should soon know whether there is such a gene.

DISCUSSION

Questioner: Does the fact you can find certain sets of haplotypes localised suggest a certain period of time for the change to have taken place?

Tyler-Smith: We can estimate the times at which the various haplogroups appeared from the amounts of variation we find within them when we use other markers like microsatellites, and in some cases these estimates are of just a few hundred to a few thousand years. In other cases they can be 10,000 or a few 10,000 years. The one that I showed with the most extreme localisation in China is a quite diverse set of chromosomes where we estimated a time of perhaps 10,000 years. So I think the distribution we see will depend on the population history, and if a chromosome arose in a sedentary population then it may remain quite localised. But some that are widespread, like the one that was spread across the north of Asia and Europe, have a younger estimated time.

Questioner: Does the distribution of Y variants tell us anything about the peopling of the Americas?

Tyler-Smith: Quite a lot of work has been done in that area that shows a high proportion of the native American populations have a single Y type that is not found outside America except in a few rare instances that are explained by back migration. So since that type is found through most of the populations including the different linguistic groups then that supports the idea that there was only one migration to the Americas and only one peopling, and goes against some hypotheses of multiple waves of migration. If we try and trace back the origins of that lineage it traces back to central Siberia: northern parts of Asia rather than southern China/Asia. What we would really like to know is the timing: did it occur perhaps just before 12–15 thousand years ago or did it occur 30–40,000 years ago? In that case the evidence is not just good enough to give a definitive answer.

Questioner: I am puzzled about the whole of your theory: it seems that it hinges on the fact that there's a later change in a part of the Y chromosome and it's transferred and then it splits up. This is what Tim Crow realised and related to lateralisation and the rest. Then you go at great length to show that there is considerable variation in the Y chromosome, not only between individuals but in areas itself. Why is it that one of them has a very significant change and another not at all?

Tyler-Smith: That depends on whether gene expression is affected. Some of the gross chromosomal rearrangements that you see, like in the movements of the centromere or an inversion of a large part of the chromosomes, if they do not affect a gene or interrupt a gene then they may have no phenotypic consequences whatever. Whereas a single base change in a crucial gene that inactivates it may be lethal, so I think that the magnitude of the genetic change is not a measure of how phenotypically significant it is. So far as Tim's hypothesis is concerned, it is essential for that hypothesis that there are genes or a gene within the region that transposed from the X to the Y.

Questioner: So you think there is a gene there?

Tyler-Smith: It is a hypothesis that there is a gene in that region and this is a testable hypothesis. And since these entire regions are now being sequenced and there are reasonably efficient ways of detecting genes, if there is such a gene it will become apparent. If there is not such a gene, then the hypothesis is wrong. I think Nabeel Affara will tell us that.

Questioner: I think my question is a follow-up on that one. You particularly did not tell us about the homologous regions and your haplotypes are clearly all very different. Are all your haplotypes contained within the homologous region, or do some not have this region?

Tyler-Smith: One of the deletions that we characterised removes the proximal part of that homology region. But as far as we know, all of the other haplotypes contain the homology region and particularly all of them contain the large block of homology.

Questioner: Can you explain how the translocation and inversion are related? Specifically what is the calibration that you used to obtain its age?

Tyler-Smith: The translocation was dated by comparing the sequence and the X and Y chromosomes. Also we know it occurred after the divergence of human and chimp Y chromosomes. So that provides broad limits. The more precise calibration that Tim mentioned of 3 million years or so is obtained from comparing sequences. That will have broad confidence limits. As for the timing of the inversion, I don't know how it can be dated except to say that it must have been later than the translocation. I wouldn't like to say what the 95% confidence limits are, but I would suspect that they are pretty broad and I would like to say that I don't know how to date the inversion.

References

Archidiacono, N., Storlazzi, C.T., Spalluto, C., Ricco, A.S., Marzella, R. & Rocchi, M. (1998) Evolution of chromosome Y in primates. *Chromosoma*, **107**, 241–6.

Arnason, U., Gullberg, A., Janke, A. & Xu, X. (1996) Pattern and timing of evolutionary divergences among hominoids based on analyses of complete mtDNAs. *Journal of Molecular Evolution*, **43**, 650–61.

Ciccodicola, A., D'Esposito, M., Esposito, T., Gianfrancesco, F., Migliaccio, C., Miano, M.G., Matarazzo, M.R., Vacca, M., Franze, A., Cuccurese, M., Cocchia, M., Curci, A., Terracciano, A., Torino, A., Cocchia, S., Mercadante, G., Pannone, E., Archidiacono, N., Rocchi, M., Schlessinger, D. & D'Urso, M. (2000) Differentially regulated and evolved genes in the fully sequenced Xq/Yq pseudoautosomal region. *Human Molecular Genetics*, **9**, 395–401.

Dunham, I., Shimizu, N., Roe, B.A., Chissoe, S., Hunt, A.R., Collins, J.E., Bruskiewich, R., Beare, D.M., Clamp, M., Smink, L.J., Ainscough, R., Almeida, J.P., Babbage, A., Bagguley, C., Bailey, J., Barlow, K., Bates, K.N., Beasley, O., Bird, C.P., Blakey, S., Bridgeman, A.M., Buck, D., Burgess, J., Burrill, W.D., O'Brien, K.P., *et al.* (1999) The DNA sequence of human chromosome 22. *Nature*, **402**, 489–95.

Freije, D., Helms, C., Watson, M.S. & Donis-Keller, H. (1992) Identification of a second pseudoautosomal region near the Xq and Yq telomeres. *Science*, **258**, 1784–7.

Hammer, M.F. (1995) A recent common ancestry for human Y chromosomes. *Nature*, **378**, 376–8.

Hammer, M.F., Karafet, T., Rasanayagam, A., Wood, E.T., Altheide, T.K., Jenkins, T., Griffiths, R.C., Templeton, A.R. & Zegura, S.L. (1998) Out of Africa and back again: nested cladistic analysis of human Y chromosome variation. *Molecular Biology and Evolution*, **15**, 427–41.

Harding, R.M., Fullerton, S.M., Griffiths, R.C., Bond, J., Cox, M.J., Schneider, J.A., Moulin, D.S. & Clegg, J.B. (1997) Archaic African and Asian lineages in the genetic ancestry of modern humans. *American Journal of Human Genetics*, **60**, 772–89.

Heyer, E., Puymirat, J., Dieltjes, P., Bakker, E. & de Knijff, P. (1997) Estimating Y chromosome specific microsatellite mutation frequencies using deep rooting pedigrees. *Human Molecular Genetics*, **6**, 799–803.

Jobling, M.A. & Tyler-Smith, C. (2000) New uses for new haplotypes: the human Y chromosome, disease and selection. *Trends in Genetics*, **16**, 356–62.

Kaessmann, H., Heissig, F., von Haeseler, A. & Paabo, S. (1999) DNA sequence variation in a non-coding region of low recombination on the human X chromosome. *Nature Genetics*, **22**, 78–81.

Karafet, T.M., Zegura, S.L., Posukh, O., Osipova, L., Bergen, A., Long, J., Goldman, D., Klitz, W., Harihara, S., de Knijff, P., Wiebe, V., Griffiths, R.C., Templeton, A.R. & Hammer, M.F. (1999) Ancestral Asian source (s) of new world Y-chromosome founder haplotypes. *American Journal of Human Genetics*, **64**, 817–31.

Lahn, B.T. & Page, D.C. (1997) Functional coherence of the human Y chromosome. *Science*, **278**, 675–80.

Page, D.C., Harper, M.E., Love, J. & Botstein, D. (1984) Occurrence of a transposition from the X-chromosome long arm to the Y-chromosome short arm during human evolution. *Nature*, **311**, 119–23.

Pritchard, J.K., Seielstad, M.T., Perez-Lezaun, A. & Feldman, M.W. (1999) Population growth of human Y chromosomes: a study of Y chromosome microsatellites. *Molecular Biology and Evolution*, **16**, 1791–8.

Santos, F.R., Pandya, A. & Tyler-Smith, C. (1998) Reliability of DNA-based sex tests. *Nature Genetics*, **18**, 103.

Sarich, V.M. & Wilson, A.C. (1967) Immunological time scale for hominid evolution. *Science*, **158**, 1200–3.

Schwartz, A., Chan, D.C., Brown, L.G., Alagappan, R., Pettay, D., Disteche, C., McGillivray, B., de la Chapelle, A. & Page, D.C. (1998) Reconstructing hominid Y evolution: X-homologous block, created by X–Y transposition, was disrupted by Yp inversion through LINE–LINE recombination. *Human Molecular Genetics*, **7**, 1–11.

Shimmin, L.C., Chang, B.H. & Li, W.H. (1993) Male-driven evolution of DNA sequences. *Nature*, **362**, 745–7.

Sinclair, A.H., Berta, P., Palmer, M.S., Hawkins, J.R., Griffiths, B.L., Smith, M.J., Foster, J.W., Frischauf, A.M., Lovell-Badge, R. & Goodfellow, P.N. (1990) A gene from the human sex-determining region encodes a protein with homology to a conserved DNA-binding motif. *Nature*, **346**, 240–4.

Tavare, S., Balding, D.J., Griffiths, R.C. & Donnelly, P. (1997) Inferring coalescence times from DNA sequence data. *Genetics*, **145**, 505–18.

Thomson, R., Pritchard, J.K., Shen, P., Oefner, P.J. & Feldman, M.W. (2000) Recent common ancestry of human Y chromosomes: evidence from DNA sequence data. *Proceedings of the National Academy of Sciences of the USA*, **97**, 7360–5.

Underhill, P.A., Jin, L., Lin, A.A., Mehdi, S.Q., Jenkins, T., Vollrath, D., Davis, R.W., Cavalli-Sforza, L.L. & Oefner, P.J. (1997) Detection of numerous Y chromosome biallelic polymorphisms by denaturing high-performance liquid chromatography. *Genome Research*, **7**, 996–1005.

Vigilant, L., Stoneking, M., Harpending, H., Hawkes, K. & Wilson, A.C. (1991) African populations and the evolution of human mitochondrial DNA. *Science*, **253**, 1503–7.

Whitfield, L.S., Sulston, J.E. & Goodfellow, P.N. (1995) Sequence variation of the human Y chromosome. *Nature*, **378**, 379–80.

Yunis, J.J. & Prakash, O. (1982) The origin of man: a chromosomal pictorial legacy. *Science*, **215**, 1525–30.

Do the Hominid-Specific Regions of X–Y Homology Contain Candidate Genes Potentially Involved in a Critical Event Linked to Speciation?

CAROLE A. SARGENT, PATRICIA BLANCO & NABEEL A. AFFARA

Summary. It has been postulated that the critical events leading to major differences between humans and the great apes (such as language and lateralisation of the brain) are associated with major changes on sex chromosomes. Regions of homology between the human sex chromosomes have arisen at different points during mammalian evolution. The two largest blocks are specific to hominids, having appeared at some time after the divergence of humans and chimpanzees. These are the second pairing region found at the telomeres of the sex chromosome long arms and a region of homology between Xq21.3 (X chromosome long arm) and Yp11 (Y chromosome short arm). Questions arise as to whether (1) these regions of the sex chromosomes contain functional genes and (2) these genes might be candidates for the differences in cognitive function that distinguish modern humans from their ancestors. Furthermore, divergence between functional sequences on the X and the Y, and the alteration of the immediate environment around a locus by rearrangements, may allow for the acquisition and evolution of genes conferring more subtle sexually dimorphic variation.

INTRODUCTION

THE MAMMALIAN SEX chromosomes have evolved from an ancestral pair of autosomes (Ohno, 1967). The dominant male-determining gene evolved on the proto-Y, and became genetically sequestered from the rest of the genome through the suppression of recombination over most of the length of the proto-Y with the proto-X. In all mammals, a small region of strict X–Y

Proceedings of the British Academy, **106**, 231–250, © The British Academy 2002.

homology has been maintained in order to permit pairing and correct segregation of the sex chromosomes during male gametogenesis. This segment is known as the pseudoautosomal region (PAR). However, the PAR shows considerable variation in gene content between different groups of species. The evolution of the sex chromosomes and the establishment of distinct PARs is believed to have occurred both by the acquisition of regions from autosomes, and the loss of material from the male-determining Y chromosome: the addition–attrition model of sex chromosome evolution (Graves, 1995). Within the non-recombining portions of the sex chromosomes, several other blocks of homology and X–Y homologous genes have been described. By studying the patterns of homology in different species, either by Southern hybridisation or fluorescence *in situ* hybridisation (FISH), the events leading to discrete blocks of sequence conservation can be reconstructed (Lambson *et al.*, 1992; Affara & Ferguson-Smith, 1994; Vogt *et al.*, 1997; Perry *et al.*, 1998; Graves *et al.*, 1998; Glaser *et al.*, 1999; Lahn & Page, 1999). Consequently, regions of X–Y homology defined on the modern human sex chromosomes may represent either the ancient remnants of the ancestral pair of autosomes, or reflect more recent exchanges of material. Figure 1 summarises our present understanding of the patterns of homology between the human sex chromosomes.

One of the consequences of a heterogamous method of sex determination is that females with an XX karyotype have twice the number of copies of X-linked genes as males with an XY karyotype. Hence those genes outside the PAR on the X chromosome are subject to inactivation on one of the X chromosomes in each female cell. The exceptions to this rule are genes found to be homologous between the X and Y chromosomes, where both copies are functional. These genes escape X inactivation and are, therefore, expressed in diploid dose in both males and females. Deficits of genes in this category lead to features of Turner's syndrome (XO females, and females and males with partial deletions of the X and Y), suggesting that they are required in diploid dose in both males and females (Ferguson-Smith, 1965). The association of cognitive deficits and male/female differences in cognitive function with sex chromosome aneuploidies has given rise to the idea that a gene or genes in the X–Y homologous category may be implicated in these aspects of brain phenotype (Crow, 1993). This proposition can be tested by searching X–Y homologous sequences for genes that are likely to have a function(s) within the brain.

There are two major homology blocks on the human sex chromosomes that must represent recent additions to the Y chromosome through transposition events, because they are not found on the Y in the great apes. These regions are the Xq21.3/Yp11 homology and part of the strictly homologous PAR 2 located at the tips of the X and Y chromosome long arms. The PAR 2 is known to contain a gene related to synaptobrevin (*SYBL1*) that may have a function in the

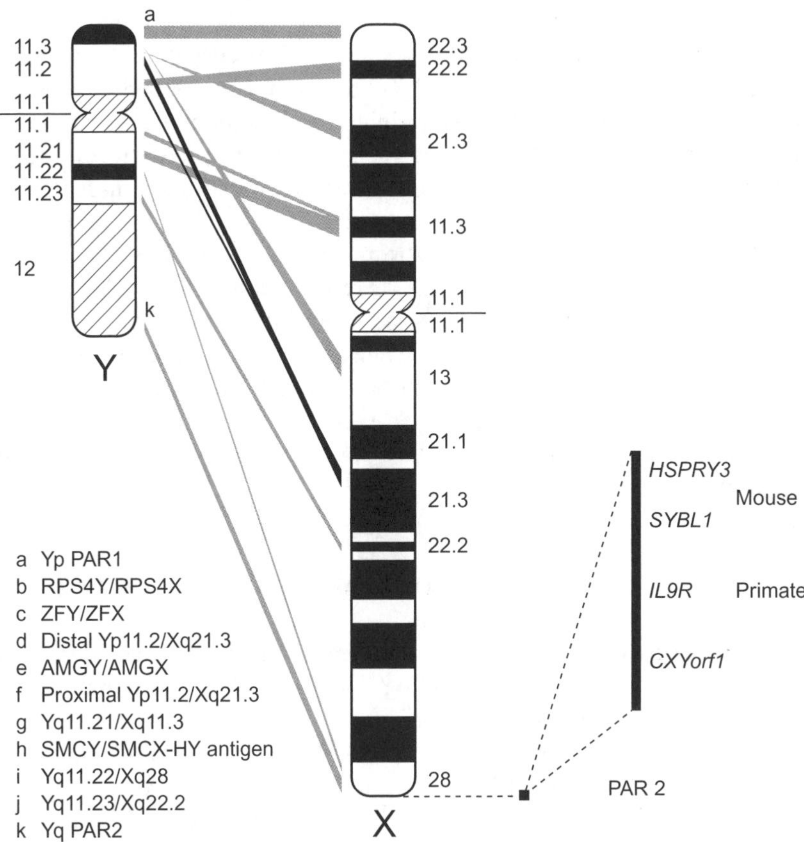

Figure 1. Homologies between the human X and Y chromosomes. Homologies between the sex chromosomes are highlighted and assigned according to the extent of conservation data available. Numerous single genes are shared on the X and Y chromosomes: for some of these genes the location within the non-recombining portion of the sex chromosomes is unique to humans, although these genes are X–Y homologous within the PAR 1 of other primate and non-primate species. An expansion of the PAR 2 is shown to the right. Homologues of *HSPRY3* and *SYBL1* are found in the mouse, *IL9R* has been added more recently in primate evolution, and *CXYorf1* is human specific. The diagram shown here is simplified, and a more detailed version can be found in Vogt *et al.* (1997).

brain (Ciccodicola *et al.*, 2000). However, it is also worth noting that the gene is ubiquitous in its expression (D'Esposito *et al.*, 1996) and may have a role in other tissues. The synaptobrevin class of proteins is involved in synaptic vesicle docking and membrane transport and may also play a role in dorsoventral patterning in the nervous system. It has been shown that *SYBL1* is inactivated on both the inactive X and the Y chromosome (D'Esposito *et al.*, 1996), but its presence in the ancestral primate PAR 2 makes it a less attractive candidate gene. It is possible that mutations leading to reactivation of either gene may have an impact on brain phenotype as a result of inappropriate expression. The organisation and gene content of the PAR 2 is shown as an expanded inset in Figure 1.

The Xq21.3/Yp11 region is approximately 3.5 Mb of DNA and almost 10-fold larger than the PAR 2; it may therefore contain many more genes. In this chapter we discuss the structure and evolution of this block. The overall structure is well conserved between the human X and Y chromosomes, and the X chromosomes from different primates. Although the sequence data reveal high homology for the human X and Y, there are regions of significant divergence, presumably as the consequence of multiple rearrangements during evolution.

STRUCTURE OF THE Xq21.3/ Yp11 HOMOLOGY BLOCK

Previous work has established extensive yeast artificial chromosome (YAC) contigs from both the X and Y homology blocks, and has shown that the gross order of DNA markers is highly conserved (Vollrath *et al.*, 1992; Jones *et al.*, 1994; Sargent *et al.*, 1996; Mumm *et al.*, 1997). The major difference in the organisation of the sequences on the sex chromosomes is the isolation on the Y chromosome of the interval defined by DXYS34 and DXYS1 by means of a LINE (long interspersed nuclear element)-mediated recombination (Schwartz *et al.*, 1998). This results in a paracentric inversion leading to the juxtaposition of this block with the *AMELY* locus, creating a small (<300 kb) proximal Yp/Xq homology block and a larger (3 Mb) distal Yp/Xq block (DXYS42–DXYS31). Figure 2 compares the organisation of the homologous blocks on the X and the Y chromosomes.

Through the isolation of bacterial artificial chromosome (BAC), P1 artificial chromosome (PAC) and cosmid genomic clones from human male and human female DNA libraries, X-specific and Y-specific clone sets have been developed for sequencing by the genome centres as part of the Human Genome Project. Analysis of these sequence data allows a more detailed study of the level of identity between the chromosomes, and the identification of candidate genes. Comparison of non-repetitive DNA sequence from homologous

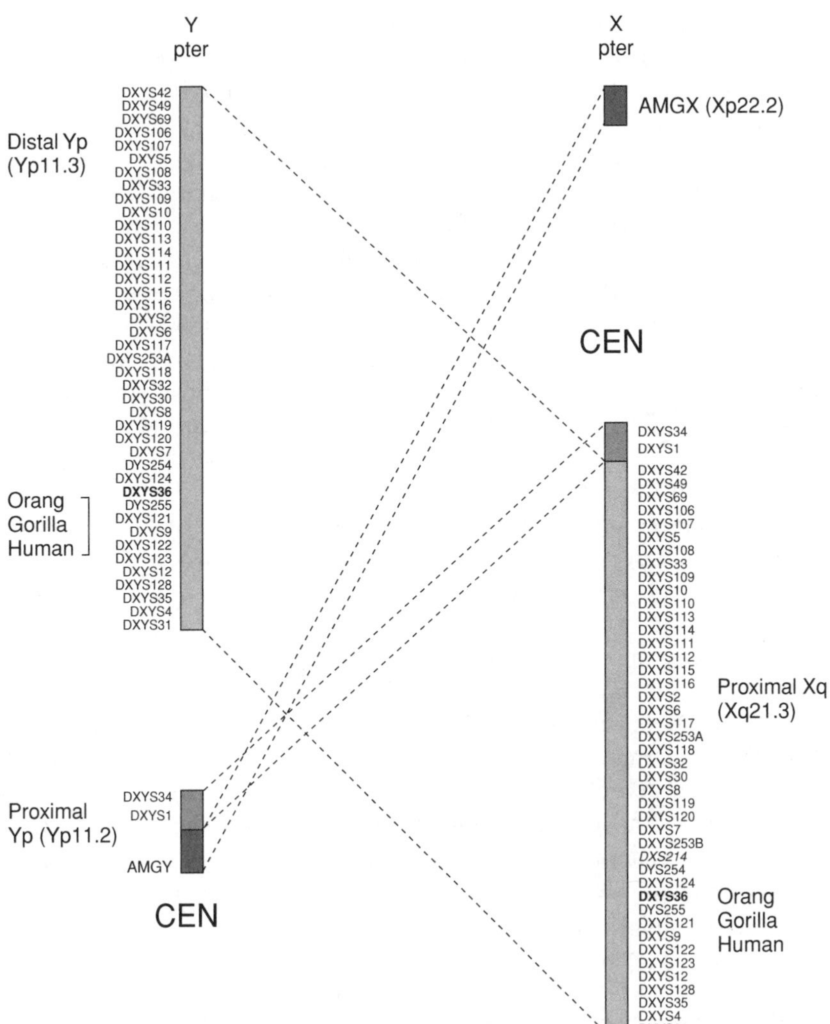

Figure 2. Transposition of the Xq21.3 block to Yp11. Diagram to show the transfer of the Xq21.3 block to the Y chromosome has occurred as a single event. Subsequent inversion of the region containing markers DXYS1 and DXYS34 have split the block on Yp11 into two. The proximal portion is sited alongside *AMGY*, a gene with an X homologue in Xp22.2. The centromeric to telomeric orientation of the Xq21.3/Yp block is reversed as a result of the transposition. DXS214 is shown in italics, and is conserved on the ancestral primate X chromosome. The marker DXYS36, in a bold text, is found on both the X and Y chromosomes of orang-utan, gorilla and human.

segments of the X and Y gives a level of identity of between 98.8% and 99.1%. The G + C content of the Xq21.3/Yp11 block is low, with a range of 35–37%. This figure is similar for both the X and the Y chromosome, and does not currently show significant local variation, although this can only be fully assessed once the sequences have been fully completed and assembled. In addition, the percentage of repetitive DNA elements detected is high, with a range of 57–87% for different stretches of the homologous blocks. The majority of repeats fall into the LINE and/or LTR (long terminal repeat) category, with very few SINE (short interspersed nuclear element) elements. This high LINE and low Alu content, combined with the paucity of G + C, suggests that the X–Y homology block belongs to the GC-poorest subfraction of bulk genomic DNA, the L isochore. Generally, this class of DNA has a lower gene content than G + C-enriched isochores, and very few CpG islands (associated with the transcriptional start sites of many genes). In addition, most genes found in regions of low G + C content have a larger than average size, and tend to be tissue-specific genes expressed at low abundance rather than ubiquitous transcripts (reviewed by Bernardi, 2000).

The general composition of the block is reflected in the analysis of coding potential. Although expressed sequence tag (EST; sequences derived from cloned fragments of RNA molecules) matches have been defined through BLAST (basic local alignment search tool) searches, it is unclear how many of these represent fragments of true X–Y homologous genes, and how many are pseudogenes. This is also true for the gene structures predicted from computer algorithms. Of the confirmed pseudogenes identified the vast majority are uninterrupted, reminiscent of retroposition. To date, only three regions corresponding to clusters of cDNA clone sequences have been identified that potentially represent functional genes. One of these (a protocadherin gene expressed in the brain) has been analysed in detail (see below) and is a promising candidate gene for brain lateralisation. Figure 3 summarises the current information on the gene content of this homology block.

From a comparison of the sequence derived from the X and Y chromosomes, there appears to be one major block of divergence covering some 200 kb of DNA. This is characterised by the marker DXS214 on the X, and NR11 on the Y. It was unclear from initial experiments whether the differences between the human sex chromosomes were either the result of rearrangement between the flanking X–Y homologous markers or the consequence of sequence divergence. More detailed analysis of the interval shows that the pattern of X–Y homology/ divergence observed is complex, and probably reflects more than one rearrangement during evolution. However, through FISH and other molecular techniques such as polymerase chain reaction (PCR) amplification of DNA, we have found that different ancestral Y chromosome types show the same marker distribution: i.e. we would predict that there are no major

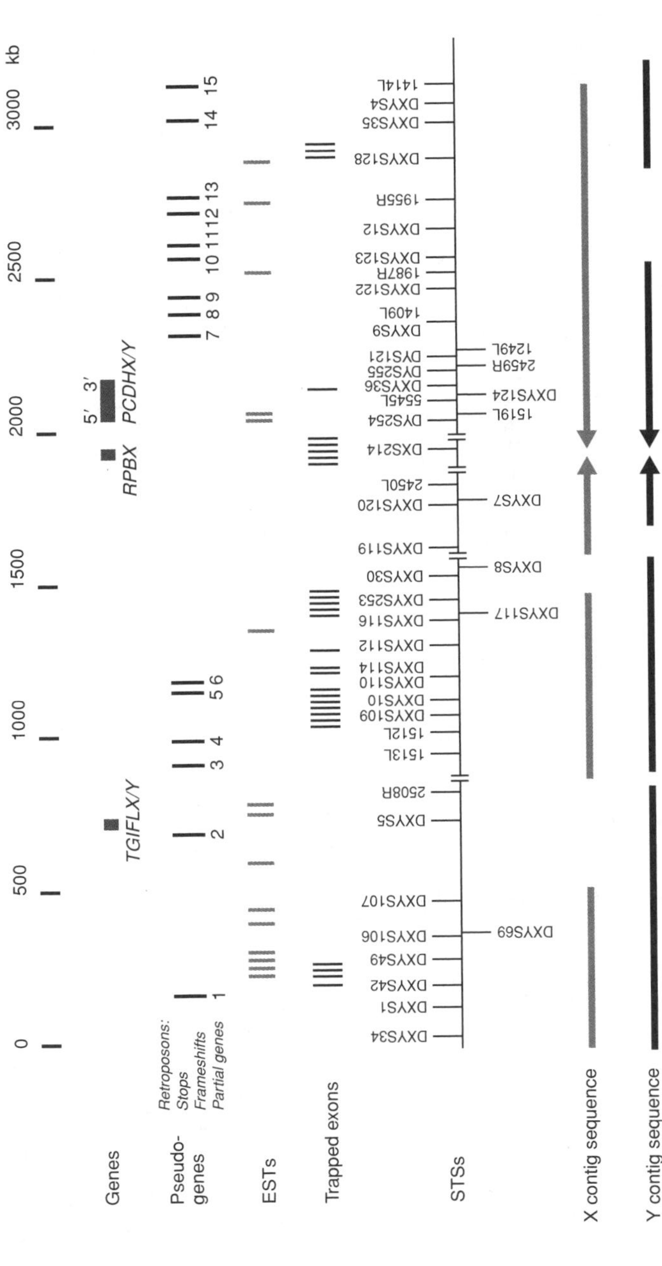

Figure 3. Coding potential of the Xq21.3/Yp11 homology block. This figure summarises the data from exon trapping experiments, and sequence analysis using the NIX program at HGMP. The extent of available X and Y genomic clones in a near completed state is shown below the STS (marker) content: the arrows indicate that the sequence extends into regions of X or Y unique sequence. Pseudogenes are 1, Sorcin; 2, Polypro dUTPase; 3, Staufen; 4, *UBH1*; 5, *CROC1B*; 6, XY body protein; 7, *VDAC1*; 8, *SAP62*; 9, *EIF4A1*; 10, *ZNF127*; 11, Transposon 10; 12, Keratin 18; 13, *SNX3*; 14, *RPL26*; 15, *FUS1*. The only EST matches that appear to be representative of genes to date are those for *TGIF*-like (*TGIFLX/Y*), an X-linked ribosomal binding domain protein (*RBPX*), and the protocadherin genes (*PCDHX/Y*).

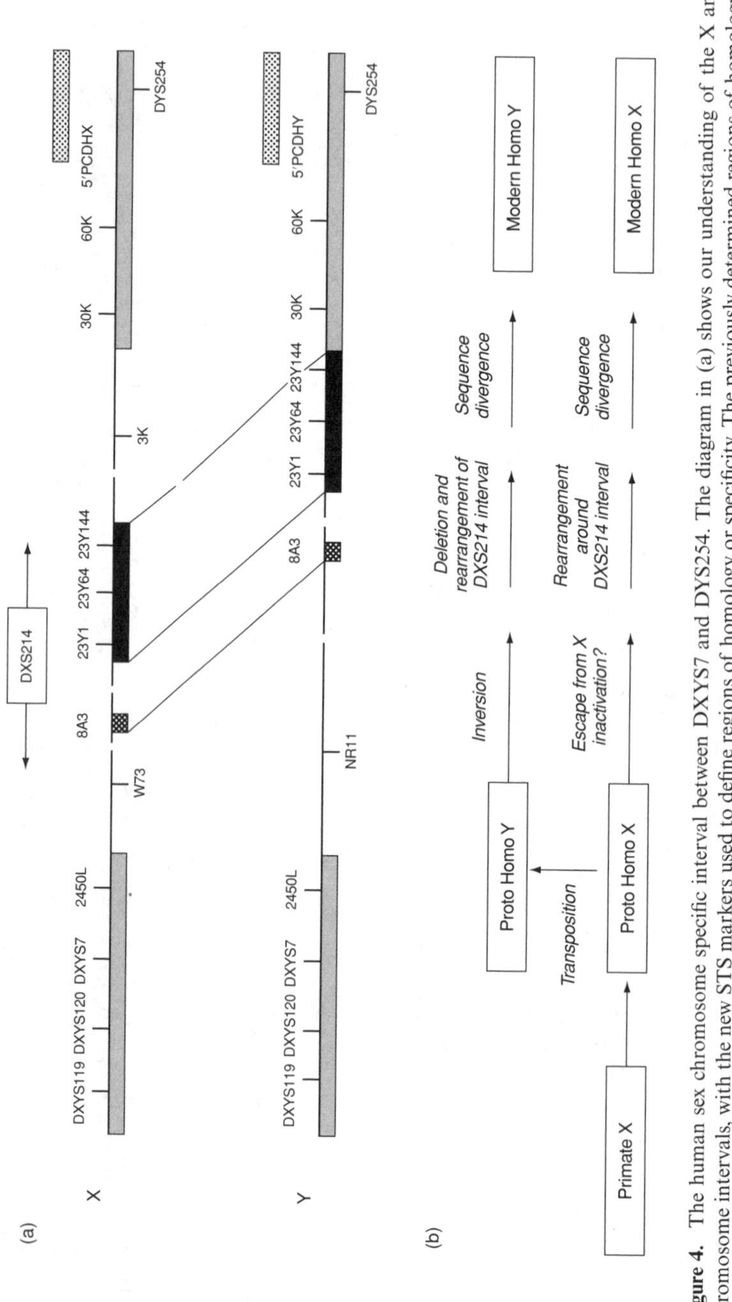

Figure 4. The human sex chromosome specific interval between DXYS7 and DYS254. The diagram in (a) shows our understanding of the X and Y chromosome intervals, with the new STS markers used to define regions of homology or specificity. The previously determined regions of homology are defined by the shaded boxes; new homology is defined by the solid box and the single exon 8A3 (white dots on black); the location of the 5′ ends of the protocadherin genes is shown by the black dots on white. The exact order of sequences in the central portion is not known (broken lines). DXS214 lies on the X chromosome between W73 and 3K. In (b), a proposed set of events leading to the current mosaic structure of this interval is outlined. The precise order and date of each event is unknown.

variations of the interval in modern *Homo sapiens* populations. Figure 4a shows the organisation of this segment and demonstrates the mosaic interspersion of X–Y homologous and X- or Y-specific sequences.

A second region of potential sequence divergence lies between the markers DXYS5 and DXYS109, as defined by the divergence of the DNA sequence DXYS108 between the human sex chromosomes (Mumm *et al.*, 1997). As yet, the comparative sequence data for both the X and Y are not available for a more detailed commentary on possible candidate genes.

CONSERVATION OF THE HUMAN XQ21.3 INTERVAL IN THE GREAT APES

Previous failure to find homology to the probe DXS214 in primates by Southern blotting led to the supposition that this marker was representative of an insertion on the X chromosome that occurred after the transposition of the homologous block to the human Y chromosome. However, as we (Vogt *et al.*, 1997), and others (Vacca *et al.*, 1999), have demonstrated by FISH, these X-specific sequences occur on the ancestral primate X chromosome. Thus substantial homology to this DNA segment has been lost from the human Y either through chromosomal rearrangements or by sequence divergence.

For a more detailed study, genomic clones spanning the Xq21.3 interval were selected for FISH analysis of male and female normal human, male chimpanzee (*Pan troglodytes*), female gorilla (*Gorilla gorilla*), male orang-utan (*Pongo pygmaeus*), female gibbon (*Hylobates lar*) and male and female lemur (*Eulemur macaco macaco*). In the human male signals were observed on the long arm of the human X chromosome, and the short arm of the Y chromosome for all the clones, with the exception of a clone containing DXS214. The signal for this clone was restricted to the human X chromosome, as expected. In the primates and the lemur lines, all hybridisation signals were observed on the long arms of the X chromosome. With a two-colour FISH approach the results suggest that the region from DXYS42 to DXYS31, including the DXS214 interval, has a very similar gross structural arrangement between the great apes.

Additional results from Southern restriction fragment length polymorphism (RFLP) (Lambson *et al.*, 1992) and sequencing of genomic DNA allow estimation of the level of identity with the human sequence as *c*. 97% for gorilla and *c*. 98% for chimpanzee. This is concordant with other data for the evolutionary distance between these species, and supports previous claims that the transfer of the Xq21 block to the Y chromosome in hominid evolution occurred after the divergence of the great apes (Page *et al.*, 1984; Lambson *et al.*, 1992).

A schematic view of possible events leading to the arrangement of the sex chromosomes in modern humans is shown in Figure 4b.

GENE CONTENT OF THE Xq21.3/Yp11 BLOCK

As noted above, only three potential candidate genes for brain lateralisation have been identified within this region of the X chromosome: (1) a homeodomain-related transcribed sequence (*TGIF*-like), (2) a transcribed sequence containing four RNA binding motifs that is expressed predominantly in brain and testis, and (3) a gene (*PCDHX*) belonging to the protocadherin family, expressed predominantly in the brain (Yoshida & Sugano, 1999; Blanco *et al.*, 2000). For the *TGIF*-like transcript, the functional status is unclear, as analysis is incomplete. The RNA-binding domain transcript does not have a functional Y homologue in humans, but encodes a potentially functional protein from the X locus. The absence of a Y homologue makes this gene a less favourable candidate, despite expression within the brain. The third gene appears to be a promising candidate in several respects. The X homologue has an open reading frame; it has a closely related Y homologue also with an open reading frame; both genes are predominantly expressed within different regions of fetal and adult brain, but are virtually undetectable in mRNA from other tissues. This is consistent with the general observation that protocadherins are predominantly expressed in the brain (Sano *et al.*, 1993). Redies & Takeichi (1996) have argued forcefully that these genes play a fundamental role in the cell–cell recognition essential for the segmental development and function of the central nervous system. Thus the *PCDHX* and *PCDHY* genes potentially have a role not only in the establishment of neuronal networks in the developing brain, but also in the maintenance of functions in adult structures.

THE PCDHX AND PCDHY GENES

General structure and classification of the genes

Figure 5 illustrates the predicted structures of the X- and Y-linked protocadherin genes and their products, as determined from nucleotide sequence analysis. The proteins conform closely to the general organisation of other members of the protocadherin subfamily: a signal peptide to permit secretion of the protein; seven extracellular cadherin motifs that mediate cell interactions; a transmembrane domain achieving insertion into the cell membrane; and a cytoplasmic domain that interacts with intracellular proteins. In addition to the arrangement of the protein domains, the gene organisation of *PCDHX*/*PCDHY* strongly supports their inclusion in the protocadherin gene subfamily. Typically, the protocadherin subfamily genes consist of four to five exons, with most of the coding region contained in a single large exon (Wu & Maniatis, 1999). Similarly, the majority of the open reading frame of *PCDHX*/*PCDHY* is encoded in exon 5.

Three further features of *PCDHX*/*PCDHY* indicate that they belong to a specific subgroup of the protocadherins known as the CNR (cadherin-related neuronal receptor) genes. First, the cytoplasmic region of *PCDHX*/*PCDHY* contains a lysine-rich segment that is characteristic of the CNR genes (Kohmura *et al.*, 1998; Wu & Maniatis, 1999) and may be important for interactions with other proteins within the cell. Secondly, the sequence of the cadherin domains of *PCDHX* and *PCDHY* is most similar to those found in the mouse CNR genes encoded on chromosome 18, and the cluster of human CNR genes on chromosome 5q31 (Kohmura *et al.*, 1998; Wu & Maniatis, 1999; Sugino *et al.*, 2000). Thirdly, like the CNR genes, the first cadherin domain of *PCDHX* and *PCDHY* possesses a variant (TGD) RGD (putative integrin binding domain) motif. Proteins of the CNR protocadherin subgroup show enriched or specific expression in the brain and subpopulations of synapses. In the mouse, the extracellular domain of CNR proteins has been found to associate with secreted Reelin proteins (Senzaki *et al.*, 1999), while the cytoplasmic region has been shown to interact with Fyn, a non-receptor tyrosine kinase (Kohmura *et al.*, 1998). It is thought that these molecular associations may act as a positive selection mechanism during development of the cortex, and by triggering secondary signalling pathways may also regulate the morphogenesis of the neuronal cells. It is interesting to note that mice carrying mutations of Fyn show impaired long-term potentiation, spatial learning and hippocampal development (Grant *et al.*, 1992).

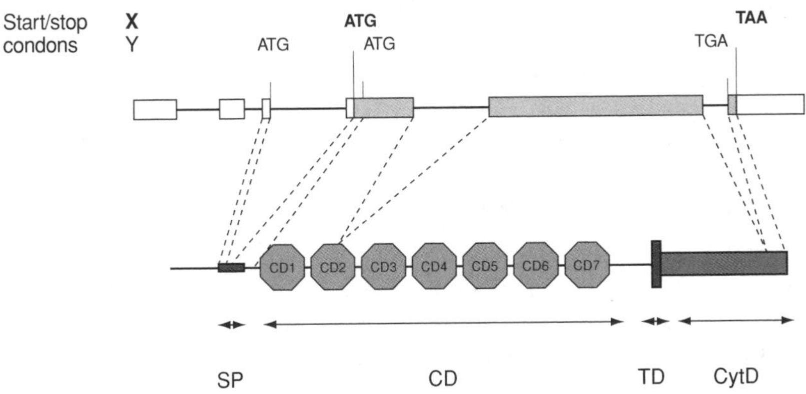

Figure 5. Comparison of the protocadherin genes. The genomic structure of *PCDHX* and *PCDHY* consists of at least six exons; only three of them contain the coding sequence (shaded). The different start and stop codons of both genes are highlighted above. The corresponding protein structure is depicted below. SP, signal peptide; CD, cadherin domains; TD, transmembrane domain; CytD, cytoplasmic domain.

Comparison of the X and Y gene sequences

Detailed sequence analysis of the X and Y genes has shown that they share some 98% identity at the nucleotide level over their respective open reading frames (Table 1). This suggests that they may possess closely related functions. However, examination of the coding regions reveals two notable differences between the X and Y transcripts. (1) The Y gene has a 13-base pair deletion that removes the region equivalent to the initiator methionine of the X gene and the first amino acid (aspartic acid) of the putative signal peptide. A single base change in an upstream exon of the Y gene creates a new initiator methionine in-frame with the remaining open reading frame, and retains an enlarged signal peptide sequence. This initiator methionine has a weaker match than the X gene initiator to the Kozak consensus. There is a second initiator methionine with a better Kozak consensus further downstream, but if this were used there would be no signal peptide in the resulting protein, and it would not be correctly transported to the intracellular apparatus required for secretion and insertion into the cell membrane. However, it is possible that both initiator codons are used in the translation of Y transcripts and may represent a means of regulating the level of Y protein that is capable of being secreted. (2) The Y gene encodes a cytoplasmic domain that stops nine amino acids before the termination of the *PCDHX* polypeptide. This nine-amino acid difference may be very significant in terms of differentiation of function of the X and Y genes. Currently, we do not know if the respective cytoplasmic domains can interact with different cytoplasmic proteins, and whether the nine residues represent part or the whole of a specific binding domain. If these genes are involved in inter- or intracellular signalling (as are other CNR family genes), then it is possible that unique X-only or Y-only interactions will influence distinct pathways.

Expression of *PCDHX*/*PCDHY*

The expression of *PCDHX* and *PCDHY* was examined using reverse transcriptase (RT)-PCR with a range of tissue-specific mRNAs. The PCR primers were designed to amplify across exons 2–4 and produce a fragment from both the X and Y transcripts. From the tissue data, expression is predominantly in brain mRNA and varies from one region of the brain to the other. The protocadherin transcripts are found in fetal brain, adult cortex amygdala, hippocampus, caudate nucleus, corpus callosum, substantia nigra and thalamus and are barely detectable in the adult cerebellum. Transcription cannot be detected in mRNAs from other adult tissues (heart, liver, kidney and skeletal muscle) except at a low level in testis.

Table 1. *PCDHX* and *PCDHY* gene structure

	Size (bp)		
	X	Y	Identity (%)
EXON 1	380	380	99
Intron 1	30,788	31,300	96.7
EXON 2	106	106	97
Intron 2	658	658	98.9
EXON 3	59	59	96
Intron 3	24,165	24,160	96.5
EXON 4	584	571	98.8
Intron 4	40,736	40,755	97.1
EXON 5	2493	2493	99.1
Intron 5	3633	3636	98.9
EXON 6	1100	1100	98.8

Using a two-step nested PCR approach that allows detection of mRNA present at very low levels, the X or Y origin of the transcripts was investigated in all the tissues. The PCR primers were designed to amplify across a coding region that contains a restriction enzyme site difference between the X- and Y-linked sequences. This allows an estimate of the relative transcription levels from the X- and Y-linked genes. The mRNA samples used in the study comprised pools from male and female individuals, with the exceptions of whole brain, heart, liver and testis (all male only). Both X and Y transcripts are evident in the subregions of the brain, except cerebellum (predominantly *PCDHX*) and the heart, whereas kidney, liver, muscle and testis are predominantly *PCDHY* (Table 2).

Additional studies into the regulation of *PCDHX*/*PCDHY* gene expression were carried out using the testis tumour-derived pluripotential cell line NTERA. This cell line can be induced with retinoic acid to differentiate along the spermatogenic pathway and into neuronal cells. The NTERA cells were cultured either in the presence of retinoic acid, hydrocortisone or DMSO, or left in an untreated state. The hydrocortisone is used to mimic the effect of steroid hormones upon the cells and, as the retinoic acid is prepared in DMSO solution, DMSO-treated cells act as an experimental control. The nested set of primers was used to amplify the relatively low level of gene expression in NTERA cells as described above. In the uninduced state and with the DMSO-treated cells, only the X-linked transcripts are detectable, whereas hydrocortisone-treated cells contain both X and Y transcripts at similar levels. In contrast, treatment with retinoic acid increases the level of Y-linked transcripts and appears to depress the X-linked mRNA. These data seem to indicate that the Y-linked gene is regulated differently to the X-linked gene at the mRNA

Table 2. Summary of expression data for *PCDHX* and *PCDHY*

	PCDHX	*PCDHY*
Amygdala	+	+
Caudate nucleus	+	+
Cerebellum	+	–
Corpus callosum	+	+
Hypocampus	+	+
Substantia nigra	+	+
Thalamus	+	+
Total brain	+	+
Fetal brain	+	+
Heart	+	+
Liver	–	+
Kidney	–	+
Muscle	–	+
Testis	+	+

level, and although both of the promoters may contain retinoic acid response elements the effects are modified by other transcriptional control elements.

Currently the promoters of these genes have not been isolated or characterised. However, the 5′ ends of both the X and Y genes lie close to the boundary between the region of strict X–Y homology and the X or Y chromosome-specific regions. The sequences that have been inserted or deleted upstream of the translational start sites are likely to have an influence upon the promoter structures, and the position of long-range elements such as enhancers.

Conservation of *PCDHX*

cDNA probes for *PCDHX* were hybridised to Southern blot filters of male and female pairs of DNA from a range of different species. No bands were detected in kangaroo (representative of metatherian mammals) or in the armadillo, mouse or rat. However, sex-linked dosage was seen for rabbit, cow, goat, sheep, pig, horse, lemur, a New World monkey (squirrel monkey) and two Old World monkeys (rhesus macaque and stump-tailed macaque), suggesting that this gene is only on the X chromosome of these species (Table 3). Previous blotting and hybridisation studies used a set of probes from across the Xq21/Yp11 homology block (Lambson *et al.*, 1992). The observation that all of these markers, with the exception of DXYS36 (contained within an intron of the *PCDHX*/*PCDHY* genes), are confined to the X chromosome in primates and monkeys but appear absent from rodents is entirely consistent with the pattern seen for the *PCDHX* cDNA. Probe DXYS36, in contrast, is X–Y homologous in higher primates (orang-utan, gorilla and human) but not in lower primates or chimpanzee. This raises the question as to whether there is a functional *PCDHY* gene copy on the Y chromosome of these additional species of great apes. So far, there is no conclusive evidence of a Y gene from PCR and sequence analysis of non-human male primate DNA: sequences do not differ from those of female DNA. Either the transfer of material encompassing DXYS36 does not include a complete protocadherin gene, or the X and Y copies are identical, or the Y copy has diverged significantly from the X copy such that it no longer amplifies with the primers used in these experiments. Additional studies are in progress to resolve this issue.

Does *PCDHX/PCDHY* fit the criteria for a gene involved in brain lateralisation and/or language development?

The current hypothesis under investigation is that a gene (or genes) on the human sex chromosomes is important for brain lateralisation, and that this is fundamental for language development. It is also predicted that relative hand skill is influenced in part by the dominant hemisphere. The evidence for this is summarised as follows.

1. Brain asymmetry and directional handedness are features not observed in chimpanzees (Marchant & McGrew, 1996; Buxhoeveden & Casanova, 2000)
2. The cognitive deficits seen in individuals with sex chromosome aneuploidies can be correlated with deficits in brain lateralisation, and

Table 3. Conservation of *PCDHX/Y*

		PCDHX	*PCDHY*
Hominoid	Human	+	+
Old World	Rhesus monkey	+	–
	Stamp-tailed macaque	+	–
New World	Squirrel monkey	+	–
	Marmoset	+	–
Prosimian	Lemur	+	–
Carnivore	Dog	+	–
Artiodactyl	Cow	+	–
	Goat	+	–
	Sheep	+	–
	Pig	+	–
	Horse	+	–
	Rabbit	+	–
Rodent	Mouse	–	–
	Rat	–	–
Marsupial	Kangaroo	–	–

suggest that a gene for relative hemispheric development is located on the X chromosome. Thus, XO patients have a relative deficit of the non-dominant hemisphere, and patients with an excess of the X chromosome (XXX, XXY) have a relative deficit of the dominant hemisphere. Males do not have hemispheric deficits equivalent to those in XO patients and, therefore, the gene must fall into the X–Y homologous category (Crow, 1993).

3 Although the gene(s) is X–Y homologous, there are sex-related differences in brain development and verbal ability, with females maturing faster than males. Individuals of either sex who show equal hand skill are delayed in developing language skills, including reading (Crow *et al.*, 1998).

To fulfil the requirements demanded of the above hypothesis, any potential candidate gene must therefore: (1) be found on the human X chromosome; (2) have a homologue on the human Y chromosome; (3) show appropriate spatial and temporal patterns of expression during embryonic and post-natal development to support a role in lateralisation; (4) have acquired sufficient differences in the nucleotide sequence of the open reading frame, or in the mode of transcriptional control, to distinguish it from the homologous gene in other great apes.

The criteria defined by the first two points are met by the protocadherin genes, but also by *SYBL1* and by the partially characterised *TGIF*-like locus. However, *SYBL1* does not have an expressed Y homologue (see below). So far, in relation to expression patterns, we know that *PCDHX* and *PCDHY* transcripts are both expressed in the fetal brain during the second half of pregnancy (mRNA sample pool 21–30 weeks). Currently, we do not have detailed information about the precise onset of gene expression, or the cell types involved. The *TGIF*-like gene is not expressed in any of the brain samples investigated to date, and thus would appear to be a less attractive candidate.

The presence of a candidate cDNA at the appropriate point in development must be validated with protein data, showing that the transcripts are actually translated to give rise to functional proteins. These proteins need to have a proven role in the initiation of asymmetric development. Recently, experiments designed to dissect the pathways that establish left–right asymmetry in the developing chick embryo identified another gene of the cadherin superfamily, N-cadherin, as a key early but transiently expressed molecule required for normal morphogenesis (Garcia-Castro *et al.*, 2000).

With respect to those changes that define the unique status of *H. sapiens*, it would be useful to compare the expression patterns of candidate genes in modern humans with those of our closest extant relatives, such as the chimpanzee. Differences between the species could then be related either to variations in the potential protein product encoded by the genes, or to the alterations in the sequences involved in the control of mRNA production both at the temporal and spatial levels. To this end, it is of added importance to define the promoter regions of the human *PCDHX* and *PCDHY* genes, as there are already sufficient data from the expression analysis to support the hypothesis that the transcripts are regulated by distinctive promoter and/or enhancer elements.

Other predictions arise from the criteria listed. As the brains of XY individuals are not phenotypically like those of XO individuals, either the Y gene is functionally equivalent to the X gene, or there is a mechanism in males (such as differential expression levels in response to hormonal differences) to compensate for X haploinsufficiency. From this it also follows that if the X and Y genes are equivalent for critical stages in development, the X gene must escape from X-inactivation. For the protocadherin gene, where we should expect the ancestral primate homologue to be subject to X inactivation, such an escape from the established controls would represent an evolutionary significant event. Subtle differences in male/ female verbal performance could relate to secondary functions of the diverged regions of the *PCDHY* protein. Additionally, this sex chromosome-related dimorphism might help to explain some of the gender bias observed in sets of patients with learning difficulties related to language, such as dyslexia. Traditionally, these disorders are classified as autosomal recessive with reduced female penetrance. Variants of the *PCDHX*/*PCDHY* genes make attractive candidates for modifying effects.

The above characteristics of a gene are all testable with the appropriate materials and experimental design. Patient, anthropological and other primate samples can be used to collect data either to strengthen the case for or to exclude a given candidate. The greatest problem for molecular biology is to try to date the events leading to the genomic organisation observed in modern humans. As methodology for dating based upon the observed rates of nucleotide variation on the X and Y chromosomes improves, it may become possible to predict more precisely the date of the transposition event that placed the copy of the Xq21.3 block onto the Y chromosome. The dating of subsequent independent rearrangements of the sex chromosomes will prove a more formidable challenge in the absence of an archaeological DNA database.

NOTE ADDED IN PROOF

The protocadherin genes on the X and Y chromosomes are also known as *PCDH22* (*PCDHX*) and *PCDH11* (*PCDHY*). Alternative transcript analysis shows that each gene is much larger than originally estimated, and currently there are 17 defined exons spanning 850 kb. *RBPX* has now received the official gene symbol *PABPC5*. Additional details are published in Blanco *et al.* (2001).

ACKNOWLEDGEMENTS

We wish to thank Dr R.A. Furlong and Dr C.A. Boucher for their critical reading of the manuscript. P. Blanco is funded by the Fundacion Pedro Barrie de la Maza and the British Council, and C. Sargent is currently funded by PIC and The Isaac Newton Trust. Part of this work was supported by an MRC programme grant.

DISCUSSION

Questioner: Is this protcadherin gene actually absent in primates and mice?

Affara: It's present on the X chromosome in primates but not on the Y and we have not been able to detect cross-hybridisation with mouse genomic DNA.

Questioner: Have you found that the gene is inactivated on the X chromosome?

Affara: No we haven't. This is something obvious that we want to do. But its not such an easy thing to do for a gene that is expressed specifically in nervous tissue because a lot of the experiments looking at inactivation of genes on the X are done on cell lines in culture. If we can get a polymorphism for one of those neuroepithelial cell lines from a female then we can do that.

Questioner: Was that expression in human or non-human tissue?

Affara: It was human tissue. We haven't looked at non-human tissue.

Questioner: Were they adult brains?

Affara: Most were adult brain except some were whole fetal brain.

References

Affara, N.A. & Ferguson-Smith, M.A. (1994) DNA sequence homology between the human sex chromosomes. In: *Molecular Genetics of Sex Determination* (ed. S. Wachtel, pp. 225–66. San Diego: Academic Press.

Bernardi, G. (2000) Isochores and the evolutionary genomics of vertebrates. *Gene*, **241**, 3–17.

Blanco, P., Sargent, C.A., Boucher, C.A. & Affara, N.A. (2000) Conservation of PCDHX in mammals: expression of human X–Y genes predominantly in the brain. *Mammalian Genome*, **11**, 906–14.

Blanco, P., Sargent, C.A., Boucher, C.A., Howell, G., Ross, M., Affara, N.A. (2001) A novel poly(A)-binding protein gene (PABPC5) maps to an X-specific subinterval in the Xq21.3/Yp11.2 homology block of the human sex chromosomes. *Genomics*, **74**, 1–11.

Buxhoeveden, D. & Casanova, M. (2000) Comparative lateralization patterns in the language area of normal human, chimpanzee, and rhesus monkey brain. *Laterality*, **5**, 315–30.

Ciccodicola, A., D'Esposito, M., Esposito, T., Gianfrancesco, F., Migliaccio, C., Miano, M.G., Matarazzo, M.R., Vacca, M., Franze, A., Cuccurese, M., Cocchia, M., Curci, A., Terracciano, A., Torino, A., Cocchia, S., Mercadante, G., Pannone, E., Archidiacono, N., Rocchi, M., Schlessinger, D. & D'Urso, M. (2000) Differentially regulated and evolved genes in the fully sequenced Xq/Yq pseudoautosomal region. *Human Molecular Genetics*, **9**, 395–4012.

Crow, T.J. (1993) Sexual selection, Machiavellian intelligence and the origins of psychosis. *Lancet*, **342**, 594–8.

Crow, T.J., Crow, L.R., Done, D.J. & Leask, S. (1998) Relative hand skill predicts academic ability: global deficits at the point of hemispheric indecision. *Neuropsychologia*, **36**, 1275–82.

D'Esposito, M., Ciccodicola, A., Gianfrancesco, F., Esposito, T., Flagiello, L., Mazzarella, R., Schlessinger, D. & D'Urso, M. (1996) A synaptobrevin-like gene in the Xq28 pseudoautosomal region undergoes X inactivation. *Nature Genetics*, **13**, 227–9.

Ferguson-Smith, M.A. (1965) Karyotype–phenotype correlations in gonadal dysgenesis and their bearing on the pathogenesis of malformations. *Journal of Medical Genetics*, **2**, 142–55.

Garcia-Castro, M.I., Vielmetter, E. & Bronner-Fraser, M. (2000) N-Cadherin, a cell adhesion molecule involved in establishment of embryonic left–right asymmetry. *Science*, **288**, 1047–51.

Glaser, B., Myrtek, D., Rumpler, Y., Schiebel, K., Hauwy, M., Rappold, G.A. & Schempp, W. (1999) Transposition of SRY into the ancestral pseudoautosomal region creates a new pseudoautosomal boundary in a progenitor of simian primates. *Human Molecular Genetics*, **8**, 2071–8.

Grant, S.G., O'Dell, T.J., Karl, K.A., Stein, P.L., Soriano, P. & Kandel, E.R. (1992) Impaired long-term potentiation, spatial learning, and hippocampal development in fyn mutant mice. *Science*, **258**, 1903–10.

Graves, J.A. (1995) The origin and function of the mammalian Y chromosome and Y-borne genes: an evolving understanding. *Bioessays*, **17**, 311–20.

Graves, J.A.M., Wakefield, M.J. & Toder, R. (1998) The origin and evolution of the pseudoautosomal regions of the human sex chromosomes. *Human Molecular Genetics*, **7**, 1991–6.

Jones, M.H., Khwaja, O.S., Briggs, H., Lambson, B., Davey, P.M., Chalmers, J., Zhou, C.Y., Walker, E.M., Zhang, Y., Todd, C., Ferguson-Smith, M.A. & Affara, N.A. (1994) A set of ninety-seven overlapping yeast artificial chromosome clones spanning the human Y chromosome euchromatin. *Genomics*, **24**, 266–75.

Kohmura, N., Senzaki, K., Hamada, S., Kai, N., Yasuda, R., Watanabe, M., Ishii, H., Yasuda, M., Mishina, M. & Yagi, T. (1998) Diversity revealed by a novel family of cadherins expressed in neurons at a synaptic complex. *Neuron*, **20**, 1137–51.

Lahn, B.T. & Page, D.C. (1999) Four evolutionary strata on the human X chromosome. *Science*, **286**, 964–7.

Lambson, B., Affara, N.A., Mitchell, M. & Ferguson-Smith, M.A. (1992) Evolution of DNA sequence homologies between the sex chromosomes in primate species. *Genomics*, **14**, 1032–40.

Marchant, L.F. & McGrew, W.C. (1996) Laterality of limb function in wild chimpanzees of Gombe National Park: comprehensive study of spontaneous activities. *Journal of Human Evolution*, **30**, 427–43.

Mumm, S., Molini, B., Terrell, J., Srivastava, A. & Schlessinger, D. (1997) Evolutionary features of the 4-Mb Xq21.3 XY homology region revealed by a map at 60-kb resolution. *Genome Research*, **7**, 307–14.

Ohno, S. (1967) Sex chromosomes and sex linked genes. In: *Monographs on Endocrinology* (eds A. Londhardt, L. Samuels & J. Zander). New York: Springer-Verlag.

Page, D.C., Harper, M.E., Love, J. & Botstein, D. (1984) Occurrence of a transposition from the X chromosome long arm to the Y chromosome short arm during human evolution. *Nature*, **311**, 119–23.

Perry, J., Feather, S., Smith, A., Palmer, S. & Ashworth, A. (1998) The human FXY gene is located within Xp22.3: implications for evolution of the mammalian X chromosome. *Human Molecular Genetics*, **7**, 299–305.

Redies, C. & Takeichi, M. (1996) Cadherins in the developing central nervous system: an adhesive code for segmental and functional subdivisions. *Developmental Biology*, **180**, 413–23.

Sano, K., Tanihara, H., Heimark, R.L., Obata, S., Davidson, M., St John, T., Taketani, S. & Suzuki, S. (1993) Protocadherins: a large family of cadherin-related molecules in the central nervous system. *EMBO Journal*, **12**, 2249–56.

Sargent, C.A., Briggs, H., Chalmers, I.J., Lambson, B., Walker, E. & Affara, N.A. (1996) The sequence organization of Yp/proximal Xq homologous regions of the human sex chromosomes is highly conserved. *Genomics*, **32**, 200–9.

Schwartz, A., Chan, D.C., Brown, L.G., Alagappan, R., Pettay, D., Disteche, C., McGillivray, B., de la Chapelle, A. & Page, D.C. (1998) Reconstructing hominid Y evolution; X-homologous block, created by X–Y transposition, was disrupted by Yp inversion through LINE–LINE recombination. *Human Molecular Genetics*, **7**, 1–11.

Senzaki, K., Ogawa, M. & Yagi, T. (1999) Proteins of the CNR family are multiple receptors for Reelin. *Cell*, **99**, 635–47.

Sugino, H., Hamada, S., Yasuda, R., Tuji, A., Matsuda, Y., Fujita, M. & Yagi, T. (2000) Genomic organization of the family of CNR cadherin genes in mice and humans. *Genomics*, **63**, 75–87.

Vacca, M., Matarazzo, M.R., Jones, J., Spalluto, C., Archidiacono, N., Ma, P., Rocchi, M., D'Urso, M., Chen, E.Y., D'Esposito, M. & Mumm, S. (1999) Evolution of the X-specific block embedded in the human Xq21.3/Yp11.1 homology region. *Genomics*, **62**, 293–6.

Vogt, P.H., Affara, N., Davey, P., Hammer, M., Jobling, M.A., Lau, Y.F., Mitchell, M., Schempp, W., Tyler-Smith, C., Williams, G., Yen, P. & Rappold, G.A. (1997) Report of the Third International Workshop on Y Chromosome Mapping 1997. Heidelberg, Germany, April 13–16, 1997. *Cytogenetics and Cell Genetics*, **79**, 1–20.

Vollrath, D., Foote, S., Hilton, A., Brown, L.G., Beer-Romero, P., Bogan, J.S. & Page, D.C. (1992) The human Y chromosome: a 43 interval map based on naturally occurring deletions. *Science*, **258**, 52–9.

Wu, Q. & Maniatis, T. (1999) A striking organization of a large family of human neural cadherin-like cell adhesion genes. *Cell*, **97**, 779–90.

Yoshida, K. & Sugano, S. (1999) Identification of a novel protocadherin gene (*PCDH11*) on the human X–Y homology region in Xq 21.3. *Genomics*, **62**, 540–3.

Preferential Sex Linkage of Sexually Selected Genes: Evidence and a New Explanation

KLAUS REINHOLD

Summary. Here, I review the evidence showing that the X chromosome has a disproportional share concerning the inheritance of sexually selected traits in animals with heterogametic males, and suggest a new explanation that relates this X bias with female choice of heterozygotic males. With numeric simulations I show that female choice of heterozygotic males is usually disadvantageous. Because this disadvantage cannot occur when females prefer X-linked male traits, preferential X linkage of sexually selected traits can be expected. As an alternative to fluctuating selection on sex-limited traits the disadvantage of heterozygotic choice may thus explain the X bias observed for sexually selected traits.

INTRODUCTION

INITIALLY BASED ON the inheritance of sexually selected traits in single species, some authors have proposed that genes coding for traits involved in speciation (Ewing, 1969) and sexual selection (Reinhold, 1994) are biased towards the X chromosome. In a qualitative literature review, however, Charlesworth *et al.* (1987) found no evidence of a special role of sex-linked genes in species recognition. Recently, two quantitative literature reviews have shown that there is a substantial X bias for sex- and reproduction-related genes in humans (Saifi & Chandra, 1999) and for sexually selected traits in animals with heterogametic males (Reinhold, 1998). Based on a database search, Saifi & Chandra (1999) concluded that in humans the proportion of loci related to sex or reproduction is significantly higher on the X chromosome than on the autosomes. For three groups of animals, *Drosophila*, other insects with heterogametic males and mammals, Reinhold (1998) compared the influence of X chromosomal genes on sexually selected male traits and on traits supposedly not under sexual

Proceedings of the British Academy, **106**, 251–265, © The British Academy 2002.

selection. Regarding sexually selected traits, about one-third of the difference between closely related taxa is coded by X chromosomal genes, whereas the X chromosomal contribution is negligible regarding traits classified as not under sexual selection. Those studies not included in these two reviews, for example because they appeared after these reviews or because some data were not included in the analysis, are also generally in accordance with these findings. The X chromosome has a disproportional effect regarding the inheritance of sexually selected song traits in the bushcricket *Ephippiger ephippiger* (Ritchie, 2000) and various *Drosophila* hybrids (Noor & Aquadro, 1998; Hoikkala *et al.*, 2000; but see Ritchie & Kyriacou, 1996; Boake *et al.*, 1998). Other sexually selected traits in *Drosophila*, the sex combs and cuticular hydrocarbon pheromones, also seem to be strongly sex linked (Khadem & Krimbas, 1997; Blows & Allan, 1998; Scott & Richmond, 1998). Male agonistic behaviour in the desert spider *Agelenopsis aperta* and male display behaviour in the fiddler crab *Uca*, traits likely to be under sexual selection, are largely determined by sex-linked loci (Salmon & Hyatt, 1979; Riechert & Maynard Smith, 1989). In accordance with the above reviews, no disproportional effects of X chromosomal genes on sexually selected male traits were detected in birds (*Philomachus pugnax*: Lank *et al.*, 1999) where males are homogametic.

An X bias such as the one observed by Saifi & Chandra (1999) and Reinhold (1998) can be expected for several theoretical reasons, for example sexually antagonistic selection of fluctuating selection. When traits are under sexually antagonistic selection (Rice, 1984), i.e. when one sex would benefit from an increased and the other sex from a decreased trait size, trait expression already differs between the sexes when X chromosomal genes are involved. Under such a selection regime, X chromosomal genes therefore provide the raw material for selection to work with and will therefore be particularly likely to contribute to traits under sexually antagonistic selection. However, most of the examined traits are sex limited in their expression, i.e. females do not express those sexually selected male traits, and the antagonistic selection explanation thus cannot easily be applied. Another possible explanation relates to the difference in expression of autosomal and X chromosomal sex-limited genes. As in heterogametic males one-third of the X chromosomal genes and one-half of the autosomal genes are exposed to selection, fluctuating selection should favour the invasion of X chromosomal mutations (Reinhold, 1999).

Here, I have tried to show that the observed X-bias might also result from selection against female choice of heterozygotic males. If heterozygote advantage is frequent and if female choice is disadvantageous, there will be selection against female choice of males when attractiveness of males is correlated with heterozygote advantage. As heterozygote advantage cannot occur for X chromosomal traits, this disadvantage will not occur for X chromosomal traits and under this scenario one can thus expect to find an X bias for sexually selected traits.

In many animal species, females choose healthy males with high growth rates and developmental stability (reviewed by Andersson, 1994). On the other hand, heterozygosity has been shown to be associated with superior survival, disease resistance, high growth rates and developmental stability (Allendorf & Leary, 1986). Inspired by this pattern, Borgia (1979) and Brown (1997) suggested that females might benefit from choosing heterozygotic males. One problem for this hypothesis is that, at equilibrium, offspring viability does not increase when females preferentially mate with heterozygotic males having superior viability (Partridge, 1983). Offspring viability does not increase at equilibrium because heterozygosity is, as Brown (1997) correctly noted, not directly heritable (but see Mitton *et al.*, 1993). In accordance with the hypothesis that choice of heterozygotic males increases offspring fitness, a theoretical analysis by Charlesworth (1988) suggested that female choice for heterozygotic males might evolve when environmental conditions fluctuate. In these simulations underdominance of viability was assumed and heterozygotes were modelled to be almost as viable as the better one of the homozygotes. Viability values were switched between the homozygotes at regular intervals to mimic temporal environmental fluctuations, and the relative viability of the heterozygotes was kept constant (Charlesworth, 1988). As a result of the higher geometric mean fitness of heterozygotes, an increase in the frequency of the allele causing females to choose heterozygotic males was observed in most but not all simulations. In similar simulations, Mitton (1997) assumed overdominance of heterozygotic males and showed that female choice for heterozygotic males can evolve under temporally fluctuating selection. An allele causing female choice for heterozygotic males is, in contrast, unlikely to increase in frequency when constant viabilities are assumed and when the allele occurs in initially low frequency (Heisler & Curtsinger, 1990). In a recent analytical model, Irwin & Taylor (2000) showed that fluctuating selection is a necessary condition for the evolution of heterozygotic choice.

With the following model I also examined whether a female choice allele can increase in frequency when it causes females to choose heterozygotic males that have superior viability due to overdominance. The simulations show that female choice for heterozygotic males with superior viability will only evolve under restricted environmental conditions.

METHODS

The evolution of female choice for superior heterozygotic males was examined by a population genetic model for a diploid organism with discrete generations and a population of infinite size. Two alleles, A and B at one locus, were assumed to influence male viability before mating, and heterozygotic males were assumed to have higher viability than both types of homozygotic males. The relative viability of AA males was assumed to be $V_{AA} = 1 - s$, the relative viability of BB males was assumed to be $V_{BB} = 1 - t$ $(0 < s, t \leq 1)$, and the relative viability of heterozygotic males was assumed to be $V_{AB} = 1$. Female viability was assumed to be independent from the male viability allele. A female choice allele, C, that causes absolute choice of heterozygotic males was introduced in low frequency in linkage equilibrium with the viability alleles so that 1% of all females showed preference for heterozygotic males. All other females were assumed to mate at random with the available males. The choice allele and the trait alleles were assumed to be unlinked, and the allele causing female choice was both modelled to be dominant and recessive against the no-choice allele. The evolution of female choice of superior heterozygotic males was examined for three different types of environmental variability: (1) stable environmental conditions leading to stable viability disadvantages of the homozygotic males; (2) environmental fluctuations leading to random fluctuations of the viability disadvantages of homozygotic males within a given interval; (3) the size of the viability disadvantage of homozygotic males was assumed to vary randomly as an effect of environmental fluctuations.

Stable environment

The evolution of a rare choice allele was modelled for a range of different values $(0.01 \leq s \leq 0.9)$ for the viability of homozygotic AA males, while the viability of homozygotic BB males was assumed to be 0.5 or 0.8 times the viability of heterozygotic males. The relative change in the frequency of the C allele was calculated as the frequency of the C allele at generation 150 divided by the frequency of C at generation 50. The logarithm of this relative change in the frequency of the choice allele is given as the result. Linkage disequilibrium between the choice and viability alleles was calculated as D/D_{max} (Maynard Smith, 1989).

Environmental fluctuations within a given interval

In these simulations, viability of homozygotic AA males was assumed to vary as an effect of temporal fluctuations in environmental conditions. Viability was modelled to vary randomly within an interval of ± 0.1 around a given mean (i.e. values were chosen from a uniform distribution each generation). Viability of homozygotic BB males was assumed to vary in the same way between 0.4 and 0.6 times (or 0.7 and 0.9 times) the viability of heterozygotic males. The influence of these fluctuations on the frequency of the choice allele was examined. Viabilities of AA males were assumed to vary independently from the viability of BB males. The logarithm of the relative change in the frequency of the choice allele between generations 50 and 1050 is given as the result. In comparison to a stable environment, a 10-fold number of generations was used, otherwise chance would have a large influence on the frequency of the choice allele due to the stochastic nature of the modelled environmental fluctuations.

Environmental fluctuations of different variance

Viabilities of homozygotic AA males were assumed to vary around a mean value of $1 - s$. The influence of environmental fluctuations on the viability of AA males was modelled by $V_{AA} = (1 - s) + s*X*R_1$, where R_1 is a number between -1 and 1 chosen randomly each generation, and where X, a value between 0 and 1, gives the strength of temporal fluctuations influencing the viability disadvantages of the homozygotes. With a maximum X of 1 the viability of homozygotic AA males was assumed to vary between 1 and 2 s, and the viability of heterogametic males was assumed to be 1. With a minimum X equal to 0, the viability of AA males was assumed to be constant at $1 - s$. Viabilities of homozygotic BB males were accordingly assumed to vary around a mean value by $V_{BB} = (1 - t) + t*X*R_2$. For the simulations shown in Figure 3, AA males were assumed to have a mean viability of 0.5 and BB males were assumed to have a mean viability of 0.8. The viabilities of AA males were assumed to vary independently from the viability of BB males. The logarithm of the relative change in the frequency of the choice allele between generations 50 and 1050 is given as the result. To account for the stochastic nature of the modelled environmental fluctuations, 1000 generations were again used to estimate the change in the frequency of the choice allele.

RESULTS

The female choice allele was neutrally stable when both types of homozygotic males were assumed to have the same viability (Figure 1). For all other viability values the frequency of the female choice allele causing mate choice of heterozygotic males decreased markedly (Figure 1). The observed decrease was largest for extreme differences in viability between the two types of heterozygotic males. The size of the linkage disequilibrium between the choice allele and the viability alleles showed a similar dependence on the difference between homozygote viabilities. Linkage disequilibrium was zero when the homozygotes had identical viabilities, and increased monotonically when larger differences between homozygote viabilities were assumed (Figure 2). For all assumed viabilities of males, the observed decrease in the frequency of the female choice allele was larger for a dominant choice allele than for a recessive choice allele (Figure 1).

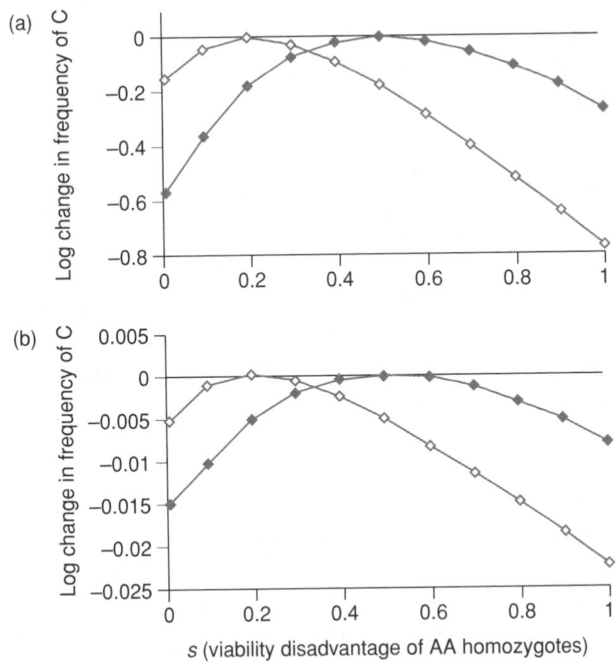

Figure 1. Effect of viability of homozygotic AA males on the frequency of the choice allele C under the assumption of stable viabilities (viability of BB males: open diamonds, $1 - t = 0.8$; closed diamonds, $1 - t = 0.5$). The relative change in the frequency of C is given as logarithm of the ratio between the frequency at generation 150 and the frequency at generation 50, (a) for a dominant choice allele, (b) for a recessive choice allele.

Assuming random variation in the viability of homozygotic males within a given interval, a small increase in the frequency of the choice allele could be observed when the average viabilities of the two types of homozygotic males were assumed to be similar (Figure 3). The frequency of the choice allele decreased when the average viabilities of the two types of males differed by about 0.1 or more. The disadvantage of the choice allele increased with increasing difference between the mean viabilities of AA and BB males (Figure 3). Given mean viabilities of AA and BB males differing by 0.2 or more, the observed decrease in the frequency of the choice allele was large compared with the observed increase in the frequency of the choice allele when the viabilities of the homozygotes were assumed to be similar. This pattern, with a small increase in the frequency of the choice allele when homozygotes had similar viabilities and a large decrease in the frequency of the choice allele when homozygotes had dissimilar viabilities, occurred both with a dominant and with a recessive choice allele. The change in the frequency of the choice allele was smaller when the choice allele was assumed to be recessive against the no-choice allele.

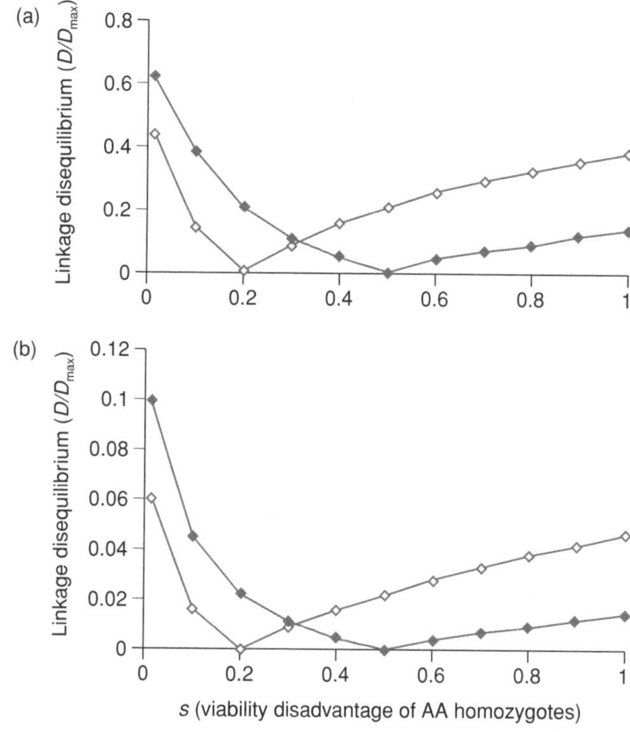

Figure 2. Effect of the viability of homozygotic AA males on the linkage disequilibrium between choice allele and viability alleles under the assumption of stable viabilities (viability of BB males: open diamonds, $V_{BB} = 1 - t = 0.8$; closed diamonds, $V_{BB} = 1 - t = 0.5$), (a) for a dominant choice allele, (b) for a recessive choice allele.

The outcome of the simulations was similar when temporal fluctuations of different strength were assumed. When mean viabilities of homozygotic males were assumed to differ substantially, the frequency of the choice allele decreased even under maximum variability of homozygotic viabilities (Figure 4). However, the disadvantage of the choice allele decreased with increasing strength of temporal fluctuations (Figure 4). As in the other simulations the decrease in the frequency of C was smaller for a recessive choice allele than for a dominant choice allele.

CONCLUSIONS

It was suggested that females should be able to increase offspring fitness by choosing heterozygotic males with superior viability for mating (Borgia, 1979;

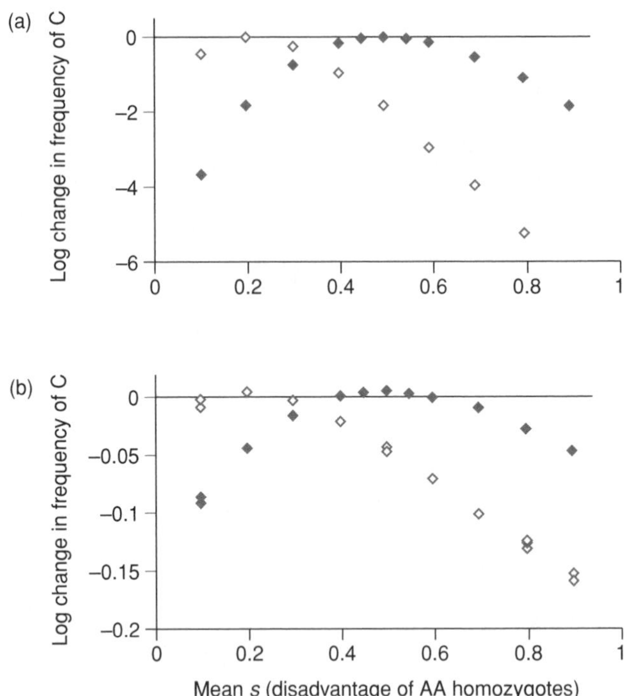

Figure 3. Effect of mean viability of homozygotic AA males on the frequency of the choice allele C under the assumption that viabilities vary randomly by ±0.1 around the mean (mean viability of BB males: open diamonds, $V_{BB} = 1 - t = 0.8$; closed diamonds, $V_{BB} = 1 - t = 0.5$). The relative change in the frequency of C is given as the logarithm of the ratio between the frequency at generation 1050 and the frequency at generation 50. For each value of the parameter s the results of three replicate simulations are shown, (a) for a dominant choice allele, (b) for a recessive choice allele.

Charlesworth, 1988; Brown, 1997). With overdominance and constant relative viabilities of homozygotic ($V_{AA} = 1 - s$; $V_{BB} = 1 - t$) and heterozygotic ($V_{AB} = 1$) males, there is an equilibrium frequency for the viability alleles. For any values of $s > 0$ and $t > 0$, maximum average viability of males is $V_{max} = 1 - (st/(s + t))$ when the frequency of the A allele is equal to $t/(s + t)$ (Partridge, 1983). At equilibrium the viability of male offspring of females choosing heterozygotic males is equal to $1 - (st/(s + t))$ (Partridge, 1983). Thus, offspring viability does not change when females choose heterozygotic males for mating, and female choice seems to be neutral under the assumption of a constant environment. However, these simulations show a disadvantage for females choosing heterozygotic males with superior viability. This disadvantage is caused by a linkage disequilibrium between the choice allele and the less frequent viability allele that is built up by female choice for heterozygotic males. Female choice of

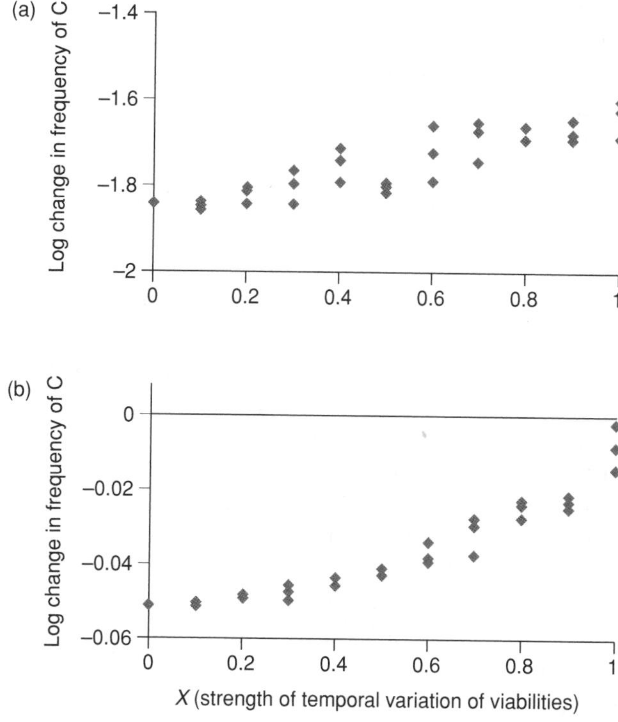

Figure 4. Effect of temporal fluctuations of different strength on the frequency of the choice allele under the assumption that viabilities of homozygotic males are influenced by these fluctuations. The extent of temporal fluctuation of homozygote viabilities is given by X, and homozygote viabilities are assumed to vary randomly by $(1 - s) \pm s^*X$ and $(1 - t) \pm t^*X$ around the mean ($s = 0.5$, $t = 0.2$). The relative change in the frequency of the choice allele is given as logarithm of the ratio between the frequency at generation 1050 and the frequency at generation 50. For each value of X the results of three replicate simulations are shown, (a) for a dominant choice allele, (b) for a recessive choice allele.

heterozygotic males gives an advantage to the less frequent viability allele, because the frequency of the rare allele is higher in the preferred males than in the overall population. As a result, the female choice allele becomes associated with the less frequent allele. This association, together with the reduced viability of the offspring homozygous for the less frequent viability allele, leads to a selection disadvantage of the choice allele compared with the no-choice allele.

When females choose heterozygotic males, the frequency of the less frequent viability allele increases because heterozygotic males have one copy of this allele. When the choice allele occurs at substantial frequency, female choice will cause a deviation of the viability alleles away from the equilibrium frequency towards a 1 : 1 ratio. This is an additional disadvantage for female choice of heterozygotic males because any deviation in the frequency of the A allele by ε from equilibrium frequency reduces mean male viability by $\varepsilon^2/(s + t)$.

When s and t are assumed to be equal, the equilibrium frequency of allele A is 0.5 and female choice of heterozygous males cannot cause a linkage equilibrium or a deviation from the equilibrium allele frequencies. For this reason the decrease of the choice allele was smallest for those cases where s and t were similar. An increase in the frequency of the choice allele could not be observed for any combination of homozygote viabilities when a stable environment was assumed.

The pathway for the evolution of female choice for heterozygotic males suggested by Brown (1997) thus seems to be impossible for temporally stable viabilities. Charlesworth (1988) showed that female choice for heterozygotic males may evolve when environmental conditions fluctuate. However, he only used symmetrical viabilities of the homozygotes in his simulations and the long-term fitness of the two types of homozygotes was thus modelled to be identical. According to the simulations presented here, female choice for heterozygotic males may only be advantageous when homozygote viabilities are similar. Inferred from data on allele frequencies, similar viabilities of the homozygotes seem to be rare in cases with overdominance (Mitton *et al.*, 1993). If temporal fluctuations in the viabilities of homozygotic males were assumed together with similar average viabilities of the two types of homogametic males, the allele causing female choice of heterozygous males increased in frequency. But even in these cases the increase in the frequency of the choice allele was small compared with the decrease that was observed when viabilities of the two types of homozygotes differed. Thus, the evolution of female choice for heterozygotic males with superior viability seems to be restricted to special environmental conditions.

According to the above theoretical analysis, female choice of heterozygotic males should be rare because it is restricted to special environmental conditions. Does the empirical evidence concerning female choice for heterozygous males fit this expectation? There are several studies that are often cited as evidence for female choice of heterozygous males (*Littorina* snails: Rolán-Alvarez *et al.*, 1995; *Danaus* butterflies: Smith, 1981; *Colias* butterflies: Watt *et al.*, 1986; *Artemia* shrimps: Zapata *et al.*, 1990). All these studies have shown that heterozygous males have a mating advantage but they have not shown that females prefer heterozygous males. The mating advantage of heterozygous males shown in these studies can also be explained by an advantage of heterozygous males in finding or acquiring mating partners due to their superior physiology or behaviour resulting from heterosis (Brncic & Koref-Santibañez, 1964). There is thus no clear evidence that the observed mating advantage involved female choice for heterozygotic males. Empirical evidence for female choice of heterozygotic males is therefore at least equivocal and female choice for heterogametic males may be rare, in accordance with the results of the model presented here.

Due to the simplistic conditions used in the simulations, some outcomes of the model may not hold for the more complicated conditions that might occur in nature. The main assumption of the model examined here, choosy females mating only with males that are heterozygotic at a single locus, is clearly unrealistic with respect to female choice in nature. Females have been proposed to select males for their overall heterozygosity at a large number of loci (but see Watt *et al.*, 1986). One should be careful to extrapolate the results of a one-locus simulation to a multilocus system. However, the cause of the disadvantage of female choice in the analysed one-locus system (the linkage disequilibrium) should also occur in a multilocus system. Female choice of heterozygotic males favours rare alleles because rare alleles occur in higher frequency among heterozygotic males than in the overall population. A linkage disequilibrium will consequently build up between choice allele and rare trait alleles. In multilocus systems, the size of the linkage disequilibrium and therefore the disadvantage of female choice of heterozygotic males might, however, be much smaller than in one-locus two-allele systems. When females have a less strong preference for heterozygotic males (for example because they make mistakes in identifying heterozygotic males) or have some intermediate preference for one type of homozygotic males, this will change the size of the linkage disequilibrium and therefore the strength of selection against female choice. But, it will not alter the direction of selection, and female choice can, at best, be expected to be neutral when the chosen males have on average the same allele frequencies as the whole population. It should also be noted that the model rests on the assumption that the population is at equilibrium for the alleles determining male viability, and the results of the simulations will not be applicable when this condition is not met. Considering the limitations of the simulations used, I conclude that the selective disadvantage of female choice of heterozygotic males is also likely to occur under more realistic conditions.

Even when not actively choosing heterozygotic males, females might for some other reason prefer healthy and viable males, in accordance with the good genes hypothesis (for references see Andersson, 1994). If some part of the above-average viability of these attractive males is due to heterozygote advantage, selection against heterozygotic choice can decrease the benefit of choosing viable males. For the good genes mechanism to work, either the benefits of choosing viable males have to be larger than the cost of choosing heterozygotic males, or fluctuating selection is necessary to render heterozygotic choice adaptive. Let us assume that female choice of viable males is adaptive, heterozygotic choice disadvantageous, and that there is variation between females in the male traits they use to recognise male viability. Let us further assume that some females base their preference on traits that are largely determined by X chromosomal genes, and some other females use autosomal traits as viability indicators. If heterozygote advantage is frequent and if the X chromosomal traits are as good as viability indicators as the autosomal traits, selection against heterozygotic choice should lead to an increase in the frequency of females preferring X chromosomal male traits. For those traits, the disadvantage of heterozygotic choice cannot occur because in heterogametic males there is only one X chromosome. Selection against heterozygotic choice should thus lead to an increased influence of X chromosomal genes on sexually selected traits, as has been observed empirically (Reinhold, 1998). Together with sex differential selection (Rice, 1984) and fluctuating selection on sex-limited traits (Reinhold, 1999), the disadvantage of heterozygotic choice thus provides an additional possible explanation of the X-bias observed for sex- and reproduction-related traits (Reinhold, 1998; Saifi & Chandra, 1999).

ACKNOWLEDGEMENTS

I thank the Institute for Advanced Study, Berlin, for support and Leif Engqvist, Thomas Gerken, Bernhard Misof and Andrew Pomiankowski for helpful comments on a previous version of the manuscript.

DISCUSSION

Questioner: Does assortative mating come into this story? Humans have all sorts of assortative mating patterns, things that don't look like sexual selection at first sight, but clearly each sex is looking for something in the other.

Reinhold: Why should this be related to the X chromosome?

Comment: It may have nothing to do with the X chromosome and could be on autosomes.

Questioner: Could it be that there are two copies of the X in the female?

Reinhold: Yes, and therefore the expression is different in females than in males.

Questioner: What are the sex chromosomes in *Drosophila*. Do they have an X and Y as in humans?

Reinhold: Males are X0 and females are XX and there is no X inactivation. In the male; the single X is upregulated.

Comment: I wonder if you have considered imprinting, that is genes would be expressed when they are inherited from the mother, but not from the father. In the latter they would be silent.

Reinhold: There has not been any modelling of imprinting. It should be done. Up to now, the information is limited to mammals. However, I don't think this is likely to be an overall explanation for sexual selection.

Questioner: What is the relationship between speciation and sexual selection? Do you think the evidence is substantial?

Reinhold: Many characteristics are the same between species except for those that are sexually selected. These appear to be species specific. There is much evidence showing that sexual selection and speciation come together. For example in the cichlids, in African lakes, if there is a different colour pattern they don't mate. But if the colour can't be seen because the water is too turbulent, then they mate between the species and the species disappear. So sexual signals are very important in keeping species apart.

Comment: One consequence of having a gene on the X chromosome is that it is dominant in the males and recessive in females. So that it produces greater variation in the males than females. Is this advantageous for sexual selection?

Reinhold: Yes, definitely.

Comment: This difference in sex could have to do with the unequal distribution of language disorders, i.e. more often occurring in males. There are all sorts of sex differences in other disorders such as autism and schizophrenia.

References

Allendorf, F.W. & Leary, R.F. (1986) Heterozygosity and fitness in natural populations of animals. In: *Conservation Biology: The Science of Scarcity and Diversity* (ed. M. E. Soule), pp. 57–76. Massachusetts: Sinauer.

Andersson, M. (1994) *Sexual Selection.* New Jersey: Princeton University Press.

Blows, M.W. & Allan, R.A. (1998) Levels of mate recognition within and between two *Drosophila* species and their hybrids. *American Naturalist*, **152**, 826–37.

Boake, C.R.B., Price, D.K. & Andreadis, D.K. (1998) Inheritance of behavioural difference between two infertile, sympatric species, *Drosophila silvestris* and *D. heteroneura*. *Heredity*, **80**, 642–50.

Borgia, G. (1979) Sexual selection and the evolution of mating systems. In: *Sexual Selection and Reproductive Competition in Social Insects* (eds M.S. Blum & N.A. Blum), pp. 19–80. New York: Academic Press.

Brncic, D. & Koref-Santibañez, S. (1964) Mating activity of homo- and heterokaryotypes in *Drosophila pavani*. *Genetics*, **49**, 585–91.

Brown, J.L. (1997) A theory of mate choice based on heterozygosity. *Behavioral Ecology*, **8**, 60–5.

Charlesworth, B. (1988) The evolution of mate choice in a fluctuating environment. *Journal of Theoretical Biology*, **130**, 191–204.

Charlesworth, B., Coyne, J.A. & Barton, N. (1987) The relative rates of evolution of sex chromosomes and autosomes. *American Naturalist*, **130**, 113–46.

Ewing, A.W. (1969) The genetic basis of sound production in *Drosophila pseudobscura* and *D. persimilis*. *Animal Behaviour*, **1**, 555–60.

Heisler, I.L. & Curtsinger, J.W. (1990) Dynamics of sexual selection in diploid populations. *Evolution*, **44**, 1164–7.

Hoikkala, A., Paallysaho, S., Aspi, J. & Lumme, J. (2000) Localization of genes affecting species differences in male courtship song between *Drosophila virilis* and *D. littoralis*. *Genetical Research*, **75**, 37–45.

Irwin, A.J. & Taylor, P.D. (2000) Heterozygous advantage and the evolution of female choice. *Evolutionary Ecology Research*, **2**, 119–28.

Khadem, M. & Krimbas, C.B. (1997) Studies of the species barrier between *Drosophila subobscura* and *D. madeirensis*. IV. A genetic dissection of the X chromosome for speciation genes. *Journal of Evolutionary Biology*, **10**, 909–20.

Lank, D.B., Coupe, M. & Wynne-Edwards, K.E. (1999) Testosterone-induced male traits in female puffs (*Philomachus pugnax*): autosomal inheritance and gender differentiation. *Proceedings of the Royal Society of London B*, **266**, 2323–30.

Maynard Smith, J. (1989) *Evolutionary Genetics*. Oxford: Oxford University Press.

Mitton, J.B. (1997) *Selection in Natural Populations*. Oxford: Oxford University Press.

Mitton, J.B., Schuster, W.S.F., Cothran, E.G. & De Fries, J.C. (1993) Correlation between the individual heterozygosity of parents and their offspring. *Heredity*, **71**, 59–63.

Noor, M.A.F. & Aquadro, C.F. (1998) Courtship songs of *Drosophila pseudoobscura* and *D. persimilis*: analysis of variation. *Animal Behaviour*, **56**, 115–25.

Partridge, L. (1983) Non random mating and offspring fitness. In: *Mate Choice* (ed. P. Bateson), pp. 227–55. Cambridge: Cambridge University Press.

Reinhold, K. (1994) Inheritance of body and testis size in the bushcricket *Poecilimon veluchianus* (Orthoptera: Tettigoniidae) examined by means of subspecies hybrids. *Biological Journal of the Linnean Society*, **52**, 305–16.

Reinhold, K. (1998) Sex linkage among genes controlling sexually selected traits. *Behavioural Ecology and Sociobiology*, **44**, 1–7.

Reinhold, K. (1999) Evolutionary genetics of sex-limited traits under fluctuating selection. *Journal of Evolutionary Biology*, **12**, 897–902.

Rice, W.R. (1984) Sex chromosomes and the evolution of sexual dimorphism. *Evolution*, **38**, 735–42.

Riechert, S.E. & Maynard Smith, J. (1989) Genetic analyses of two behavioural traits linked to individual fitness in the desert spider *Agelenopsis aperta*. *Animal Behaviour*, **37**, 624–37.

Ritchie, M.G. (2000) The inheritance of female preference functions in a mate recognition system. *Proceedings of the Royal Society of London B*, **267**, 327–32.

Ritchie, M.G. & Kyriacou, C.P. (1996) Artificial selection for a courtship signal in *Drosophila melanogaster*. *Animal Behaviour*, **52**, 603–11.

Rolán-Alvarez, E., Zapata, C. & Alvarez, G. (1995) Multilocus heterozygosity and sexual selection in a natural population of the marine snail *Littorina mariae* (Gastropoda: Prosobranchia). *Heredity*, **75**, 17–25.

Saifi, G.M. & Chandra, H.S. (1999) An apparent excess of sex- and reproduction-related genes on the human X chromosome. *Proceedings of the Royal Society of London B*, **266**, 203–9.

Salmon, M. & Hyatt, G.W. (1979) The development of acoustic display in the fiddler crab *Uca pugilator*, and its hybrids with *U. panacea*. *Marine Behaviour and Physiology*, **6**, 197–209.

Scott, D. & Richmond, R.C. (1998) A genetic analysis of male-predominant pheromones in *Drosophila melanogaster*. *Genetics*, **119**, 639–46.

Smith, D.A.S. (1981) Heterozygous advantage expressed through sexual selection in a polymorphic African butterfly. *Nature*, **289**, 174–5.

Watt, W.B., Carter, P.A. & Donohue, K. (1986) Females' choice of 'good genotypes' as mates is promoted by an insect mating system. *Science*, **233**, 1187–90.

Zapata, C., Gajardo, G. & Beardsmore, J.A. (1990) Multilocus heterozygosity and sexual selection in the brine shrimp *Artemia franciscana*. *Marine Ecology Progress Series*, **62**, 211–17.